AF616362

ADVANCED SUMMER SCHOOL IN PHYSICS 2006

To learn more about the AIP Conference Proceedings, including the Conference Proceedings Series, please visit the webpage
http://proceedings.aip.org/proceedings

ADVANCED SUMMER SCHOOL IN PHYSICS 2006

Frontiers in Contemporary Physics

EAV06

Cinvestav, México City, México *10 – 14 July 2006*

EDITORS

Omar Miranda

Mauricio Carbajal

Luis Manuel Montaño

Oscar Rosas-Ortiz

Sergio A. Tomás Velázquez

Cinvestav

México City, México

SPONSORING ORGANIZATIONS

Centro de Investigación y de Estudios Avanzados del IPN - (Cinvestav)

Consejo Nacional de Ciencia y Tecnología - (CONACyT)

Academia Mexicana de Ciencias - (AMC)

Fundación México-Estados Unidos para la Ciencia - (FUMEC)

Melville, New York, 2007

AIP CONFERENCE PROCEEDINGS ■ VOLUME 885

Editors

Omar Miranda
Mauricio Carbajal
Luis Manuel Montaño
Oscar Rosas-Ortiz
Sergio A. Tomás Velázquez

Departamento de Física
Cinvestav
Av. IPN 2508, Col San Pedro Zacatenco
GAM 07360, México D.F.
México

E-mail: omr@fis.cinvestav.mx
mdct@fis.cinvestav.mx
lmontano@fis.cinvestav.mx
orosas@fis.cinvestav.mx
stomas@fis.cinvestav.mx

L.C. Catalog Card No. 2006939766
ISBN 978-0-7354-0385-7
ISSN 0094-243X

Printed in the United States of America

CONTENTS

I. MATHEMATICAL PHYSICS AND GRAVITATION

II. PARTICLES AND FIELDS

III. STATISTICAL PHYSICS

IV. MEDICAL PHYSICS

V. SOLID STATE PHYSICS

FOREWORD

Science cannot be learned just from reading books or scientific journals, the direct contact with working scientists is essential for the students.

A. Rosenblueth, April 1961.

These words synthesize the central idea of Professor Arturo Rosenblueth, the founder director of Cinvestav (acronym in Spanish for Center of Research and Advanced Studies) about education at graduate level. This philosophy has been the cornerstone of the way Cinvestav operates its graduate programs since its beginning in 1961. The advanced summer school hosted by the Physics Department is the most important example of this way of thinking, since it allows the direct interactions between leading scientists and students in an intense environment during a short period of time. One of the advantages of this school is not only the chance that students have to become familiar with the state of the art in several areas of physics, but the actual possibility that students have to discuss their own work with leading scientists from all over the world.

The ADVANCED SUMMER SCHOOL 2006 was part of the activities celebrating the forty fifth anniversary of Cinvestav, a period of time in which this Center has evolved from an original staff of 10 faculty members working in 4 departments to 575 faculty members working in 11 campuses. The only constant during this period has been to maintain the high quality of research and education at all times.

It is important to reflect about the development of Physics in Cinvestav during this period of time. In 1964, when the first report of activities was published, the Physics Department had only three visiting professors and research assistants. Today, the Physics Department is the largest Department in Cinvestav with 48 professors covering the areas of High Energy Physics, Mathematical Physics, Solid State and Condensed Matter Physics, Statistical Physics and Medical Physics and attending more than 100 graduate students. This Department has graduated almost 200 Ph.D. students, it being one of the main contributors to the generation of new scientists in Mexico.

Maybe more important than the figures just mentioned is the fact that Cinvestav, most noticeable in Physics, has become a reference of how to decentralize science and create new successful research groups. The Physics Department has been the seed of several groups in different regions of Mexico. Specifically, in Cinvestav besides our Physics Department in Cinvestav-Zacatenco we now have an Applied Physics Department in Cinvestav-Mérida, a Materials Science Department in Cinvestav-Querétaro and our newly founded Medical Physics group in Cinvestav-Monterrey.

The ADVANCED SUMMER SCHOOL 2006 hosted students from 19 different Mexican institutions and students from Guatemala and Peru showing that these summer schools are now a reference in Mexico for students who want to become familiar with the state of the art in different areas of Physics.

We are sure that the Summer School in the future will be a mirror of the strong development of the Physics community in Cinvestav and we vow to make it an important event that will attract students from several countries.

México City, August 2006
José Mustre
Provost (Secretario Académico)
Cinvestav

PREFACE

The ADVANCED SUMMER SCHOOL 2006 (EAV06) was held at Cinvestav, Mexico City, on July 10-14, 2006. As in the previous year, the aim of the School was bring together graduate and advanced undergraduate students and researchers from the different areas developed at the Physics Department of Cinvestav. At the School a general overview of the current state of the art was presented on topics related with mathematical physics and gravitation, high energy physics, statistical, medical and solid state physics. The School was attended by 33 researches from Germany, Spain, U.S.A. and Mexico, and by about 120 students from Guatemala, Peru, and nineteen different institutions in Mexico. The courses in the School consisted of lectures starting from a general introduction and followed by a series of sessions at increasing level to reach applications and special topics.

During the School, seven courses were presented. The first one was introduced by Stephen Parke who lectured on Neutrinos in the Standard Model and beyond. A didactic overview on Rigged Hilbert spaces was covered by Rafael de la Madrid. David Aspnes taught different aspects of the Electrodynamic Properties of Nanoscopically Inhomogeneous Materials. Joe Greene lectured on the Nucleation, Growth, and the Evolution of Self-Organized Nanostructures. Special emphasis was given to understanding the interplay of surface and bulk diffusion processes on morphological evolution of compound surfaces. Albert Díaz-Guilera presented results of the last 8 years concerning complex networks which are far from being regular or completely random. Also in this field, Moisés Santillán discussed on gene regulatory dynamics and the mathematical techniques as well as procedures required to tackle complex genetic networks. Finally, Frank Scheffold presented some multiple light scattering techniques used to study turbid colloidal systems.

As a complement to the courses, nine invited talks were given by colleagues from Cinvestav. There were also twenty short talks given mainly by students. We thank to all of them for the professional attitude in preparing their speeches; their enthusiasm was contagious and produced a comfortable hard-working environment in the School.

The present volume of the AIP Conference Proceedings includes the lecture notes of all the courses, as well as the positively refereed contributions of the participants in the EAV06. As in the previous School, EAV05, most of the contributors submitted original results to be considered for publication. The contributions published in this volume were accepted in agreement with the decision of the anonymous referees. The proceedings are organized in five chapters according with our Physics Department's groups of research. In all cases, the chapter starts with the lecture notes of the courses while contributions are arranged according to similarity in subject.

The School was part of the activities celebrating the forty fifth anniversary of Cinvestav and the Physics Department. As it has been a tradition since the first Advanced

School in 1969, the high academic quality of the lectures was the primordial goal to celebrate the EAV06. Such a condition would be not possible without the financial and academic support of our Institution. The support of the Mexican Council for Science and Technology (CONACyT) is also acknowledged. Professors David Aspnes, Joe Greene and Stephen Parke were also partially supported by The United States-Mexico Foundation for Science and the Academia Mexicana de Ciencias. We are indebted to Professor José Mustre, Cinvestav's Provost, for his help in obtaining the SEP grants for the students who were supported by the EAV06. The encourage and assistance of Professor Gerardo Herrera, Chair of the Physics Department, was invaluable in solving some organizational and financial difficulties.

We are also indebted to all the people who dedicated many hours of their time to support the organization of the School. Our support team included Juan Barranco, Luz del Carmen Cortés, Sara Cruz, Nicolás Fernández, Elvia Lomelí and Israel Lomelí. Daniel Pérez gave important support to speakers, by solving all the problems that usually appears in any presentation. Beatriz Garrido helped us with some administrative and broadcasting tasks. We would like to especially thank Miriam Lomelí, secretary of the School, her experience in all the administrative duties was indispensable for a successful organization of the School, right from the early stages to the very end of the event. We acknowledge her effort and patience along the organization

Omar Miranda
Mauricio Carbajal
Luis Manuel Montaño
Oscar Rosas-Ortiz
Sergio A. Tomás Velázquez

Mexico City, 2006

LECTURERS

David E Aspnes
NC State University
USA

Albert Diaz-Guilera
Universitat de Barcelona
Spain

Joe E Greene
University of Illinois
USA

Rafael de la Madrid
University of California at San Diego
USA

Stephen Parke
Fermi National Accelerator Laboratory
USA

Moisés Santillán Zerón
Cinvestav-Unidad Monterrey
Mexico

Frank Scheffold
University of Fribourg
Switzerland

SPONSORS

Centro de Investigación y de Estudios Avanzados del IPN (Cinvestav)

Consejo Nacional de Ciencia y Tecnología (CONACyT)

Academia Mexicana de Ciencias (AMC)

Fundación México-Estados Unidos para la Ciencia (FUMEC)

ORGANIZING COMMITTEE

Mauricio D Carbajal-Tinoco — *mdct@fis.cinvestav.mx*
Departamento de Física, Cinvestav
Mexico

Luis Manuel Montaño — *lmontano@fis.cinvestav.mx*
Departamento de Física, Cinvestav
Mexico

Oscar Rosas-Ortiz — *orosas@fis.cinvestav.mx*
Departamento de Física, Cinvestav
Mexico

Sergio A Tomás Velázquez — *stomas@fis.cinvestav.mx*
Departamento de Física, Cinvestav
Mexico

Omar Miranda (*Chairman*) — *omr@fis.cinvestav.mx*
Departamento de Física, Cinvestav
Mexico

I. MATHEMATICAL PHYSICS AND GRAVITATION

Description of Resonances within the Rigged Hilbert Space

Rafael de la Madrid

Department of Physics, University of California at San Diego, La Jolla, CA 92093 USA,
rafa@physics.ucsd.edu

Abstract. The spectrum of a quantum system has in general bound, scattering and resonant parts. The Hilbert space includes only the bound and scattering spectra, and discards the resonances. One must therefore enlarge the Hilbert space to a rigged Hilbert space, within which the physical bound, scattering and resonance spectra are included on the same footing. In this work, I will explain how this is done.

INTRODUCTION

In Quantum Mechanics, observable quantities are represented by linear operators. The eigenvalues of an operator represent the possible values of the measurement of the corresponding observable. These eigenvalues, which mathematically correspond to the spectrum of the operator, can be discrete (as the energies of a particle in a box), continuous (as the energies of a free, unconstrained particle), resonant (as in α decay), or a combination thereof.

The Hilbert space includes only the bound and scattering spectra, because the Hilbert space spectrum of an observable is real, thereby discarding the resonance spectrum as unphysical. However, radioactive nuclei and unstable elementary particles are physical objects that ought to have a place in the quantum mechanical formalism. This is why we need to extend the Hilbert space to a rigged Hilbert space, within which the resonance spectrum has a place.

The purpose of this paper is to explain how one should use the rigged Hilbert space in quantum mechanics and, in particular, how to incorporate the resonance spectrum into the quantum mechanical formalism by using the rigged Hilbert space.

When the spectrum of an observable A is discrete and A is bounded, then A is defined on the whole of the Hilbert space $\mathscr{H}$ and the eigenvectors of A belong to $\mathscr{H}$. In this case, A can be essentially seen as a matrix. This means that, as far as discrete spectrum is concerned, there is no need to extend $\mathscr{H}$. However, quantum mechanical observables are in general unbounded and their spectrum has in general a continuous part. In order to deal with continuous spectrum, we use Dirac's bra-ket formalism. This formalism does not fit within the Hilbert space alone, but within the rigged Hilbert space.

Loosely speaking, a rigged Hilbert space (also called a Gelfand triplet) is a triad of spaces

$$\Phi \subset \mathscr{H} \subset \Phi^{\times} \tag{1}$$

CP885, *Advanced Summer School in Physics 2006, Frontiers in Contemporary Physics—EAV06,* edited by O. Miranda, M. Carbajal, L. M. Montaño, O. Rosas-Ortiz, and S. A. Tomás Velázquez

such that $\mathscr{H}$ is a Hilbert space, Φ is a dense subspace of $\mathscr{H}$, and $\Phi^{\times}$ is the space of antilinear functionals over Φ. Mathematically, Φ is the space of test functions, and $\Phi^{\times}$ is the space of distributions. The space $\Phi^{\times}$ is called the antidual space of Φ. Associated with the rigged Hilbert space (1), there is always another rigged Hilbert space,

$$\Phi \subset \mathscr{H} \subset \Phi' , \tag{2}$$

where Φ' is called the dual space of Φ and contains the linear functionals over Φ.

The basic reason why we need the spaces Φ' and $\Phi^{\times}$ is that the bras and kets associated with the elements in the continuous spectrum of an observable belong, respectively, to Φ' and $\Phi^{\times}$ rather than to $\mathscr{H}$. The basic reason reason why we need the space Φ is that unbounded operators are not defined on the whole of $\mathscr{H}$ but only on dense subdomains of $\mathscr{H}$ that are not invariant under the action of the observables. Such non-invariance makes expectation values, uncertainties and commutation relations not well defined on the whole of $\mathscr{H}$. The space Φ is the largest subspace of the Hilbert space on which such expectation values, uncertainties and commutation relations are well defined.

Besides accommodating resonances and Dirac's bra-ket formalism, the rigged Hilbert space seems to capture the physical principles of quantum mechanics better than the Hilbert space. For example, assuming that the Hilbert space provides the whole mathematical framework for quantum mechanics leads to the conclusion that Heisenberg's uncertainty relations are not physical, since they cannot be defined on the whole of the Hilbert space [1]. Using the rigged Hilbert space, one overcomes this difficulty after realizing that the commutation relations are well defined on Φ.

The completeness relation is a good place to appreciate the added value of the rigged Hilbert space. Consider, for example, the Hamiltonian H of a system. In the Hilbert space, one writes the completeness relation as

$$1 = \int_{\mathrm{Sp}(H)} d\mathsf{E}_E , \tag{3}$$

where E_E are the spectral projections of H and $\mathrm{Sp}(H)$ is its spectrum. However, within the rigged Hilbert space one can write[1]

$$1 = \sum_n |E_n\rangle\langle E_n| + \int_0^{\infty} dE\, |E\rangle\langle E| , \tag{4}$$

where $|E_n\rangle$ and $|E\rangle$ are the bound and scattering states of H, respectively. In addition to (4), the rigged Hilbert space gives you an additional completeness relation in which the resonance states participate:

$$1 = \sum_n |E_n\rangle\langle E_n| + \sum_n |z_n\rangle\langle z_n| + \int_{-\infty}^{0} dE\, |E\rangle\langle E| , \tag{5}$$

where $|z_n\rangle$ are the Gamow (resonance) states of H and the last integral, called the background, is performed in the complex plane right below the negative real axis of the

[1] In the Hilbert space, one can actually write something close to, although not the same as (4), by means of direct integral decompositions.

second sheet.[2] Thus, the completeness relation (5) substitutes the scattering states contribution by the resonance contribution plus a background, thereby putting the resonance spectrum on the same footing as the bound and scattering spectra.

It is important to note that the integrals in (4) and (5) are different, and that the resonance contribution does not appear in (4), because resonances are not asymptotic states. Also important is to note that the resonance states, and therefore expansion (5), need a different rigged Hilbert space from that needed by the scattering states and expansion (4).

There are dangers in using the rigged Hilbert space, though. For instance, A. Bohm and collaborators have been using a rigged Hilbert space (of Hardy class) to construct a quantum theory of resonances, see review [2] and references therein. However, such theory is inconsistent with quantum mechanics [3].

The structure of this work is as follows. After an introductory Section 1, I will explain in Section 2 how to construct the rigged Hilbert space of the one-dimensional rectangular barrier potential. This will show that the rigged Hilbert space is already needed to provide the mathematical support of the most basic quantum systems. In Section 3, I will construct the rigged Hilbert space of the Lippmann-Schwinger equation, which equation governs quantum scattering. In Section 4, I will construct the rigged Hilbert space of the analytic continuation of the Lippmann-Schwinger equation. The rigged Hilbert space of Section 4 will be needed in the final Section 5 to provide the mathematical support for the resonance states. The PDF file of the talks corresponding to each section will be posted at `http://www.ucsd.edu/~rafa`.

Since space prevents a full account, at each section I will refer the reader to the appropriate papers where further details can be found. The essentials of functional analysis needed to understand those papers can be found in [4, Chapter 2].

CONSTRUCTION OF A SIMPLE RIGGED HILBERT SPACE

If resonances are not included, the way to construct the rigged Hilbert space of a quantum system is as follows:

1. We identify the observables of the system. Their expressions are usually given by linear, differential operators.
2. We identify the Hilbert space, whose scalar product is used to calculate probability amplitudes.
3. We identify the domains, spectra and eigenfunctions of the observables. If the observables have a discrete spectrum, we need not go beyond the Hilbert space. However, if at least one of the observables has continuous spectrum, we need to enlarge the Hilbert space to the rigged Hilbert space. (If position and momentum are among the observables, we will always need the rigged Hilbert space.)

---

[2] As we will see in Section 5, the expansion (5) must be either regulated or understood in a time-dependent way.

4. We construct the space Φ in which physical quantities such as expectation values, uncertainties and commutation relations are well defined. When resonances are not included, the space Φ is usually given by the maximal invariant subspace of the algebra of observables.
5. We construct the dual Φ' and antidual $\Phi^{\times}$ spaces. We construct the bras and kets of the observables and check that they respectively belong to Φ' and $\Phi^{\times}$.
6. The completeness relations and all the features of Dirac's bra-ket formalism now follow.

Let's see how the above steps are carried out in the case of a spinless particle moving in one dimension and impinging on a rectangular barrier. The observables relevant to this system are the position Q, the momentum P, and the Hamiltonian H:

$$Qf(x) = xf(x) \,, \tag{6}$$

$$Pf(x) = -i\hbar \frac{d}{dx} f(x) \,, \tag{7}$$

$$Hf(x) = \left(-\frac{\hbar^2}{2m} \frac{d^2}{dx^2} + V(x) \right) f(x) \,, \tag{8}$$

where

$$V(x) = \begin{cases} 0 & -\infty < x < a \\ V_0 & a < x < b \\ 0 & b < x < \infty \end{cases} \tag{9}$$

is the 1D rectangular barrier potential. These observables satisfy the following commutation relations:

$$[Q,P] = i\hbar I \,, \tag{10}$$

$$[H,Q] = -\frac{i\hbar}{m} P \,, \tag{11}$$

$$[H,P] = i\hbar \frac{\partial V}{\partial x} \,. \tag{12}$$

Since our particle can move in the full real line, the Hilbert space on which the differential operators (6)-(8) should act is

$$L^2 = \{ f(x) \,|\, \int_{-\infty}^{\infty} dx |f(x)|^2 < \infty \} \,. \tag{13}$$

The corresponding scalar product is

$$(f,g) = \int_{-\infty}^{\infty} dx \overline{f(x)} g(x) \,, \qquad f,g \in L^2 \,. \tag{14}$$

The differential operators (6)-(8) induce three linear operators on the Hilbert space L^2. These operators cannot be defined on the whole of L^2, but only on the following subdomains of L^2:

$$\mathscr{D}(Q) = \left\{ f \in L^2 \,|\, xf \in L^2 \right\} \,, \tag{15}$$

$$\mathscr{D}(P) = \left\{ f \in L^2 \mid f \in AC,\, Pf \in L^2 \right\} , \tag{16}$$

$$\mathscr{D}(H) = \left\{ f \in L^2 \mid f \in AC^2,\, Hf \in L^2 \right\} , \tag{17}$$

where, essentially, AC is the space of functions whose derivative exists, and AC^2 is the space of functions whose second derivative exists. On these domains, the operators Q, P and H are self-adjoint, and their spectra are

$$\mathrm{Sp}(Q) = \mathrm{Sp}(P) = (-\infty, \infty) , \quad \mathrm{Sp}(H) = [0, \infty) , \tag{18}$$

which spectra coincide with those we would expect on physical grounds.

To obtain the eigenfunctions corresponding to each eigenvalue, we have to solve the eigenvalue equation for each observable:

$$x \langle x | x' \rangle = x' \langle x | x' \rangle , \tag{19}$$

$$-i\hbar \frac{d}{dx} \langle x | p \rangle = p \langle x | p \rangle , \tag{20}$$

$$\left(-\frac{\hbar^2}{2m} \frac{d^2}{dx^2} + V(x) \right) \langle x | E \rangle = E \langle x | E \rangle . \tag{21}$$

The eigenfunctions of Q are delta functions,

$$\langle x | x' \rangle = \delta(x - x') , \tag{22}$$

those of P are plane waves,

$$\langle x | p \rangle = \frac{e^{ipx/\hbar}}{\sqrt{2\pi\hbar}} , \tag{23}$$

and those of H are given by

$$\langle x | E^+ \rangle_{\mathrm{r}} = \left(\frac{m}{2\pi k \hbar^2} \right)^{1/2} \times \begin{cases} T(k) e^{-ikx} & -\infty < x < a \\ A_{\mathrm{r}}(k)^{i\kappa x} + B_{\mathrm{r}}(k) e^{-i\kappa x} & a < x < b \\ R_{\mathrm{r}}(k) e^{ikx} + e^{-ikx} & b < x < \infty , \end{cases} \tag{24}$$

$$\langle x | E^+ \rangle_{\mathrm{l}} = \left(\frac{m}{2\pi k \hbar^2} \right)^{1/2} \times \begin{cases} e^{ikx} + R_{\mathrm{l}}(k) e^{-ikx} & -\infty < x < a \\ A_{\mathrm{l}}(k) e^{i\kappa x} + B_{\mathrm{l}}(k) e^{-i\kappa x} & a < x < b \\ T(k) e^{ikx} & b < x < \infty , \end{cases} \tag{25}$$

where

$$k = \sqrt{\frac{2m}{\hbar^2} E} , \quad \kappa = \sqrt{\frac{2m}{\hbar^2} (E - V_0)} , \tag{26}$$

and where the coefficients that appear in Eqs. (24)-(25) can be easily found by the standard matching conditions at the discontinuities of the potential. Physically, $\langle x | E^+ \rangle_{\mathrm{r}}$ ($\langle x | E^+ \rangle_{\mathrm{l}}$) represents a particle of energy E impinging on the barrier from the right (left).

The eigenfunctions (22)-(25) are not square integrable, that is, they do not belong to L^2. Mathematically speaking, this is the reason why they are to be dealt with as distributions (note that all of them except for the delta function are also proper functions).

We now start the construction of the rigged Hilbert space by constructing Φ. The space Φ is given by

$$\Phi = \bigcap_{\substack{n,m=0 \\ A,B=Q,P,H}}^{\infty} \mathscr{D}(A^n B^m) \,. \tag{27}$$

In view of expressions (6)-(8), Φ is simply

$$\Phi = \{\varphi \in L^2 \mid \varphi \in C^\infty(\mathbb{R}),\ \varphi^{(n)}(a) = \varphi^{(n)}(b) = 0\,,\ n = 0,1,\ldots\,, \\ P^n Q^m H^l \varphi(x) \in L^2\,,\ n,m,l = 0,1,\ldots\}\,, \tag{28}$$

where $C^\infty(\mathbb{R})$ is the collection of infinitely differentiable functions, and $\varphi^{(n)}$ denotes the nth derivative of φ. From the last condition in Eq. (28), we deduce that the elements of Φ satisfy the following estimates:

$$\|\varphi\|_{n,m,l} \equiv \sqrt{\int_{-\infty}^{\infty} dx \left|P^n Q^m H^l \varphi(x)\right|^2} < \infty\,, \quad n,m,l = 0,1,\ldots\,. \tag{29}$$

These estimates mean that the action of any combination of any power of the observables remains square integrable. For this to happen, the functions $\varphi(x)$ must be infinitely differentiable and must fall off at infinity faster than any polynomial. Hence, Φ is a Schwartz-like space.

Because Φ is invariant under the action of the observables,

$$A\Phi \subset \Phi\,, \qquad A = P,Q,H\,, \tag{30}$$

the expectation values

$$(\varphi, A^n \varphi)\,, \quad \varphi \in \Phi\,,\ A = P,Q,H\,,\ n = 0,1,\ldots \tag{31}$$

are finite, and the commutation relations (10)-(12) are well defined. In particular, Heisenberg's uncertainty principle makes sense on Φ.

The spaces Φ' and $\Phi^\times$ are simply the collection of linear and antilinear functionals over Φ, respectively. By combining the spaces Φ, $\mathscr{H}$, $\Phi^\times$ and Φ', we obtain the rigged Hilbert spaces of our system,

$$\Phi \subset \mathscr{H} \subset \Phi^\times\,, \tag{32}$$

$$\Phi \subset \mathscr{H} \subset \Phi'\,. \tag{33}$$

The space $\Phi^\times$ accommodates the eigenkets $|p\rangle$, $|x\rangle$ and $|E^+\rangle_{1,\mathrm{r}}$ of P, Q and H, whereas Φ' accommodates the eigenbras $\langle p|$, $\langle x|$ and ${}_{1,\mathrm{r}}\langle^+E|$.

Mathematically, the bras and kets are distributions defined as follows. Given a function $f(x)$ and a space of test functions Φ, the antilinear functional F that corresponds to the function $f(x)$ is an integral operator whose kernel is precisely $f(x)$:

$$F(\varphi) \equiv \int dx\, \overline{\varphi(x)} f(x)\,, \tag{34}$$

and the linear functional $\tilde{F}$ generated by the function $f(x)$ is an integral operator whose kernel is the complex conjugate of $f(x)$:

$$\tilde{F}(\varphi) \equiv \int dx\, \varphi(x)\overline{f(x)} \,. \tag{35}$$

In Dirac's notation, these two equations become

$$\langle \varphi | F \rangle = \int dx \langle \varphi | x \rangle \langle x | f \rangle \,, \tag{36}$$

$$\langle F | \varphi \rangle = \int dx \langle f | x \rangle \langle x | \varphi \rangle \,. \tag{37}$$

Note that these definitions are very similar, except that the complex conjugation affects either $f(x)$ or $\varphi(x)$, which makes the corresponding functional either linear or antilinear.

Definitions (34) and (35) provide the link between the quantum mechanical formalism and the theory of distributions. In practical applications, what one obtains from the quantum mechanical formalism is the distribution $f(x)$ (in this section, the plane waves $\frac{1}{\sqrt{2\pi\hbar}}e^{ipx/\hbar}$, the delta function $\delta(x-x')$ and the eigenfunctions $\langle x|E^{+}\rangle_{\mathrm{l,r}}$). Once $f(x)$ is given, one can use definitions (34) and (35) to generate the functionals $|F\rangle$ and $\langle F|$. Then, the theory of distributions can be used to obtain the properties of the functionals $|F\rangle$ and $\langle F|$, which in turn yield the properties of the distribution $f(x)$.

By using prescription (34), we can define for each eigenvalue p the eigenket $|p\rangle$ associated with the eigenfunction (23):

$$\langle \varphi | p \rangle \equiv \int_{-\infty}^{\infty} dx\, \overline{\varphi(x)} \frac{1}{\sqrt{2\pi\hbar}} e^{ipx/\hbar} \,, \tag{38}$$

which, using Dirac's notation for the integrand, becomes

$$\langle \varphi | p \rangle \equiv \int_{-\infty}^{\infty} dx \langle \varphi | x \rangle \langle x | p \rangle \,. \tag{39}$$

Similarly, for each x, we can define the ket $|x\rangle$ as

$$\langle \varphi | x \rangle \equiv \int_{-\infty}^{\infty} dx'\, \overline{\varphi(x')} \delta(x-x') \,, \tag{40}$$

which, using Dirac's notation for the integrand, becomes

$$\langle \varphi | x \rangle \equiv \int_{-\infty}^{\infty} dx' \langle \varphi | x' \rangle \langle x' | x \rangle \,. \tag{41}$$

The definition of the kets $|E^{+}\rangle_{\mathrm{l,r}}$ that correspond to the Hamiltonian's eigenfunctions (24)-(25) follows the same prescription:

$$\langle \varphi | E^{+} \rangle_{\mathrm{l,r}} \equiv \int_{-\infty}^{\infty} dx\, \overline{\varphi(x)} \langle x | E^{+} \rangle_{\mathrm{l,r}} \,, \tag{42}$$

that is,

$$\langle\varphi|E^+\rangle_{\mathrm{l,r}} \equiv \int_{-\infty}^{\infty} dx\,\langle\varphi|x\rangle\langle x|E^+\rangle_{\mathrm{l,r}}\,. \tag{43}$$

One can now show that the definition of the kets $|p\rangle$, $|x\rangle$ and $|E^+\rangle_{\mathrm{l,r}}$ makes sense, and that these kets indeed belong to the space of distributions $\Phi^\times$.

By using prescription (35), we can also define for each eigenvalue p the eigenbra $\langle p|$ associated with the eigenfunction (23):

$$\langle p|\varphi\rangle \equiv \int_{-\infty}^{\infty} dx\,\varphi(x)\frac{1}{\sqrt{2\pi\hbar}}e^{-ipx/\hbar} \equiv \int_{-\infty}^{\infty} dx\,\langle p|x\rangle\langle x|\varphi\rangle\,. \tag{44}$$

Comparison with Eq. (38) shows that the action of $\langle p|$ is the complex conjugate of the action of $|p\rangle$,

$$\langle p|\varphi\rangle = \overline{\langle\varphi|p\rangle}\,, \tag{45}$$

and that

$$\langle p|x\rangle = \overline{\langle x|p\rangle} = \frac{1}{\sqrt{2\pi\hbar}}e^{-ipx/\hbar}\,. \tag{46}$$

The bra $\langle x|$ is defined as

$$\langle x|\varphi\rangle \equiv \int_{-\infty}^{\infty} dx'\,\varphi(x')\delta(x-x') \equiv \int_{-\infty}^{\infty} dx'\,\langle x|x'\rangle\langle x'|\varphi\rangle\,. \tag{47}$$

Comparison with Eq. (40) shows that the action of $\langle x|$ is complex conjugated to the action of $|x\rangle$,

$$\langle x|\varphi\rangle = \overline{\langle\varphi|x\rangle}\,, \tag{48}$$

and that

$$\langle x|x'\rangle = \langle x'|x\rangle = \delta(x-x')\,. \tag{49}$$

Analogously, the eigenbras of the Hamiltonian are defined as

$${}_{\mathrm{l,r}}\langle^+E|\varphi\rangle \equiv \int_{-\infty}^{\infty} dx\,\varphi(x)\,{}_{\mathrm{l,r}}\langle^+E|x\rangle \equiv \int_{-\infty}^{\infty} dx\,{}_{\mathrm{l,r}}\langle^+E|x\rangle\langle x|\varphi\rangle\,, \tag{50}$$

where

$${}_{\mathrm{l,r}}\langle^+E|x\rangle = \overline{\langle x|E^+\rangle}_{\mathrm{l,r}}\,. \tag{51}$$

Comparison of Eq. (50) with Eq. (42) shows that the actions of the bras ${}_{\mathrm{l,r}}\langle^+E|$ are the complex conjugates of the actions of the kets $|E^+\rangle_{\mathrm{l,r}}$:

$${}_{\mathrm{l,r}}\langle^+E|\varphi\rangle = \overline{\langle\varphi|E^+\rangle}_{\mathrm{l,r}}\,. \tag{52}$$

Now, by using the rigged Hilbert space, one can show that the definitions of $\langle p|$, $\langle x|$ and ${}_{\mathrm{l,r}}\langle^+E|$ make sense and that $\langle p|$, $\langle x|$ and ${}_{\mathrm{l,r}}\langle^+E|$ belong to Φ'.

It is important to keep in mind the difference between eigenfunctions and kets. For instance, $\langle x|p\rangle$ is an eigenfunction of a differential equation, Eq. (20), whereas $|p\rangle$ is a functional, the relation between them being given by Eq. (39). A similar relation holds between $\langle x'|x\rangle$ and $|x\rangle$, and between $\langle x|E^+\rangle_{\rm l,r}$ and $|E^+\rangle_{\rm l,r}$. It is also important to keep in mind that "scalar products" like $\langle x|p\rangle$, $\langle x'|x\rangle$ or $\langle x|E^+\rangle_{\rm l,r}$ do not represent an actual scalar product of two functionals; these "scalar products" are simply solutions to differential equations.

The kets $|p\rangle$, $|x\rangle$ and $|E^+\rangle_{\rm l,r}$ are indeed eigenvectors of P, Q and H, respectively:

$$P|p\rangle = p|p\rangle\,, \quad p \in \mathbb{R}\,, \tag{53}$$

$$Q|x\rangle = x|x\rangle\,, \quad x \in \mathbb{R}\,, \tag{54}$$

$$H|E^+\rangle_{\rm l,r} = E|E^+\rangle_{\rm l,r}\,, \quad E \in [0,\infty)\,. \tag{55}$$

Similarly, the bras $\langle p|$, $\langle x|$ and ${}_{\rm l,r}\langle^+E|$ are left eigenvectors of P, Q and H, respectively:

$$\langle p|P = p\langle p|\,, \quad p \in \mathbb{R}\,, \tag{56}$$

$$\langle x|Q = x\langle x|\,, \quad x \in \mathbb{R}\,, \tag{57}$$

$${}_{\rm l,r}\langle^+E|H = E\,{}_{\rm l,r}\langle^+E|\,, \quad E \in [0,\infty)\,. \tag{58}$$

Note that these equations are to be understood in the distributional way, that is, as "sandwiches" with elements of Φ. For example, Eq. (53) should be understood as

$$\langle\varphi|P|p\rangle = p\langle\varphi|p\rangle\,, \quad p \in \mathbb{R},\ \varphi \in \Phi\,, \tag{59}$$

and the same for (54)-(58). Usually, the "sandwiching" is implicit and therefore omitted.

Now that we have constructed the Dirac bras and kets, we can see how other aspects of Dirac's bra-ket formalism hold within the rigged Hilbert space. For example, the completeness relations

$$\int_{-\infty}^{\infty} dp\,|p\rangle\langle p| = I\,, \tag{60}$$

$$\int_{-\infty}^{\infty} dx'\,|x'\rangle\langle x'| = I\,, \tag{61}$$

$$\int_0^{\infty} dE\,|E^+\rangle_{\rm l\,l}\langle^+E| + \int_0^{\infty} dE\,|E^+\rangle_{\rm r\,r}\langle^+E| = I\,, \tag{62}$$

and the action of P, Q and H,

$$P = \int_{-\infty}^{\infty} dp\,p|p\rangle\langle p|\,, \tag{63}$$

$$Q = \int_{-\infty}^{\infty} dx x|x\rangle\langle x|\,, \tag{64}$$

$$H = \int_0^{\infty} dE\,E|E^+\rangle_{\rm l\,l}\langle^+E| + \int_0^{\infty} dE\,E|E^+\rangle_{\rm r\,r}\langle^+E|\,, \tag{65}$$

all hold within the rigged Hilbert space as a "sandwich" with elements of Φ. Other expressions such as the delta normalization of eigenfunctions, e.g.,

$$\frac{1}{2\pi\hbar}\int_{-\infty}^{\infty} dx\,e^{i(p-p')x/\hbar} = \delta(p-p')\,, \tag{66}$$

or the "matrix elements" of the observables, e.g.,

$$\langle x|Q|x'\rangle = x'\,\delta(x-x')\,, \tag{67}$$

$$\langle x|P|x'\rangle = -i\hbar\frac{d}{dx}\,\delta(x-x')\,, \tag{68}$$

$$\langle x|H|x'\rangle = \left(-\frac{\hbar^2}{2m}\frac{d^2}{dx^2}+V(x)\right)\delta(x-x')\,. \tag{69}$$

are interpreted the same way.

To conclude this section, I would like to refer the reader to [5, 6] for a detailed account of the above results.

THE RIGGED HILBERT SPACE OF THE LIPPMANN-SCHWINGER EQUATION

The Lippmann-Schwinger equation is one of the cornerstones of scattering theory. It is written as

$$|E^{\pm}\rangle = |E\rangle + \frac{1}{E-H_0\pm i\varepsilon}V|E^{\pm}\rangle\,, \tag{70}$$

where $|E^{\pm}\rangle$ are the "in" and "out" Lippmann-Schwinger kets, $|E\rangle$ is an eigenket of the free Hamiltonian H_0,

$$H_0|E\rangle = E|E\rangle\,, \tag{71}$$

and V is the potential. The Lippmann-Schwinger kets are, in particular, eigenvectors of H:

$$H|E^{\pm}\rangle = E|E^{\pm}\rangle\,. \tag{72}$$

To the kets $|E^{\pm}\rangle$, there correspond the bras $\langle^{\pm}E|$, which satisfy

$$\langle^{\pm}E| = \langle E| + \langle^{\pm}E|V\frac{1}{E-H_0\mp i\varepsilon}\,. \tag{73}$$

The bras $\langle^{\pm}E|$ are left eigenvectors of H,

$$\langle^{\pm}E|H = E\langle^{\pm}E|\,, \tag{74}$$

and the bras $\langle E|$ are left eigenvectors of H_0,

$$\langle E|H_0 = E\langle E|\,. \tag{75}$$

The Lippmann-Schwinger equation (70) for the "in" $|E^+\rangle$ and "out" $|E^-\rangle$ kets has the scattering "in" and "out" boundary conditions built into the $\pm i\varepsilon$, since Eq. (70) is equivalent to the time-independent Schrödinger equation (72) subject to those "in" $(+i\varepsilon)$ and "out" $(-i\varepsilon)$ boundary conditions. In the position representation, the $\pm i\varepsilon$ prescriptions yield the following asymptotic behaviors:

$$\langle \mathbf{x}|E^+\rangle \xrightarrow[r\to\infty]{} e^{ikz} + f(k,\theta)\frac{e^{ikr}}{r}\,, \tag{76}$$

$$\langle \mathbf{x}|E^-\rangle \xrightarrow[r\to\infty]{} e^{ikz} + \overline{f(k,\theta)}\,\frac{e^{-ikr}}{r}\,, \tag{77}$$

where $\mathbf{x} \equiv (x,y,z) \equiv (r,\theta,\phi)$ are the position coordinates, k is the wave number of Eq. (26) and $f(k,\theta)$ is the scattering amplitude.

Likewise any bra and ket, the Lippmann-Schwinger bras and kets do not have a place in the Hilbert space. In this section, we will construct the rigged Hilbert space to which they belong. We will use the example of the spherical shell potential.

The radial Lippmann-Schwinger equation

For the spherical shell potential,

$$V(\mathbf{x}) \equiv V(r) = \begin{cases} 0 & 0<r<a \\ V_0 & a<r<b \\ 0 & b<r<\infty\,, \end{cases} \tag{78}$$

the Lippmann-Schwinger equation can be solved explicitly. Because the potential (78) is spherically symmetric, we will work in the radial position representation and restrict ourselves to angular momentum $l=0$.

In the radial representation, the Lippmann-Schwinger equation (70) becomes

$$\langle r|E^\pm\rangle = \langle r|E\rangle + \langle r|\frac{1}{E-H_0\pm i\varepsilon}V|E^\pm\rangle\,. \tag{79}$$

The procedure to solve Eq. (79) is well known. Since Eq. (79) is an integral equation, it is equivalent to a differential equation subject to the boundary conditions that are built into it. In our case, for $l=0$, Eq. (79) is equivalent to the Schrödinger differential equation,

$$\left(-\frac{\hbar^2}{2m}\frac{d^2}{dr^2}+V(r)\right)\langle r|E^\pm\rangle = E\langle r|E^\pm\rangle\,, \tag{80}$$

subject to the following boundary conditions:

$$\langle r|E^\pm\rangle = 0\,, \tag{81}$$

$$\langle r|E^\pm\rangle \text{ is continuous at } r=a,b\,, \tag{82}$$

$$\frac{d}{dr}\langle r|E^\pm\rangle \text{ is continuous at } r=a,b\,, \tag{83}$$

$$\langle r|E^+\rangle \sim e^{-ikr} - S(E)\,e^{ikr} \quad \text{as } r\to\infty\,, \tag{84}$$

$$\langle r|E^-\rangle \sim e^{ikr} - \overline{S(E)}\,e^{-ikr} \quad \text{as } r\to\infty\,, \tag{85}$$

where $S(E)$ is the S matrix in the energy representation. The boundary conditions (84) and (85) originate from the $\pm i\varepsilon$ conditions of Eq. (79). The asymptotic behaviors (84) and (85) are the $l=0$, radial counterparts of the asymptotic behaviors (76) and (77).

If we insert (78) into (80), and solve (80) subject to (81)-(85), we obtain

$$\langle r|E^\pm\rangle \equiv \chi^\pm(r;E) = N(E)\,\frac{\chi(r;E)}{\mathscr{J}_\pm(E)}\,, \qquad E\in[0,\infty)\,, \tag{86}$$

where $N(E)$ is a delta-normalization factor,

$$N(E) = \sqrt{\frac{1}{\pi}\frac{2m/\hbar^2}{\sqrt{2m/\hbar^2 E}}}, \tag{87}$$

$\chi(r;E)$ is the so-called regular solution of Eq. (80),

$$\chi(r;E) = \begin{cases} \sin(\sqrt{\frac{2m}{\hbar^2}E}\, r) & 0<r<a \\ \mathscr{J}_1(E)e^{i\sqrt{\frac{2m}{\hbar^2}(E-V_0)}\, r} + \mathscr{J}_2(E)e^{-i\sqrt{\frac{2m}{\hbar^2}(E-V_0)}\, r} & a<r<b \\ \mathscr{J}_3(E)e^{i\sqrt{\frac{2m}{\hbar^2}E}\, r} + \mathscr{J}_4(E)e^{-i\sqrt{\frac{2m}{\hbar^2}E}\, r} & b<r<\infty, \end{cases} \tag{88}$$

and $\mathscr{J}_\pm(E)$ are the Jost functions,

$$\mathscr{J}_+(E) = -2i\,\mathscr{J}_4(E), \tag{89}$$

$$\mathscr{J}_-(E) = 2i\,\mathscr{J}_3(E). \tag{90}$$

The explicit expressions for $\mathscr{J}_1$-$\mathscr{J}_4$ can be obtained by matching the values of $\chi(r;E)$ and of its derivative at the discontinuities of the potential. In terms of the Jost functions, the S matrix is given by

$$S(E) = \frac{\mathscr{J}_-(E)}{\mathscr{J}_+(E)}, \qquad E \in [0,\infty). \tag{91}$$

Similarly to the "right" ones, we can obtain the "left" Lippmann-Schwinger eigenfunctions by writing Eq. (73) in the radial position representation,

$$\langle {}^\pm E|r\rangle = \langle E|r\rangle + \langle {}^\pm E|V\frac{1}{E-H_0 \mp i\varepsilon}|r\rangle. \tag{92}$$

Solving Eq. (92) is analogous to solving Eq. (79). The solutions to (92) are the complex conjugates of the solutions to (79):

$$\langle {}^\pm E|r\rangle = \overline{\chi^\pm(r;E)} = \chi^\mp(r;E). \tag{93}$$

The rigged Hilbert space of the Lippmann-Schwinger bras and kets

By using (34), we associate "in" and "out" kets $|E^\pm\rangle$ with the eigenfunctions $\chi^\pm(r;E)$ for each $E \in [0,\infty)$:

$$\langle \varphi^\pm|E^\pm\rangle \equiv \int_0^\infty dr\, \overline{\varphi^\pm(r)}\chi^\pm(r;E) \equiv \int_0^\infty dr\, \langle \varphi^\pm|r\rangle\langle r|E^\pm\rangle, \tag{94}$$

where $\varphi^\pm$ belong to the space Φ that will be constructed below. (Note that the Φ of this section is different from the Φ of Section 2.) Similarly, by using (35), we define the bras $\langle {}^\pm E|$ as

$$\langle {}^\pm E|\varphi^\pm\rangle \equiv \int_0^\infty dr\, \overline{\chi^\pm(r;E)}\varphi^\pm(r) \equiv \int_0^\infty dr\, \langle {}^\pm E|r\rangle\langle r|\varphi^\pm\rangle. \tag{95}$$

Note that even though $\chi^{\pm}(r;E) \equiv \langle r|E^{\pm}\rangle$ are also meaningful for complex energies, the energy in Eqs. (94) and (95) runs only over $\mathrm{Sp}(H) = [0,\infty)$, because in this section we restrict ourselves to bras and kets associated with energies that belong to the scattering spectrum of the Hamiltonian.

From definitions (94) and (95), it follows that the action of $\langle {}^{\pm}E|$ is complex conjugated to the action of $|E^{\pm}\rangle$:

$$\langle {}^{\pm}E|\varphi^{\pm}\rangle = \overline{\langle \varphi^{\pm}|E^{\pm}\rangle} \, . \tag{96}$$

We now need to find the subspace Φ on which the above definitions make sense. Besides making (94)-(95) well defined, the space Φ must also be invariant under the action of the observables of the system. Since in this section the only observable we are concerned with is the Hamiltonian, we will simply require invariance under H. Thus, the space Φ must satisfy the following conditions:

- The space Φ is invariant under the action of H. (97)
- The elements of Φ are such that the integrals in Eqs. (94)-(95) make sense. (98)

In order to meet requirement (97), the wave functions $\varphi^{\pm}(r)$ must at least be in the maximal invariant subspace of H:

$$\mathscr{D} = \bigcap_{n=0}^{\infty} \mathscr{D}(H^n) \, . \tag{99}$$

In order to meet requirement (98), the wave functions $\varphi^{\pm}(r)$ must behave well enough so the integrals in Eqs. (94)-(95) are well defined. From the expression for $\chi^{\pm}(r;E)$, Eq. (86), one can see that the $\varphi^{\pm}(r)$ have essentially to control purely imaginary exponentials. Therefore, the space Φ that meets the requirements (97)-(98) is

$$\Phi = \left\{ \varphi^{\pm} \in L^2([0,\infty),dr) \,|\, \varphi^{\pm} \in \mathscr{D}, \, \|\varphi^{\pm}\|_{n,m} < \infty, \; n,m = 0,1,2,\ldots \right\}, \tag{100}$$

where the $\| \; \|_{n,m}$ are given by

$$\|\varphi^{\pm}\|_{n,m} = \sqrt{\int_0^{\infty} dr \, |(1+r)^n (1+H)^m \varphi^{\pm}(r)|^2} \, , \quad n,m = 0,1,2,\ldots \, . \tag{101}$$

The space Φ is thus the collection of square integrable functions that belong to the maximal invariant subspace of H and for which the estimates (101) are finite. In particular, because $\varphi^{\pm}(r)$ satisfy the estimates (101), $\varphi^{\pm}(r)$ fall off at infinity faster than any polynomial of r:

$$\lim_{r\to\infty} (1+r)^n \varphi^{\pm}(r) = 0 \, , \quad n = 0,1,2,\ldots \, . \tag{102}$$

Thus, likewise in Section 2, we have obtained a Schwartz-like space. Obviously, the space Φ can also be seen as the maximal invariant subspace of the algebra generated by the Hamiltonian and the operator multiplication by r.

Note that we have used superscripts $\pm$ to denote the elements of one and the same space Φ. The reason is that when we use $+$, it means that the elements of Φ are acted upon by $|E^{+}\rangle$ or $\langle {}^{+}E|$, and when we use $-$, they are acted upon by $|E^{-}\rangle$ or $\langle {}^{-}E|$.

We can now construct the spaces Φ' and $\Phi^{\times}$, and see that $|E^{\pm}\rangle$ belong to $\Phi^{\times}$, whereas $\langle^{\pm}E|$ belong to Φ'. Thus, the Lippmann-Schwinger equations (70) and (73), Eqs. (72) and (74), as well as

$$e^{-iHt/\hbar}|E^{\pm}\rangle = e^{-iEt/\hbar}|E^{\pm}\rangle\,, \tag{103}$$

$$\langle^{\pm}E|e^{-iHt/\hbar} = e^{iEt/\hbar}\langle^{\pm}E|\,, \tag{104}$$

hold in the distributional sense, i.e., as "sandwiches" with elements of Φ.

For a detailed account on the results of this section, the reader may wish to refer to [7].

THE RIGGED HILBERT SPACE OF THE ANALYTIC CONTINUATION OF THE THE LIPPMANN-SCHWINGER EQUATION

This section is devoted to construct and characterize the analytic continuation of the Lippmann-Schwinger bras and kets. As in Section 3, we restrict ourselves to the spherical shell potential (78) and zero angular momentum.

The ultimate goal we want to achieve by analytically continuing the solutions of the Lippmann-Schwinger equation is to obtain, in Section 5, the resonance (decay) amplitude.

The wave number representation

The Lippmann-Schwinger eigenfunctions depend explicitly on the wave number k of Eq. (26) rather than on the energy E. It is therefore convenient to rewrite their expressions in terms of k before performing analytic continuations.

We start by writing the regular solution (88) in terms of k:

$$\chi(r;k) = \chi(r;E) = \left\{ \begin{array}{ll} \sin(kr) & 0<r<a \\ \mathscr{J}_1(k)e^{i\kappa r} + \mathscr{J}_2(k)e^{-i\kappa r} & a<r<b \\ \mathscr{J}_3(k)e^{ikr} + \mathscr{J}_4(k)e^{-ikr} & b<r<\infty\,, \end{array} \right. \tag{105}$$

where κ is given by (26). In terms of k, the Lippmann-Schwinger eigenfunctions read as

$$\chi^{\pm}(r;E) = \sqrt{\frac{1}{\pi}\frac{2m/\hbar^2}{k}}\,\frac{\chi(r;k)}{\mathscr{J}_{\pm}(k)}\,. \tag{106}$$

The eigenfunctions $\chi^{\pm}(r;E)$ are δ-normalized as functions of E. The Lippmann-Schwinger eigenfunctions that are δ-normalized as functions of k are given by

$$\chi^{\pm}(r;k) \equiv \sqrt{\frac{\hbar^2}{2m}2k}\,\chi^{\pm}(r;E) = \sqrt{\frac{2}{\pi}}\,\frac{\chi(r;k)}{\mathscr{J}_{\pm}(k)}\,. \tag{107}$$

In bra-ket notation, we will write

$$\langle r|k^{\pm}\rangle = \chi^{\pm}(r;k)\,, \quad k>0\,, \tag{108}$$

$$\langle^{\pm}k|r\rangle = \overline{\chi^{\pm}(r;k)} = \chi^{\mp}(r;k)\,, \quad k>0\,. \tag{109}$$

Because of (107), in terms of k the Lippmann-Schwinger bras and kets read as

$$\langle^{\pm}k| = \sqrt{\frac{\hbar^2}{2m}2k}\,\langle^{\pm}E|\,, \quad k>0\,, \tag{110}$$

$$|k^{\pm}\rangle = \sqrt{\frac{\hbar^2}{2m}2k}\,|E^{\pm}\rangle\,, \quad k>0\,. \tag{111}$$

The bras $\langle^{\pm}k|$ and kets $|k^{\pm}\rangle$ are, respectively, left and right eigenvectors of H with eigenvalue $\frac{\hbar^2}{2m}k^2$:

$$\langle^{\pm}k|H = \frac{\hbar^2}{2m}k^2\langle^{\pm}k|\,, \tag{112}$$

$$H|k^{\pm}\rangle = \frac{\hbar^2}{2m}k^2|k^{\pm}\rangle\,. \tag{113}$$

The analytic continuation of the Lippmann-Schwinger eigenfunctions

The analytic continuation of $\chi^{\pm}(r;k)$ is done as follows. First, one specifies the boundary values that the Lippmann-Schwinger eigenfunctions take on the positive k-axis. And second, one continues those boundary values into the whole k-plane. Since the boundary values of the Lippmann-Schwinger eigenfunctions on the positive k-axis are given by Eq. (107), and since $\chi^{\pm}(r;k)$ are expressed in terms of well-known analytic functions, the continuation of $\chi^{\pm}(r;k)$ from the positive k-axis into the whole wave-number plane is well defined.

A word on notation. Whenever they become complex, we will denote the energy E and the wave number k by respectively z and q. Accordingly, the continuations of $\chi^{\pm}(r;E)$ and $\chi^{\pm}(r;k)$ will be denoted by $\chi^{\pm}(r;z)$ and $\chi^{\pm}(r;q)$. In bra-ket notation, the analytically continued eigenfunctions will be written as

$$\langle r|q^{\pm}\rangle = \chi^{\pm}(r;q)\,, \tag{114}$$

$$\langle^{\pm}q|r\rangle = \chi^{\mp}(r;q)\,. \tag{115}$$

Note that, in distinction to (93), the analytically continued "left" eigenfunction is not the complex conjugate of the "right" eigenfunction but

$$\langle^{\pm}q|r\rangle = \overline{\langle r|\overline{q}^{\pm}\rangle}\,. \tag{116}$$

In order to characterize the analytic properties of $\chi^{\pm}(r;q)$, it is useful to define

$$Z_{\pm} \equiv \{q \in \mathbb{C} \mid \mathscr{J}_{\pm}(q) = 0\}\,. \tag{117}$$

The elements of Z_{+} are simply the resonance energies. Since $\chi(r;q)$ and $\mathscr{J}_{\pm}(q)$ are analytic in the whole k-plane, $\chi^{\pm}(r;q)$ is analytic in the whole k-plane except at $Z_{\pm}$, where its poles are located.

In order to define the analytically continued Lippmann-Schwinger bras and kets, we need to know how $\chi^{\pm}(r;q)$ grow with q. To find out, let us recall first that the growth of $\chi(r;q)$ is bounded by [8]

$$|\chi(r;q)| \leq C \frac{|q|\, r}{1+|q|\, r} e^{|\mathrm{Im}(q)|r} , \quad q \in \mathbb{C} . \tag{118}$$

>From Eqs. (107) and (118), it follows that the eigenfunctions $\chi^{\pm}(r;q)$ satisfy

$$\left|\chi^{\pm}(r;q)\right| \leq \frac{C}{|\mathscr{J}_{\pm}(q)|} \frac{|q|r}{1+|q|r} e^{|\mathrm{Im}(q)|r} . \tag{119}$$

When $q \in Z_{\pm}$, $\chi^{\pm}(r;q)$ blows up to infinity.

We can further refine the estimates (119) by characterizing the growth of $1/|\mathscr{J}_{\pm}(q)|$ in different regions of the complex plane. In the upper half of the complex k-plane, the inverse of $\mathscr{J}_{+}(q)$ is bounded:

$$\frac{1}{|\mathscr{J}_{+}(q)|} \leq C , \quad \mathrm{Im}(q) \geq 0 . \tag{120}$$

In the lower half-plane, $\frac{1}{\mathscr{J}_{+}(q)}$ is infinite whenever $q \in Z_{+}$. As $|q|$ tends to ∞ in the lower half plane, we have

$$\frac{1}{\mathscr{J}_{+}(q)} \approx \frac{1}{1-Cq^{-2}e^{2iqb}} , \quad (|q| \to \infty ,\ \mathrm{Im}(q) < 0) . \tag{121}$$

The above estimates are satisfied by $\mathscr{J}_{-}(q)$ when we exchange the upper for the lower half plane, and Z_{+} for Z_{-}.

The analytic continuation of the Lippmann-Schwinger bras and kets

The analytic continuation of the Lippmann-Schwinger bras is defined for any complex wave number q in the distributional way (35):

$$\langle^{\pm}q|\varphi^{\pm}\rangle \equiv \int_0^{\infty} dr\, \varphi^{\pm}(r)\chi^{\mp}(r;q) = \int_0^{\infty} dr\, \langle^{\pm}q|r\rangle\langle r|\varphi^{\pm}\rangle , \tag{122}$$

where the functions $\varphi^{\pm}(r)$ belong to a space of test functions Φ_{exp} that will be constructed below. Similarly to the bras, the analytic continuation of the Lippmann-Schwinger kets is defined by way of (34):

$$\langle\varphi^{\pm}|q^{\pm}\rangle \equiv \int_0^{\infty} dr\, \overline{\varphi^{\pm}(r)}\chi^{\pm}(r;q) = \int_0^{\infty} dr\, \langle\varphi^{\pm}|r\rangle\langle r|q^{\pm}\rangle . \tag{123}$$

Note that definition (122) is actually a slight generalization of (35), due to (116).

The bras (122) and kets (123) are defined for all complex q except at those q at which the corresponding eigenfunction has a pole. At those poles, one can still define bras and kets if in definitions (122) and (123) one substitutes the eigenfunctions $\chi^{\pm}(r;q)$ by their residues at the pole.

From the analytic continuation of the bras and kets into any complex wave number, one can now obtain the analytic continuation of the bras and kets into any complex energy of the Riemann surface (compare with Eqs. (110) and (111)):

$$|z^{\pm}\rangle = \sqrt{\frac{2m}{\hbar^2}\frac{1}{2q}}\,|q^{\pm}\rangle\,, \quad \langle^{\pm}z| = \sqrt{\frac{2m}{\hbar^2}\frac{1}{2q}}\,\langle^{\pm}q|\,. \tag{124}$$

Construction of the rigged Hilbert space

Likewise the bras and kets associated with real energies, the analytic continuation of the Lippmann-Schwinger bras and kets must be described within the rigged Hilbert space rather than just within the Hilbert space. We will denote the rigged Hilbert space for the analytically continued bras by

$$\Phi_{\rm exp} \subset L^2([0,\infty),dr) \subset \Phi_{\rm exp}'\,, \tag{125}$$

and the one for the analytically continued kets by

$$\Phi_{\rm exp} \subset L^2([0,\infty),dr) \subset \Phi_{\rm exp}^{\times}\,. \tag{126}$$

The functions $\varphi^{\pm} \in \Phi_{\rm exp}$ must satisfy the following conditions:

- They belong to the maximal invariant subspace $\mathscr{D}$ of H, see Eq. (99). (127)
- They are such that definitions (122) and (123) make sense. (128)

The reason why $\varphi^{\pm}$ must satisfy condition (127) is that such condition guarantees that all the powers of the Hamiltonian are well defined. Condition (127), however, is not sufficient to obtain well-defined bras and kets associated with complex wave numbers. In order for $\langle^{\pm}q|$ and $|q^{\pm}\rangle$ to be well defined, the wave functions $\varphi^{\pm}(r)$ must be well behaved so the integrals in Eqs. (122) and (123) converge. Since by Eq. (119) $\chi^{\pm}(r;q)$ grow exponentially with r, the wave functions $\varphi^{\pm}(r)$ have to, essentially, tame real exponentials. If we define

$$\|\varphi^{\pm}\|_{n,n'} \equiv \sqrt{\int_0^{\infty} dr \left|\frac{nr}{1+nr}e^{nr^2/2}(1+H)^{n'}\varphi^{\pm}(r)\right|^2}\,, \quad n,n'=0,1,2,\ldots\,, \tag{129}$$

then the space $\Phi_{\rm exp}$ is given by

$$\Phi_{\rm exp} = \left\{\varphi^{\pm} \in \mathscr{D} \,|\, \|\varphi^{\pm}\|_{n,n'} < \infty,\; n,n'=0,1,2,\ldots\right\}. \tag{130}$$

This is just the space of square integrable functions which belong to the maximal invariant subspace of H and for which the quantities (129) are finite. In particular, because $\varphi^{\pm}(r)$ satisfy the estimates (129), their tails fall off faster than Gaussians.

From Eq. (119), it is clear that the integrals in Eqs. (122) and (123) converge already for functions that fall off at infinity faster than any exponential. We have imposed Gaussian falloff because it will allow us to perform resonance expansions in Section 5.

It is illuminating to compare the space Φ of Section 3 with the space $\Phi_{\rm exp}$ of Eq. (130). Because for real wave numbers the Lippmann-Schwinger eigenfunctions behave like purely imaginary exponentials, in Section 3 we only needed to impose on the test functions a polynomial falloff, thereby obtaining a space of test functions very similar to the Schwartz space. By contrast, for complex wave numbers the Lippmann-Schwinger eigenfunctions blow up exponentially, and therefore we need to impose on the test functions an exponential falloff that damps such an exponential blowup.

One can now easily show that the kets $|q^{\pm}\rangle$ belong to $\Phi_{\rm exp}^{\times}$ and satisfy

$$H|q^{\pm}\rangle = \frac{\hbar^2}{2m}q^2\,|q^{\pm}\rangle\,, \tag{131}$$

$$e^{-iHt/\hbar}|q^{\pm}\rangle = e^{-iq^2\hbar t/(2m)}|q^{\pm}\rangle\,. \tag{132}$$

Similarly, $\langle^{\pm}q| \in \Phi'_{\rm exp}$ and

$$\langle^{\pm}q|H = \frac{\hbar^2}{2m}q^2\langle^{\pm}q|\,, \tag{133}$$

$$\langle^{\pm}q|e^{-iHt/\hbar} = e^{iq^2\hbar t/(2m)}\langle^{\pm}q|\,. \tag{134}$$

Equations (131) and (133) can be rewritten in terms of the complex energy z as

$$H|z^{\pm}\rangle = z|z^{\pm}\rangle\,, \tag{135}$$

$$\langle^{\pm}z|H = z\langle^{\pm}z|\,. \tag{136}$$

Note that (136) is not given by $\langle^{\pm}z|H = \overline{z}\langle^{\pm}z|$, as one may naively expect from formally obtaining (136) by Hermitian conjugation of (135).

For a full account of this section, the reader can refer to [9].

RESONANCE STATES AND THEIR RIGGED HILBERT SPACE

The Gamow states are the state vectors of resonances. They are eigenvectors of the Hamiltonian with a complex eigenvalue. The real (imaginary) part of the complex eigenvalue is associated with the energy (width) of the resonance.

Because self-adjoint operators on a Hilbert space can only have real eigenvalues, the Gamow states fit not within the Hilbert space but within the rigged Hilbert space. In this section, we will see that the Gamow states belong to the rigged Hilbert space of Section 4.

Like in Sections 3 and 4, we will use the spherical shell potential (78) and study the zero angular momentum case only. Unlike in Sections 4, we will write most results in terms of the energy, because they tend to be simpler than in terms of the wave number. The energy and the wave number of a resonance R will be denoted by $z_{\rm R}$ and $k_{\rm R}$.

The Gamow eigenfunctions satisfy the Schrödinger equation

$$\left(-\frac{\hbar^2}{2m}\frac{d^2}{dr^2} + V(r) \right) u(r;z_{\rm R}) = z_{\rm R}\, u(r;z_{\rm R}) \,, \tag{137}$$

subject to "purely outgoing boundary conditions,"

$$u(0;z_{\rm R}) = 0 \,, \tag{138}$$

$$u(r;z_{\rm R}) \text{ is continuous at } r = a,b \,, \tag{139}$$

$$\frac{d}{dr}u(r;z_{\rm R}) \text{ is continuous at } r = a,b \,, \tag{140}$$

$$u(r;z_{\rm R}) \sim e^{ik_{\rm R}r} \text{ as } r \to \infty \,, \tag{141}$$

where (141) is the "purely outgoing boundary condition" (POBC). Comparison of (138)-(141) with (81)-(85) shows that it is the POBC what selects the resonance energies.

For the potential (78), the only possible eigenvalues of (137) subject to (138)-(141) are the zeros of the Jost function,

$$\mathscr{J}_{+}(z_{\rm R}) = 0 \,. \tag{142}$$

The solutions of this equation come as a denumerable number of complex conjugate pairs z_n, z_n^*. The number $z_n = E_n - i\Gamma_n/2$ is the nth resonance energy, and $z_n^* = E_n + i\Gamma_n/2$ is the nth anti-resonance energy. The corresponding wave numbers are

$$k_n = \sqrt{\frac{2m}{\hbar^2}z_n} \,, \quad -k_n^* = \sqrt{\frac{2m}{\hbar^2}z_n^*} \,, \quad n = 1,2,\ldots \,. \tag{143}$$

For the potential (78), the resonance energies are simple poles of the S matrix (see [10] for an example of a potential that produces double poles). In order to write expressions for resonances and anti-resonances together, we will label the resonances by a positive integer $n = 1,2,\ldots$ and the anti-resonances by a negative integer $n = -1,-2,\ldots$.

The nth Gamow eigensolution, $n = \pm 1, \pm 2, \ldots$, reads

$$u(r;z_n) = u(r;k_n) = N_n \begin{cases} \frac{1}{\mathscr{J}_3(k_n)} \sin(k_n r) & 0 < r < a \\ \frac{\mathscr{J}_1(k_n)}{\mathscr{J}_3(k_n)} e^{iQ_n r} + \frac{\mathscr{J}_2(k_n)}{\mathscr{J}_3(k_n)} e^{-iQ_n r} & a < r < b \\ e^{ik_n r} & b < r < \infty \,, \end{cases} \tag{144}$$

where

$$Q_n = \sqrt{\frac{2m}{\hbar^2}(z_n - V_0)} \,, \tag{145}$$

and N_n is a normalization factor,

$$N_n^2 = i \,\text{res}\left[S(q)\right]_{q=k_n} \,. \tag{146}$$

The Gamow eigenfunctions (144) are related with $\chi^{\pm}(r;q)$ by

$$u(r;k_n) = -\frac{\sqrt{2\pi}}{N_n} \text{res}\left[\chi^{+}(r;q)\right]_{q=k_n} \,, \tag{147}$$

$$u(r;k_n) = i\sqrt{2\pi}N_n\chi^-(r;k_n)\,. \tag{148}$$

Because of (148) and (116), the "left" Gamow eigenfunctions read

$$\langle z_n|r\rangle = [u(r;z_n^*)]^*\,, \quad n = \pm 1, \pm 2, \ldots\,. \tag{149}$$

Thus, the "left" Gamow eigenfunction is not just the complex conjugate of the "right" eigenfunction, but the complex conjugated eigenfunction evaluated at the complex conjugated energy. Because the Gamow eigenfunctions satisfy

$$[u(r;z_n^*)]^* = u(r;z_n)\,, \quad n = \pm 1, \pm 2, \ldots\,, \tag{150}$$

the "left" and the "right" Gamow eigenfunctions are actually the same eigenfunction,

$$\langle z_n|r\rangle = [u(r;z_n^*)]^* = u(r;z_n) = \langle r|z_n\rangle\,, \quad n = \pm 1, \pm 2, \ldots\,. \tag{151}$$

In terms of the wave number, Eq. (151) reads as

$$\langle k_n|r\rangle = [u(r;-k_n^*)]^* = u(r;k_n) = \langle r|k_n\rangle\,, \quad n = \pm 1, \pm 2, \ldots\,. \tag{152}$$

The Gamow eigenfunctions $u(r;z_n)$ are not square integrable and therefore must be treated as distributions. By treating them as distributions, we will be able to generate the Gamow bras and kets. According to (34), the Gamow ket $|z_n\rangle$ associated with the eigenfunction $u(r;z_n)$ must be defined as

$$\langle\varphi|z_n\rangle \equiv \int_0^\infty dr\,[\varphi(r)]^* u(r;z_n) = \int_0^\infty dr\,\langle\varphi|r\rangle\langle r|z_n\rangle\,, \quad n = \pm 1, \pm 2, \ldots\,. \tag{153}$$

Similarly, the Gamow bra associated with the resonance (or anti-resonance) energy z_n is defined as

$$\langle z_n|\varphi\rangle \equiv \int_0^\infty dr\,\varphi(r)u(r;z_n) = \int_0^\infty dr\,\langle z_n|r\rangle\langle r|\varphi\rangle\,, \quad n = \pm 1, \pm 2, \ldots\,. \tag{154}$$

From these definitions and from Eqs. (147) and (148), it is clear that the Gamow bras and kets are accommodated by the rigged Hilbert spaces (125) and (126).

Within the rigged Hilbert spaces (125) and (126), it holds that the Gamow bras and kets are eigenvectors of the Hamiltonian:

$$H|z_n\rangle = z_n|z_n\rangle\,, \quad n = \pm 1, \pm 2, \ldots\,, \tag{155}$$

$$\langle z_n|H = z_n\langle z_n|\,, \quad n = \pm 1, \pm 2, \ldots\,. \tag{156}$$

Because their energy is complex, the time evolution of the Gamow bras and kets should be time asymmetric. For resonances, it should be that

$$\langle z_n|e^{-iHt/\hbar} = e^{iz_nt/\hbar}\langle z_n|\,, \quad \text{only for } t < 0\,,\ n = 1, 2, \ldots\,, \tag{157}$$

$$e^{-iHt/\hbar}|z_n\rangle = e^{-iz_nt/\hbar}|z_n\rangle\,, \quad \text{only for } t > 0\,,\ n = 1, 2, \ldots\,, \tag{158}$$

whereas for anti-resonances, it should be that

$$\langle z_n|e^{-iHt/\hbar} = e^{iz_nt/\hbar}\langle z_n| \,, \quad \text{only for } t>0\,,\ n=-1,-2,\ldots\,, \tag{159}$$

$$e^{-iHt/\hbar}|z_n\rangle = e^{-iz_nt/\hbar}|z_n\rangle \,, \quad \text{only for } t<0\,,\ n=-1,-2,\ldots. \tag{160}$$

If we define the complex delta function at z_n by

$$\int_0^\infty dE\, f(E)\delta(E-z_n) = f(z_n)\,, \tag{161}$$

one can use the results of Section 4 to show that the Gamow eigenfunction $u(r;z_n)$, the complex delta function (multiplied by a normalization factor) and the Breit-Wigner amplitude (multiplied by a normalization factor) are linked with each other:

$$\begin{array}{ccccc} u(r;z_n) & \leftrightarrow & i\sqrt{2\pi}\mathcal{N}_n\delta(E-z_n)\,,\ E\in[0,\infty) & \leftrightarrow & -\frac{\mathcal{N}_n}{\sqrt{2\pi}}\frac{1}{E-z_n}\,,\ E\in(-\infty,\infty) \\ \text{posit. repr.} & & \text{energy repr.} & & (-\infty,\infty)\text{-“energy” repr.} \end{array} \tag{162}$$

where

$$\mathcal{N}_n^2 = i\,\text{res}[S(z)]_{z=z_n}\,. \tag{163}$$

Physically, these links mean that the Gamow states yield a decay amplitude $\mathcal{A}(z_R \to E)$ given by the complex delta function, and that such decay amplitude can be approximated by the Breit-Wigner amplitude when we can ignore the lower bound of the energy, i.e., when the resonance is so far from the threshold that we can safely assume that the energy runs over the full real line:

$$\mathcal{A}(z_n \to E) = \langle {}^-E|z_n\rangle = i\sqrt{2\pi}\mathcal{N}_n\delta(E-z_n) \simeq -\frac{\mathcal{N}_n}{\sqrt{2\pi}}\frac{1}{E-z_n}\,. \tag{164}$$

Thus, the almost-Lorentzian peaks in cross sections are caused by intermediate, unstable particles. However, because there is actually a lower bound for the energy, the decay amplitude is never exactly given by the Breit-Wigner amplitude. This means, in particular, that the standard Gamow states are different from the so-called “Gamow vectors” of [2].

Resonance expansions

The scattering bras and kets are basis vectors that furnish the completeness relation (4). The Gamow states are also basis vectors. The completeness relation (5) generated by the Gamow states is called the resonance expansion.

Resonance expansions are almost always obtained in the same way. One starts from the expansion in terms of bound and scattering states and then, by deforming the continuum integral into the complex plane, and by Cauchy's theorem, one extracts the contributions from the resonances that are hidden in the continuum and write them in the same way as the contributions from the bound states.

For the sake of simplicity, we will focus on the resonance expansion of the transition amplitude from an "in" state φ^+ into an "out" state φ^-:

$$(\varphi^-, \varphi^+) = \int_0^{\infty} dE \, \langle \varphi^- | E^- \rangle S(E) \langle ^+E | \varphi^+ \rangle \, . \tag{165}$$

We now extract the resonance contributions out of (165) by deforming the contour of integration into the lower half plane of the second sheet of the Riemann surface, where the resonance poles are located, and by applying Cauchy's theorem. Assuming that the integrand $\langle \varphi^- | E^- \rangle S(E) \langle ^+E | \varphi^+ \rangle$ tends to zero in the infinite arc of the lower half plane of the second sheet, the resulting resonance expansion is

$$(\varphi^-, \varphi^+) = \sum_{n=1}^{\infty} \langle \varphi^- | z_n \rangle \langle z_n | \varphi^+ \rangle + \int_0^{-\infty} dE \, \langle \varphi^- | E^- \rangle S(E) \langle ^+E | \varphi^+ \rangle \, . \tag{166}$$

The integral in Eq. (166) is supposed to be done infinitesimally below the negative real semiaxis of the second sheet. By omitting the wave functions in (166), we obtain the completeness relation (5). In Eq. (166), the infinite sum contains the contribution from the resonances, while the integral is the non-resonant background. Note that in (166) bound states do not appear, since the potential (78) doesn't bind any.

In obtaining Eq. (166), we have assumed that the integrand $\langle \varphi^- | E^- \rangle S(E) \langle ^+E | \varphi^+ \rangle$ tends to zero in the infinite arc of the lower half plane of the second sheet. Since (166) makes physical sense, we are tempted to conclude that such must be the case. However, as showed in [9], $\langle \varphi^- | E^- \rangle S(E) \langle ^+E | \varphi^+ \rangle$ does not tend to zero in the infinite arc of the lower half plane of the second sheet. On the contrary, it diverges exponentially there. Therefore, Eq. (166) doesn't make sense as it stands. In order to make sense of it, one has to control the exponential blowup of $\langle \varphi^- | E^- \rangle S(E) \langle ^+E | \varphi^+ \rangle$. The way to do so is by introducing an exponentially damping regulator $e^{-i\alpha z}$, $\alpha > 0$. Thus, Eq. (166) should read

$$(\varphi^-, \varphi^+) = \lim_{\alpha \to 0} \sum_{n=1}^{\infty} e^{-i\alpha z_n} \langle \varphi^- | z_n \rangle \langle z_n | \varphi^+ \rangle + \int_0^{-\infty} dE \, e^{-i\alpha E} \langle \varphi^- | E^- \rangle S(E) \langle ^+E | \varphi^+ \rangle . \tag{167}$$

Physically, the regulator $e^{-i\alpha z}$ is simply the analytic continuation of the time evolution operator in the energy representation, $e^{-iz\alpha} \equiv e^{-izt/\hbar}$. Thus, the above regularized equation must be understood in a time-asymmetric, time-dependent fashion as

$$(\varphi^-, e^{-iH\frac{t}{\hbar}} \varphi^+) = \sum_{n=1}^{\infty} e^{-iz_n \frac{t}{\hbar}} \langle \varphi^- | z_n \rangle \langle z_n | \varphi^+ \rangle + \int_0^{-\infty} dE \, e^{-iE\frac{t}{\hbar}} \langle \varphi^- | E^- \rangle S(E) \langle ^+E | \varphi^+ \rangle \tag{168}$$

for $t > 0$ only. Equation (166) should then be seen as the (singular) limit of Eq. (168) when $t \to 0^+$. That for resonances $t \equiv \alpha\hbar$ must be positive is in accord with the time asymmetry of (158).

Expansions (167) and (168) are the reason why we chose a Gaussian falloff for the elements of $\Phi_{\rm exp}$: When the wave functions have a Gaussian falloff in the position representation, we can regularize their blowup in the energy representation and interpret the regulator as a time-asymmetric evolution.

Resonance expansions allow us to understand the deviations from exponential decay. When a particular resonance, say resonance 1, is dominant, then the Gamow state of resonance 1 will carry the exponential decay, whereas the background, which includes in this case also the contribution from other possible resonances, carries the deviations from exponential decay.

The full account of this section will appear in a forthcoming paper.

CONCLUSIONS

We have seen why the rigged Hilbert space, rather than the Hilbert space alone, is needed to formulate quantum mechanics when the observables have continuous and/or resonance spectra. The rigged Hilbert space captures the physics of continuous and resonance spectra better than the Hilbert space, because in the rigged Hilbert space physical quantities such as commutation relations, uncertainty principles and resonances have always a precise meaning.

In addition to provide the mathematical support for Dirac's bra-ket formalism, for the Lippmann-Schwinger equation and for the Gamow states, the rigged Hilbert space can be used to obtain the resonance (decay) amplitude in terms of the complex delta function. Such decay amplitude can be approximated by the Breit-Wigner amplitude when the lower bound of the energy can be ignored.

To finish, I would like to mention that there is still a long list of pending questions worth pursuing, such as the invariance properties of $\Phi_{\rm exp}$ under time evolution or a detailed proof of the asymmetry in the time evolution of the Gamow states.

ACKNOWLEDGMENTS

It is a pleasure to thank the organizers for their invitation to participate in this summer school and Oscar Rosas-Ortiz for his hospitality. This work was supported by MEC fellowship No. SD2004-0003.

REFERENCES

1. A. Peres, *Quantum Theory: Concepts and Methods*, Dordrecht, Kluwer Academic (1993).
2. A. Bohm, Int. J. Theo. Phys. **42**, 2317 (2003).
3. R. de la Madrid, J. Phys. A: Math. Gen. **39**, 9255 (2006); quant-ph/0606186.
4. R. de la Madrid, *Quantum Mechanics in Rigged Hilbert Space Language*, PhD Dissertation, Universidad de Valladolid, Spain (2001). Available at http://www.physics.ucsd.edu/~rafa.
5. R. de la Madrid, J. Phys. A: Math. Gen. **37**, 8129 (2004); quant-ph/0407195.
6. R. de la Madrid, Eur. J. Phys. **26**, 287 (2005); quant-ph/0502053.
7. R. de la Madrid, J. Phys. A: Math. Gen. **39**, 3949 (2006); quant-ph/0603176.
8. J. R. Taylor, *Scattering theory*, John Wiley & Sons, Inc., New York (1972).
9. R. de la Madrid, J. Phys. A: Math. Gen. **39**, 3981 (2006); quant-ph/0603177.
10. E. Hernández, A. Jáuregui, A. Mondragón, J. Phys. A: Math. Gen. **33**, 4507 (2000).

Time Continuity and the Positivity Problem of the Floquet Hamiltonian

Sara Cruz y Cruz[1] and Bogdan Mielnik

Departamento de Física, Cinvestav, AP 14-740 México DF 07000, Mexico.

Abstract. The Floquet scheme, if applied directly in the rotating frame, may shed false information about the radiative processes of rotating quantum systems. The requirement of some time continuity conditions on the Floquet Hamiltonian could help in the search of an energetically interpretable operator for those systems.

In quantum theory one can distinguish between two kind of systems: those described by constant Hamiltonians H_0 (stationary systems) and those described by time dependent Hamiltonians $H(t)$. The information encoded in a constant Hamiltonian is extremely narrow because it only predicts the behavior of a particle in fixed external conditions. Hence, if one looks for a complete picture of the dynamical possibilities of the system, rather than its particular behavior in a given surrounding, one has to pay more attention to the formulation of quantum mechanics with time dependent Hamiltonians.

The dynamics of quantum systems is defined then by the evolution operator $U(t,t_0)$ satisfying the equation ($\hbar = 1$)

$$\frac{d}{dt}U(t,t_0) = -iH(t)U(t,t_0), \qquad U(t_0,t_0) = 1. \tag{1}$$

In a fixed time interval $[t_0,t]$, $U(t,t_0)$ has exponential representations in terms of Hermitian operators $F(t,t_0)$ called the *effective Hamiltonians*:

$$U(t,t_0) = e^{-i\delta t F(t,t_0)}, \qquad \delta t = t - t_0. \tag{2}$$

This expression does not uniquely define $F(t,t_0)$. Indeed, there is an infinite number of effective Hamiltonians for a given system in a given time interval.

One of the fundamental elements in the study of quantum systems is the radiation spectrum. In the stationary case, the radiated frequencies are given by the eigenvalues of the Hamiltonian H_0; however, the problem is not simple at all in the time dependent case. Actually, the most known attempt to solve it, concerns the periodical systems $H(t+T) = H(t)$, for which the evolution operator in a complete period of time $U(t_0+T,t_0)$ (called the *Floquet operator*) and its corresponding effective Hamiltonian $F(t_0+T,t_0) = F(t_0)$ (the *Floquet Hamiltonian* or *quasienergy* operator) are crucial. In 1967, Ya. B.

[1] On leave of absence from: Acad. Ciencias Básicas, UPIITA-IPN, Av. IPN 2508 CP 07340 México DF, Mexico. Becaria EDI and COFAA-IPN.

CP885, *Advanced Summer School in Physics 2006, Frontiers in Contemporary Physics—EAV06,* edited by O. Miranda, M. Carbajal, L. M. Montaño, O. Rosas-Ortiz, and S. A. Tomás Velázquez

Zel'dovich formulated the hypothesis that the radiative processes of a periodical system are described by the Floquet Hamiltonian, *i.e.*, the radiation spectrum is determined by the quasienergy eigenvalues, modulo integer multiples of $\omega = 2\pi/T$ [1]. Although this *Floquet scheme* seems simple and natural, it should be cautiously applied. In his hypothesis, Zel'dovich implicitly assumes the existence of a quasienergy ground state. This is not necessarily the case; the Floquet theory admits Hamiltonians which are not bounded from below (with no energy interpretation). This fact is easily detectable for the rotating systems [2, 3, 4, 5, 6, 7, 8]

$$H(t) = e^{-i\omega\delta t\mathbf{L}\cdot\mathbf{n}} H(t_0) e^{i\omega\delta t\mathbf{L}\cdot\mathbf{n}}, \tag{3}$$

where $\mathbf{L}$ stands for the orbital angular momentum operator while $\mathbf{n}$ defines the rotation axis. As an example consider a two-dimensional harmonic oscillator $H_0 = \frac{\mathbf{p}^2}{2} + \frac{\alpha^2}{2}\mathbf{x}^2$ in presence of an electric field $\mathbf{E}(t)$ of strength ε, rotating around the z-axis ($m = e = 1$) [9]:

$$H(t) = H_0 + \varepsilon\,(x\cos\omega t + y\sin\omega t) = H_0 + \varepsilon e^{-i\omega\delta t L_z}(x\cos\omega t_0 + y\sin\omega t_0)e^{i\omega\delta t L_z}. \tag{4}$$

The evolution equation (1) is easily solved by a transformation to the rotating frame:

$$U(t,t_0) = e^{-i\omega\delta t L_z} e^{-i\delta t G(t_0)} = e^{-i\delta t F(t,t_0)} \tag{5}$$

where

$$G(t_0) = H_0 + \varepsilon(x\cos\omega t_0 + y\sin\omega t_0) - \omega L_z. \tag{6}$$

For each $\delta t = nT$, the rotational factor in the second term of (5) becomes 1, meaning that $G(t_0)$ is also one of the Floquet Hamiltonians. According to the Zel'dovich hypothesis, one would expect that the eigenvalues of $G(t_0)$ define the radiation spectrum of the system; however, the algebraic structure of this operator turns out an important obstacle. Indeed, $G(t_0)$ can be expressed as a superposition of two harmonic oscillators [9]

$$G(t_0) = \omega_+ A_+^\dagger A_+ - \omega_- A_-^\dagger A_- + \omega + \frac{1}{2}\frac{\varepsilon^2}{\omega^2 - \alpha^2}, \tag{7}$$

with $\omega_\pm = \omega \pm \alpha$, and the operators $A_\pm$ satisfying

$$[A_\pm, A_\mp] = \left[A_\pm^\dagger, A_\mp^\dagger\right] = 0 \qquad \left[A_\pm, A_\pm^\dagger\right] = \pm 1. \tag{8}$$

Whenever $\omega > \alpha$, both $\omega_\pm > 0$ and $G(t_0)$ is not bounded from below (one of the oscillators in (7) is inverted). Its eigenvalues and eigenvectors are, respectively

$$E_{n_+n_-} = \alpha + \frac{1}{2}\frac{\varepsilon^2}{\omega^2 - \alpha^2} + n_+\omega_+ - n_-\omega_-, \tag{9}$$

$$|n_+n_-\rangle = \frac{1}{\sqrt{n_+!n_-!}}\left(A_+^\dagger\right)^{n_+} A_-^{n_-}|0\rangle, \tag{10}$$

where $n_+, n_- \in \mathbb{Z}^+$, and $|0\rangle$ is the vacuum state with $n_+, n_- = 0$. If one could consider $G(t_0)$ as the energy operator, defining the spontaneous radiation, then the system should fall down eternally through the infinite number of spectral levels descending to minus infinity, emitting an infinite amount of energy. A suggestive interpretation could be that the rotating field is continuously supplying with new energy the unperturbed harmonic oscillator. This is however not the case: even when $\varepsilon = 0$, $G(t_0)$ still has the same algebraic structure and the spontaneous emission rates, evaluated according to the fundamental formulae of radiation theory [10], do not vanish.

Does it means that the system indeed "spontaneously radiates", even in absence of the rotating field? The result seems an illusion caused by the non inertial nature of the frame in which this description is carried out [9]. Notice that evaluating the radiation spectrum and the emission rates by means of $G(t_0)$ is equivalent to assume that a quantum system with the "Hamiltonian" $G(t_0)$ obeys an ordinary quantum theory no matter if applied in the rotating frame. However, the validity of quantum mechanics in non inertial frames is an open problem [11, 12], so it is not a surprise that to take it too seriously leads to incorrect results.

Now, among the infinite number of Floquet Hamiltonians, is there any one interpretable as an energy operator supplying reasonable information about the radiation processes of the system? The answer is `yes`, though the concepts split. The operator $G(t_0)$, the "Hamiltonian in the rotating frame", describes correctly the evolution of the system "in eyes of the rotating observer". Although it is one of the Floquet Hamiltonians, it contains an inverted ladder of levels and henceforth, it cannot depend continuously on the intensity ε of the perturbing (rotating) field. A natural candidate to apply the Zel'dovich hypothesis is another Floquet Hamiltonian $F(t_0)$ obtained by the continuous prolongation of H_0 to an effective Hamiltonian in the full period interval $[t_0, t_0 + T]$ (see [13]).

As obvious from (5), the Floquet Hamiltonian $G(t_0)$ was not continuously constructed. Although $G(t_0)$ is the Floquet Hamiltonian for $\delta t = T$, it is not the effective Hamiltonian for an arbitrary, intermediate δt. The continuous effective Hamiltonians can be constructed by making a transformation to the interaction frame:

$$U(t, t_0) = e^{-iH_0 \delta t} W_1(t, t_0). \tag{11}$$

The evolution equation for $W(t, t_0)$, in general, can be at least approximately solved by means of the continuous Baker-Campbell-Hausdorff formulae [14]; but for this particular case, this operator is exactly determined by using the time dependent position and momentum operators $q(t)$, $p(t)$ in the Heisenberg picture (Heisenberg trajectories). A straightforward calculation leads to ([15]):

$$\begin{aligned} F(t_0) &= \frac{1}{2}\left(p_x - \frac{\varepsilon\omega}{2}\sin\omega t_0\right)^2 + \frac{1}{2}\left(p_y + \frac{\varepsilon\omega}{2}\cos\omega t_0\right)^2 \\ &+ \frac{\alpha^2}{2}\left(x + \frac{\varepsilon}{2}\cos\omega t_0\right)^2 + \frac{\alpha^2}{2}\left(y + \frac{\varepsilon}{2}\sin\omega t_0\right)^2 - \frac{\varepsilon^2}{2}\frac{\omega^2 + \alpha^2}{\omega^2 - \alpha^2}. \end{aligned} \tag{12}$$

The operator (12) does not include inverted oscillators, infinite sequences of negative eigenvalues, or neverending spontaneous emissions sustained even by a nonexisting field. The effect of the rotating field is reduced to simple translations along the coordinate and momenta axes. In addition, $F(t_0)$ commutes with $G(t_0)$, so the former can generate real rotating states as well. Notice that, while $F(t_0)$ and $G(t_0)$ properly describe the transformation of the rotating system at the end of each full rotation interval, $F(t_0)$ is more convenient to describe the radiative phenomena. Of course, there are some additional problems the Floquet theory should face, as, *e.g.*, the time scale of radiative processes to be reported elsewhere [16, 17].

ACKNOWLEDGMENTS

SCyC is grateful to the organizers of the *Advanced Summer School in Physics 2006* for their kind hospitality. The valuable comments by L.C. Cortés-Cuautli are appreciated. The financial support of Cinvestav and CONACyT is acknowledged.

REFERENCES

1. Ya.B. Zel'dovich, "The quasi-energy of a quantum mechanical system subjected to a periodic action", *JETP* **24** (1967) 1006-1008
2. B. Mielnik and D.J. Fernández C., "Electron trapped in a rotating magnetic field", *J. Math. Phys.* **30** (1989) 537-550
3. D.J. Fernández C., "Semiclassical resonance in rotating magnetic fields", *Acta Phys. Polon. B* **21** (1990) 589-601
4. I. Białynicki-Birula, M. Kaliński, J.H. Eberly, "Lagrange equilibrium points in celestial mechanics and nonspreading wave packets for strongly driven Rydberg electrons", *Phys. Rev. Lett.* **73** (1994) 1777-1780
5. I. Białynicki-Birula, Z. Białynicka-Birula, "Nonspreading wave packets for Rydberg electrons in molecules with electric dipole moments", *Phys. Rev. Lett.* **77** (1996) 4298-4301
6. Z. Białynicka-Birula, I. Białynicki-Birula, "Radiative decay of Trojan wave packets", *Phys. Rev. A* **56** (1997) 3623-3625
7. D. Delande, J. Zakrzewski, "Spontaneous emission for nondispersive wave packets", *Phys. Rev. A* **58** (1998) 466-477
8. S. Hacyan, "Squeezed states and uncertainty relations in rotating frames and Penning traps", *Phys. Rev. A* **53** (1996) 4481-4487
9. S. Cruz y Cruz and B. Mielnik, "Illusions of quantum theory in rotating frames", preprint Cinvestav
10. H.A. Bethe and E.E. Salpeter, *Quantum mechanics of one- and two-electron atoms*, (Berlin: Springer-Verlag 1957)
11. E. Schmutzer and J. Plebański, "Quantum mechanics in non-inertial frames of reference", *Fortschritte der Physik* **25** (1977) 37-82
12. S.A. Fulling, "Nonuniqueness of canonical field quantization in Riemmanian space-time" *Phys. Rev. D* **7** (1973) 2850-2862
13. S. Cruz y Cruz and B. Mielnik, "The parity phenomenon of the Floquet spectra", *Phys. Lett. A* **352** (2006) 36-40
14. B. Mielnik and J. Plebański, "Combinatorial approach to Baker-Campbell-Hausdorff exponentials", *Ann. Inst. Henri Poincaré* **XII** (1970) 215-254
15. S. Cruz y Cruz, "Esquemas cuánticos de Floquet, espectros y operaciones", Ph.D. Thesis, Physics Department, Cinvestav, Mexico (2006)
16. S. Cruz y Cruz, "Time of events: an applicability limit for the Floquet schemes", in O. Rosas, M. Carbajal and O. Miranda (Eds.) *Advanced Summer School in Physics 2005*, AIP Conference Proceedings **809** (2006) 106-108
17. S. Cruz y Cruz and B. Mielnik, "Quantum jumps and quasienergy spectra", preprint Cinvestav

On a Class of Hairy Square Barriers and Gamow Vectors

N Fernández-García

Physics Department, Cinvestav, AP 14-740, México DF 07000, Mexico

Abstract. The second order Darboux-Gamow transformation is applied to deform square one dimensional barriers in non-relativistic quantum mechanics. The initial and the new "hairy" potentials have the same transmission probabilities (for the appropriate parameters). In general, new Gamow vectors are constructed as Darboux deformations of the initial ones.

Gamow vectors are solutions of the Schrödinger equation with complex eigenvalue $\varepsilon = \varepsilon_R - i\Gamma/2$ and fulfilling purely outgoing boundary conditions [1]. For one dimensional stationary potentials, these not square integrable functions Ψ_G represent resonant states with density probability $\rho_G(x,t) = e^{-\Gamma t}|\Psi_G(x)|^2$. The inverse of the half time life Γ and the resonance ε_R are respectively the width and the position of one of the peaks of the corresponding transmission coefficient (see [2] and references therein). In this paper we shall use the Gamow vectors as transformation functions in the context of supersymmetric quantum mechanics (Darboux transformations). In particular, as a supplement of a previous work recently reported [2] (see also [3]), we shall analyze the Schrödinger equation for the square barrier potential:

$$\left[-\frac{d^2}{dx^2} + V(x)\right]\Phi(x) = E\Phi(x), \quad \text{with} \quad V(x) = \begin{cases} 0, & x < -b/2 \\ V_0, & -b/2 \le x \le b/2 \\ 0, & b/2 < x \end{cases} \tag{1}$$

where $V_0 > 0$ and $b \ge 0$. The general solution (for particles coming from the left) is obtained as usual:

$$\Phi(x) = \begin{cases} \alpha e^{ikx} + \alpha \frac{i(q^2-k^2)}{2\Delta(k,q)} (\sin qb)\, e^{-ik(x+b)}, & x < -b/2 \\ \alpha \frac{k e^{\frac{-ikb}{2}}}{\Delta(k,q)} \Big(\left[i(k\cos\frac{qb}{2} - iq\sin\frac{qb}{2}) \sin qx \right] & \\ \quad + \left[(q\cos\frac{qb}{2} - ik\sin\frac{qb}{2}) \cos qx \right] \Big), & -b/2 \le x \le b/2 \\ \alpha \frac{kq}{\Delta(k,q)} e^{ik(x-b)}, & b/2 < x \end{cases} \tag{2}$$

with α an arbitrary constant, $k^2 = E$, $q^2 = k^2 - V_0$, and

$$\Delta(k,q) = \left(k\cos\tfrac{qb}{2} - iq\sin\tfrac{qb}{2}\right)\left(q\cos\tfrac{qb}{2} - ik\sin\tfrac{qb}{2}\right). \tag{3}$$

CP885, *Advanced Summer School in Physics 2006, Frontiers in Contemporary Physics—EAV06,* edited by O. Miranda, M. Carbajal, L. M. Montaño, O. Rosas-Ortiz, and S. A. Tomás Velázquez

This solution is analytical in all the complex plane but in the poles of (3). Let us remark that the equation $\Delta(k,q) = 0$ does not have solutions if $E \in \mathbf{R}$. Thus, necessarily $E = \varepsilon \in \mathbf{C}$. On the other hand, the transmission coefficient

$$T = \left| \frac{kq}{\Delta(k,q)} \right|^2 \tag{4}$$

does not depend on α (which is arbitrary). Therefore, the substitution $\alpha = \Delta(k,q) = 0$ in (2) fulfills the required purely outgoing boundary condition and provides the same T. In order to solve $\Delta(k,q) = 0$, one can write the function T as a superposition of Breit-Wigner distributions:

$$T \approx \omega_N(E) = \sum_{n=1}^{N} \omega(E, E_{R_n}) \tag{5}$$

where

$$\omega(E, E_R) = \frac{(\Gamma/2)^2}{(E - E_R)^2 + (\Gamma/2)^2}. \tag{6}$$

Hence, every one of the peaks of the transmission coefficient (4) coincides with one Breit-Wigner function (6) for the appropriate values of the parameters [2, 3] (see Figure 1). For this graphical method, n labels the resonances while N corresponds to the order of the approximation.

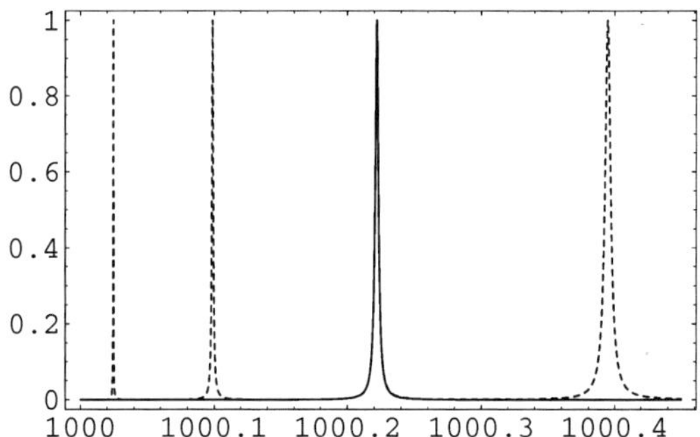

FIGURE 1. The transmission coefficient T for $V = 1000$ and $b = 20$ (dotted curve) besides the Breit-Wigner distribution (solid curve) corresponding to the third resonance.

Now, let u be solution of the Schrödinger equation

$$Hu = -u''(x) + V(x)u(x) = \varepsilon u(x) \tag{7}$$

where $\varepsilon = \varepsilon_R - i\varepsilon_I \in \mathbf{C}$; $\varepsilon_R \in \mathbf{R}$, $0 \neq \varepsilon_I \in \mathbf{R}$. The transformation $\beta(x) = -\frac{u'(x)}{u(x)}$ allows to factorize the initial Hamiltonian H as

$$H = AB + \varepsilon, \tag{8}$$

with

$$A = -\frac{d}{dx} + \beta, \quad B = \frac{d}{dx} + \beta$$

for which the following Riccati equation holds:

$$-\beta'(x,\varepsilon)+\beta^2(x,\varepsilon)+\varepsilon = V(x). \tag{9}$$

As β is a complex function, H is Hermitian although the operators A and B are not one the adjoint of the other (see [4]). A property which extends the domain of the factorization method (see also [5, 6, 7, 8, 9]). In particular, the intertwining relations $hB = BH$, $Ah = HA$ are true for a new non-Hermitian Hamiltonian

$$h = -\frac{d^2}{dx^2}+\widetilde{V}(x), \quad \widetilde{V}(x) = V(x)+2\beta'(x). \tag{10}$$

The spectrum of this Hamiltonian is either purely real or includes an extra complex eigenvalue with square integrable function. Thus, the new set of eigenfunctions is normalizable but non-orthogonal [4] (though some other possible bases have been recently discussed [10]). The iteration of the method (Sususy quantum mechanics [11]), in general, produces a complex new potential as well. However, by taking the complex conjugates u^* and ε^* as the new transformation function and factorization energy respectively, one obtains the real potential [4, 5, 8, 10]:

$$V_2(x) = V(x) - 2\mathrm{Im}\left(\frac{\varepsilon}{\beta}\right)' \tag{11}$$

for which the new solutions Ψ are given by

$$\Psi = (\varepsilon - E)\Phi - y\,\mathrm{Im}\left(\frac{\varepsilon}{\beta}\right) \tag{12}$$

where $y = \mathrm{W}(u,\Phi)/u$, $H\Phi = E\Phi$, $hy = Ey$, and $H_2\Psi = E\Psi$.

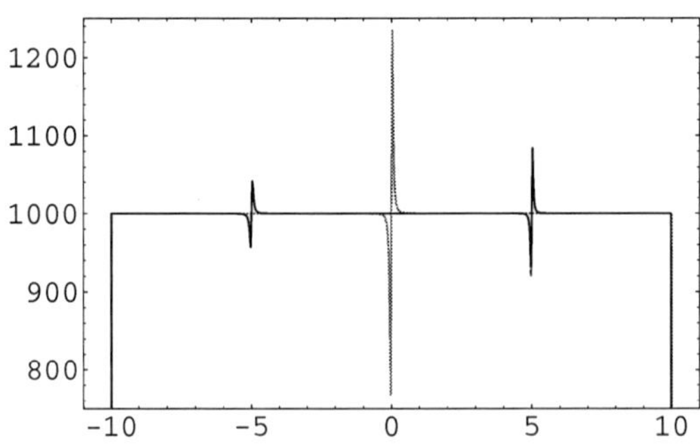

FIGURE 2. The top of the twice Darboux-deformed square barriers with $V_0 = 1000$, $b = 20$ for the first (gray curve) and second (solid curve) resonances.

Figure 2 shows the twice Darboux-deformations of the square barrier potential. These new barriers present "hair" over the top which induces stronger resonant phenomena [2]. The number of hairs depends on the excitation of the involved Gamow vector. Thereby, the lowest resonance induces only one distortion (a couple of hairs), the second one induces four hairs and so on. A graphical explanation of this behavior can be given by admitting the interpretation of resonances as quasi-bounded states. Thus, the kinetic term q grows with the energy of the resonance and the deformation becomes stronger (see equation (2)). Finally, a straightforward calculation leads to the transmission coefficient T_2 of the new potential:

$$T_2 = \left| \frac{\varepsilon - E + \varepsilon_I k/\kappa_R + \varepsilon_I(\kappa_I - i\kappa_R)/\kappa_R}{\varepsilon - E + \text{sgn}(\kappa_I)\varepsilon_I(ik/\kappa_R) + \varepsilon_I(\kappa_I - i\kappa_R)/\kappa_R} \right|^2 T \qquad (13)$$

with $sgn(\cdot)$ the sign function and $\kappa \in \mathbf{C}$, $k \in \mathbf{R}$, the kinetic terms of the transformation function u and the transformed function Φ respectively.

The approximation we are dealing with holds for peaks which are narrow compared with the distance between close resonances $\frac{\varepsilon_I}{\kappa_R} \sim 0$. Hence, equation (13) reads $T_2 \sim T$. Therefore, the twice Darboux-deformed potentials V_2 have similar transmission probabilities as their partners V. A result which is consistent with the conventional factorization (Susy) method in the limit $\varepsilon_I/\kappa_R \to 0$ (see e.g. [12]). The complex eigenvalues of the new Gamow vectors will be, in general, slightly different from the initial ones.

ACKNOWLEDGMENTS

I want to thank Prof. Oscar Rosas-Ortiz for his interesting comments about this topic. I am grateful to the organizers of the *Advanced Summer School 2006* for the opportunity to give this talk. The support of CONACyT projects 50766 and 49253-F is acknowledged.

REFERENCES

1. Gadella M and de la Madrid R, A pedestrian introduction to Gamow vectors, Am. J. Phys. **70** 626-637 (2002).
2. Fernández-García N, Darboux-deformed barriers and resonances in Quantum Mechanics, (submitted to Rev. Mex. Fis.).
3. Fernández-García N, Transformaciones de Gamow Supersimétricas en Mecánica Cuántica (Mexico: MSc dissertation, Physics Department, Cinvestav 2005).
4. Rosas-Ortiz O and Muñoz R, Non-Hermitian SUSY hydrogen-like Hamiltonians with real spectra, J. Phys. A: Math. Gen. **36** 8497-8506 (2003).
5. Fernández DJ, Muñoz R and Ramos A, Second order SUSY transformations with complex energies, Phys. Lett. A **308** 11-16 (2003).
6. Ramírez A and Mielnik B, The challenge on non hermitian structures in physics, Rev. Mex. Fis. **4952**, 130-133 (2003)
7. Mielnik B and Rosas-Ortiz O, Factorization: Little or great algorithm?, J. Phys A: Math. Gen. **37**, 10007-10035 (2004).
8. Rosas-Ortiz O, Gamow Vectors and Supersymmetric Quantum Mechanics, Rev. Mex. Fis. (at press).
9. Muñoz R, On the generation of non-Hermitian Hamiltonians with real spectra by a modified Darboux transform, Phys. Lett. A **345**, 287-292 (2005).
10. Muñoz R, Complex Supersymmetric Transformations and Exactly Solvable Potentials (Mexico: PhD dissertation, Physics Department, Cinvestav 2004).
11. Fernández DJ, SUSUSY Quantum Mechanics, Int. J. Mod. Phys. A **12**, 171-176 (1997).
12. Cooper F, Khare A, Sukhatme U, Supersymmetry and Quantum Mechanics, Phys. Rep. **251** 267-385 (1995).

Some Nonlinear Solutions of the Linear Schrödinger Equation for a Free Particle

Gabino Torres Vega

Depto. de Física, Cinvestav, Apdo. Postal 14-740, 07000 México, D.F., Mexico, gabino@fis.cinvestav.mx

Abstract. We depict a method for reducing a linear combination of circular functions with *different* frequencies in their arguments (a non eigenfunction of the linear Schrödinger equation for a free particle) to a linear combination of the same functions but with the *same* frequency (now an eigenfunction). The method is inspired in the Jacobi's form of elliptic and hyperelliptic functions.

INTRODUCTION

Trigonometric and elliptic functions [1, 2, 6, 7, 8] are very useful in non relativistic linear and nonlinear quantum mechanics, as well as in other areas [9]. For the linear Schrödinger equation, some times we need linear combinations of trigonometric functions in order to describe physical phenomena, but, for the nonlinear equation, it is not clear how to form linear combinations of elliptic functions. On the other hand, could it be possible to find nonlinearity in the linear Schrödinger equation? In this article we try to answer the last question.

After looking at Jacobi's form of elliptic functions,

$$\mathrm{sn}(u) = \sin(\phi)\,, \tag{1}$$

$$\mathrm{cn}(u) = \cos(\phi)\,, \tag{2}$$

$$\mathrm{dn}(u) = \sqrt{1 - m\sin^2\phi}\,, \tag{3}$$

where

$$u = \int_0^\phi \frac{d\theta}{\sqrt{1 - m\sin^2\theta}}\,, \tag{4}$$

one is compelled to explore the idea of adding "texture" to the real axes and use it as the argument of the trigonometric functions and see if we can obtain solutions of some differential equations of interest. Elliptic and Abel's functions are solutions of the linear and non linear Schrödinger equations among some other equations [3, 4, 5].

In particular, in this paper we add texture to the real axes in order to get *nonlinear* functions that are solutions of the *linear Schrödinger equation for a free particle.*

CP885, *Advanced Summer School in Physics 2006, Frontiers in Contemporary Physics—EAV06,* edited by O. Miranda, M. Carbajal, L. M. Montaño, O. Rosas-Ortiz, and S. A. Tomás Velázquez

THE METHOD

Let us work out a general case first. Some of the steps can be a bit different in a specific application depending upon the function and differential equation chosen. We start with a bounded, C^2, function $f = f(x(u))$ dependent of the variable $x = x(u)$ which is also a function of another variable u. The second derivative of $f(u)$ with respect to u can be written in terms of x as

$$\frac{d^2}{du^2}f(u) = \frac{1}{2}f'(x)\frac{d}{dx}j^2(x) + j^2(x)f''(x)\,, \tag{5}$$

where $j(x) \equiv dx/du$ is the Jacobian of the transformation [10], and the prime $'$ indicates the derivative of the function. Then, upon substitution of the above equality into the time-independent linear Schrödinger equation for a free particle, written as

$$f''(u) + \Delta f(u) = 0\,, \tag{6}$$

where Δ is a real constant, we obtain an equation for the unknown function $j(x)$ in the x-space. The solution of the resulting equation is

$$j(x) = \frac{\sqrt{\Delta[c - f^2(x)]}}{f'(x)}\,, \tag{7}$$

with c a real constant such that $\Delta[c - f^2(x)] > 0$ for all x. If $f(x)$ is bounded, we can chose $|c|$ large enough so that $j(x)$ is real and then there are no branch points.

By integrating the relation $du = dx/j(x)$, the relationship between x and u is found to be

$$u = \frac{1}{\sqrt{\Delta}}\left\{\tan^{-1}\left[\frac{f(x)}{\sqrt{c - f^2(x)}}\right] - \tan^{-1}\left[\frac{f(0)}{\sqrt{c - f^2(0)}}\right]\right\}\,, \tag{8}$$

i.e.,

$$f(u) = \pm\sqrt{c_1}\sin\left\{\sqrt{\Delta}u + \tan^{-1}\left[\frac{f(0)}{\sqrt{c - f(0)^2}}\right]\right\}\,. \tag{9}$$

Few properties of $f(u)$ are the derivative,

$$\frac{d}{du}f(u) = \sqrt{\Delta[c - f(u)^2]}\,, \tag{10}$$

and the derivative of the inverse function

$$\frac{d}{dy}f^{-1}(y) = \frac{1}{\sqrt{\Delta(c - y^2)}}\,. \tag{11}$$

Similar to what happens with trigonometric, elliptic and hyperelliptic functions, the f function is the inverse function of the integral

$$\int_0^x \frac{dy}{\sqrt{\Delta(c - y^2)}}\,, \tag{12}$$

i.e., these functions are of the trigonometric type.

Therefore, this method allows us to generate nonlinear functions which can be solutions of some linear differential equations of interest in physics and can be applied to other equations involving trigonometric and elliptic functions. As an application of the above ideas, in what follows we add texture to the real axes in order to get *nonlinear* functions that are solutions of the *linear* Schrödinger equation for a free particle.

A periodic function

Let us consider the linear combination

$$\text{sia}(x) \equiv a\sin(\omega_1 x) - b\cos(\omega_2 x)\,, \tag{13}$$

where ω_1, ω_2, a and b are real constants. We will take this linear combination of trigonometric functions to the nonlinear domain. This function is no longer an eigenfunction of the time independent Schrödinger equation for a free particle, but we can transform it to one of them.

For Eq. (13) to become a solution of the Schrödinger equation, $\text{sia}''(u) + \Delta\,\text{sia}(u) = 0$, the Jacobian $j(x)$ of the transformation should be (we call it $\text{dia}(x)$ for this case)

$$\text{dia}(x) = \sqrt{\Delta}\frac{\sqrt{c + a^2 - \text{sia}^2(x)}}{\text{sia}'(x)}\,, \tag{14}$$

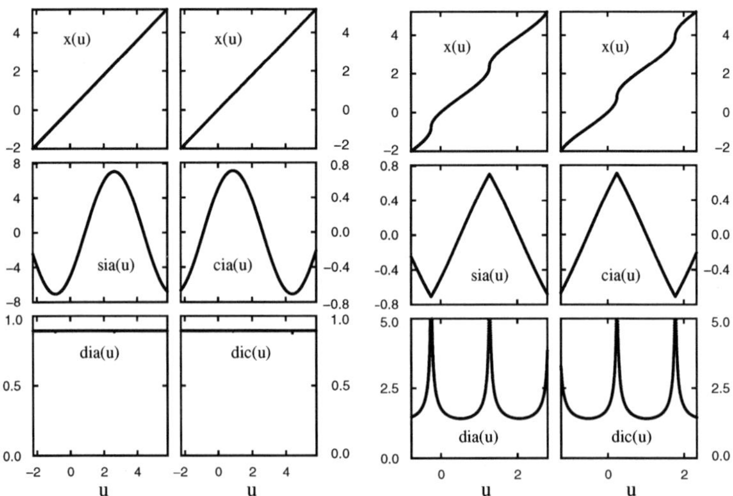

FIGURE 1. Plots of $x(u)$, $\text{sia}(u)$, $\text{cia}(u)$, $\text{dia}(u)$ and $\text{dic}(u)$ for $a = 0.5$, $b = 0.5$, $\omega_1 = 1$, $\omega_2 = 1$, $c = 0.2499999$ and $\Delta = 0.8$. On the left two columns $c = 0.2499999$ and on the right two columns $c =!$.

where c and Δ are real constants such that $\Delta[c+a^2-\mathrm{sia}^2(x)]>0$ for all values of x, avoiding branching points. Then

$$\mathrm{sia}(u)=\pm[b\,\cos(\sqrt{\Delta}u)-\sqrt{c+a^2-b^2}\sin(\sqrt{\Delta}u)]\,, \tag{15}$$

with derivatives

$$\frac{d}{du}\mathrm{sia}(u)=\sqrt{\Delta[c+a^2-\mathrm{sia}(u)^2]}\,, \tag{16}$$

and

$$\frac{d}{dy}\mathrm{sia}^{-1}(y)=\frac{1}{\sqrt{\Delta(c+a^2-y^2)}}\,. \tag{17}$$

Then, after the transformation, we have ended up with a function which is now an eigenfunction of the linear time-independent Schrödinger equation. This implies that any linear combination of eigenfunctions, with different frequencies, can be converted to an eigenfunction with only one frequency, and, then the linear combinations with different frequencies are just different manifestations of a single function, a great simplification of the problem. Plots of these functions can be seen in Fig. 1.

An hyperbolic-type function

Another functions that can be taken to the nonlinear domain are the hyperbolic functions, useful for studding tunneling through a barrier [11, 12, 13]. If we consider a negative sign in front of Δ in the Schrödinger equation, $\mathrm{sha}''(u)-\Delta\,\mathrm{sha}(u)=0$, the change of variable is given in terms of

$$\mathrm{dha}(x)=\sqrt{\Delta}\frac{\sqrt{c+a^2+\mathrm{sha}^2(x)}}{\mathrm{sha}'(x)}\,, \tag{18}$$

i.e.

$$u=\frac{1}{\sqrt{\Delta}}\log\left[\frac{\mathrm{sha}(x)+\sqrt{c+a^2+\mathrm{sha}^2(x)}}{\sqrt{c+a^2+b^2}-b}\right] \tag{19}$$

Here $\Delta[c+a^2+\mathrm{sha}^2(x)]>0$ for all x.

The function and its derivatives are

$$\begin{aligned}\mathrm{sha}(u) &= a\sinh(\omega_1 x)-b\cosh(\omega_2 x) &(20)\\ &= \sqrt{c+a^2+b^2}\sinh(\sqrt{\Delta}u)-b\,\cosh(\sqrt{\Delta}u)\,, &(21)\end{aligned}$$

$$\frac{d}{du}\mathrm{sha}(u)=\sqrt{\Delta[c+a^2+\mathrm{sha}(u)^2]}\,, \tag{22}$$

$$\frac{d}{dy}\mathrm{sha}^{-1}(y)=\frac{1}{\sqrt{\Delta(c+a^2+y^2)}}\,. \tag{23}$$

Another periodic function

A second periodic function is

$$\text{coa}(x) \equiv a\sin(\omega_1 x) + b\cos(\omega_2 x) \ . \tag{24}$$

The choice

$$\text{do}(x) = \sqrt{\Delta}\frac{\sqrt{c - a^2 + \text{coa}^2(x)}}{\text{coa}'(x)} \ , \tag{25}$$

makes that (24) be a solution of the differential equation $\text{coa}''(u) + \Delta\,\text{coa}(u) = 0$. In this case, c is such that $\Delta[c - a^2 + \text{coa}^2(x)] > 0$ for all x.

The function and its derivatives are

$$\text{coa}(u) = \pm\sqrt{c + a^2}\sin\left[\sqrt{\Delta}u + \tan^{-1}\left(\frac{b}{\sqrt{c + a^2 - b^2}}\right)\right] \ , \tag{26}$$

$$\frac{d}{du}\text{coa}(u) = \sqrt{\Delta[c + a^2 - \text{coa}(x)^2]} \ , \tag{27}$$

and

$$\frac{d}{dy}\text{coa}^{-1}(y) = \frac{1}{\sqrt{\Delta(c + a^2 - y^2)}} \ . \tag{28}$$

Another function

Finally, a second hyperbolic-type function can be defined as

$$\text{cha}(x) \equiv a\sinh(\omega_1 x) + b\cosh(\omega_2 x) \ . \tag{29}$$

where

$$\text{dhc}(x) = \sqrt{\Delta}\frac{\sqrt{c + a^2 + \text{cha}^2(x)}}{\text{cha}'(x)} \ . \tag{30}$$

Now c is large enough so that $\Delta[c + a^2 - \text{cha}^2(x)]$ is positive for all x.

The relationship between u and x is

$$u = \frac{1}{\sqrt{\Delta}}\log\left[\frac{\sqrt{c + a^2 + \text{cha}^2(x)} + \text{cha}(x)}{\sqrt{c + a^2 + b^2} + b}\right] \ . \tag{31}$$

Then

$$\text{cha}(u) = b\cosh(\sqrt{\Delta}u) + \sqrt{c + a^2 + b^2}\sinh(\sqrt{\Delta}u) \ , \tag{32}$$

and its derivatives are given by

$$\frac{d}{du}\text{cha}(u) = \sqrt{\Delta[c + a^2 + \text{cha}(u)^2]} \ , \tag{33}$$

and

$$\frac{d}{dy}\text{cha}^{-1}(y) = \frac{1}{\sqrt{\Delta(c+a^2+y^2)}} \tag{34}$$

The function Eq. (32) is a solution of the Schrödinger equation $\text{cha}''(u) - \Delta\,\text{cha}(u) = 0$.

Unfortunately, this method does not work for complex exponential functions $e^{\pm ikx}$ but it shows that one can find nonlinearity in linear systems and that it is possible to find additional solutions to ordinary differential equations from already known solutions.

REFERENCES

1. A. I. Markushevich, *The Remarkable Sine Functions, Elsevier, New York* (1966).
2. Harris Hancock, *Theory of Elliptic Functions, Dover Publications, New York* (1958).
3. I. M. Krichever, Acta Appl. Math. **36**, 7 (1994).
4. E.D. Belokolos and V.Z. Enolskii, J. Math. Sc. **18**, 295 (2002).
5. E.D. Belokolos, V.Z. Enolskii and M. Salerno, Theor. Math. Phys. **144**, 1081 (2005).
6. Arthur Cayley, *An elementary treatise on Elliptic Functions, Dover Publications, New York* (1961).
7. Derek F. Lawden, *Elliptic Functions and Applications, Springer-Verlag, New York* (1989).
8. Eric H. Neville, *Elliptic Functions: a Primer, Pergamon Press, Oxford* (1971).
9. R. K. Dodd, J. C. Eilbeck, J. D. Gibbon and H. C. Morris, *Solitons and Nonlinear Wave Equations, Academic Press, London* (1984).
10. George Arfken, *Mathematical Methods for Physicists, Academic Press, New York* (1970).
11. R. Sala, S. Brouard and J. G. Muga, J. Phys. A **28**, 6233 (1995).
12. J. G. Muga, J. Phys A: Math Gen. **24**, 2003 (1991).
13. R. Landauer, Rev. Mod. Phys. **66**, 217 (1994).

Evolution of Axisymmetric Initial Data for the Schrödinger-Poisson System

Argelia Bernal

Departamento de Física, Centro de Investigación y de Estudios Avanzados del IPN, AP 14-740, 07000 México D.F., Mexico and
Instituto de Física y Matemáticas, Universidad Michoacana de San Nicolás de Hidalgo. Edificio C-3, Cd. Universitaria, AP 2-82, 58040 Morelia, Michoacán, Mexico

Abstract. In this work it is shown that spherical and stable gravitationally-bound configurations of scalar fields are late time attractors for axisymmetric initial configurations which evolve towards those attractors via scalar field emission. These results are relevant in astrophysical scenarios because such scalar field configurations have been related to dark matter galactic halos and Newtonian boson stars.

INTRODUCTION

Recent observations are very conclusive about the existence of Dark Matter, Dark Energy and a period of inflation. Scalar fields have played an important role in describing the behavior of such exciting phenomena. Even more, there exist a big expectation that such fields could be related with those predicted by theoretical models which pretend to be fundamental. In this scenario, it is also important to study the behavior of scalar fields at astrophysical scales where they have been related with dark matter galactic halos [1] [2] [3] or boson stars [4]. In this regime, the Newtonian limit of Einstein Klein-Gordon equations (EKG), the Schrödinger-Poisson (SP) system, becomes relevant. On the other hand, the SP system is important by itself as it describes attractive solid state systems achieved in laboratories [5, 6], [7], this fact opens the exciting possibility to extrapolating results of well controlled laboratory systems to astrophysical objects.

THE SCHRÖDINGER-POISSON SYSTEM

In this section I obtain the Newtonian limit of the relativistic Einstein Klein-Gordon (EKG) equations, the SP system. Important issues about the stationary spherical solutions of it are also discussed. A complex scalar field minimally coupled to gravity in general relativity is described by the action

$$I=\frac{c^4}{16\Pi G}\int d^4x\sqrt{-g}R-\int d^4x[\sqrt{-g}\frac{1}{2}(\hbar^2g^{\mu\nu}\partial_\mu\Phi\partial_\nu\Phi^*+m^2c^2\Phi\Phi^*)] \qquad (1)$$

In this action, m is the boson mass whose field operator is $\hat{\Phi}$. At present, we are working in a semiclassical approximation, that is, Φ is the expectation value of the field operator $\Phi(\mathbf{x},t)=<\hat{\Phi}(\mathbf{x},t)>$. This action leads to the EKG equations and the

CP885, *Advanced Summer School in Physics 2006, Frontiers in Contemporary Physics—EAV06,* edited by O. Miranda, M. Carbajal, L. M. Montaño, O. Rosas-Ortiz, and S. A. Tomás Velázquez

scalar field energy momentum tensor. In spherical symmetry, the line element could be written as $ds^2 = -N(r,t)^2c^2dt^2 + g(r,t)^2dr^2 + r^2d\Omega^2$, and the KG equations, the $0,0$ and the $0,1$ components of the Einstein equations form a complete set for Φ, Φ^*, N and g. Intuitively, it is clear that the Newtonian limit of the EKG equations describes the behavior of non relativistic scalar field and weak gravity. In our case the second condition is expressed mathematically as $N(r,t)^2 = 1 + V(r,t)$ and $g(r,t)^2 = 1 + U(r,t)$, with $V \sim U \sim O(\varepsilon^2)$, where $\varepsilon \ll 1$ is an expansion parameter. The perturbations to the Minkowski metric U and V also satisfy $\partial_0 \sim \varepsilon\partial_r$, where subindex 0 corresponds to ct. This condition implies that typical velocities for this metric are non relativistic. Conditions for the scalar field are less direct. It is known that KG equation in Minkowski space for a free particle has as solution $\Phi_{flat} = exp[\frac{i}{\hbar}(\mathbf{p}\cdot\mathbf{r} - Et)]$ where the energy could be expanded as $E = \pm mc^2[1 + \frac{1}{2}\frac{p^2}{m^2c^2} - \frac{1}{8}\left(\frac{p^2}{m^2c^2}\right)^2 + ...]$ and for a non relativistic particle $p^2/(m^2c^2) \ll 1$. Substituting this expansion in Φ_{flat} give us a clue to writing the scalar field minimally coupled to gravity as $\Phi = exp[-i/\hbar(mc^2t)]\Psi(r,t)$. The exponential term accounts for the energy part associated with the particle rest mass, and we should expect to dropping it in some way as Newtonian systems do not deal with this part of the energy. Ψ instead behaves as a Newtonian field and satisfies, as the metric perturbations, the relation $\partial_0\Psi \sim \varepsilon\partial_r\Psi$, where $\varepsilon^2 \sim p^2/m^2c^2$. With such conditions for the fields, to the leading nontrivial order in ε, it is straightforward to obtain from the EKG equations the SP system which reads

$$i\hbar\partial_t\Psi = \frac{\hbar^2}{2m}\nabla^2\Psi + m\mathrm{V}\Psi \quad (2)$$

$$\nabla^2\mathrm{V} = 4\pi Gm^2\Psi\Psi^* \quad (3)$$

Stationary Spherical Solutions. Assuming $\Psi = \psi(r)e^{-i\sigma t}$, regularity at origin, asymptotic flatness and a central value $\psi(0) = \psi^c$, the SP becomes in a eigenvalue problem for σ. Given a fixed ψ^c there is a numerable set $\Psi_i = \psi_i e^{-i\sigma_i t}$, $i = 0,1,2..$, of solutions with $\psi_i(0) = \psi^c$. Each ψ_i has i number of nodes and it has been proved, by linear perturbation theory, that only Ψ_0 is stable under small perturbations [8]. Then, if Ψ_0 is slightly perturbed, its density $\rho = |\Psi_0 + \delta\Psi_0|^2$ will oscillate with a specific frequency called the *quasinormal frequency* of Ψ_0. Stable solutions are important as they are employed to test evolution numerical codes for the SP system and because they are late time attractors for non spherically symmetric initial configurations, as we will see in next section. Another important property of stationary solutions arise because the SP system obeys a scaling symmetry of the form $\{t, r, \mathrm{V}, \Psi\} \rightarrow \{\lambda^{-2}\hat{t}, \lambda^{-1}\hat{r}, \lambda^2\hat{\mathrm{V}}, \lambda^2\hat{\Psi}\}$, where λ is an arbitrary parameter, then it is possible to construct a family of solutions once we have found one. Finally, physical quantities are constructed, as in non relativistic quantum mechanics, via the operator's expectation values.

EVOLUTION OF AXISYMMETRIC INITIAL CONFIGURATIONS

In this section it is studied the collapse of non-spherical initial profiles for the SP system. It is shown that stable spherically symmetric configurations are late time attractors for initial configurations which are not spherically symmetric. As we assume axial symmetry we write the Laplacian operator as $\nabla^2 = \frac{\partial^2}{\partial r^2} + \frac{1}{r}\frac{\partial}{\partial r} + \frac{\partial^2}{\partial z^2}$, where $r = \sqrt{x^2+y^2}$. I also choose units in which $c = \hbar = 1$ and the mass has been normalized to one. It is important to notice that Poisson equation is a constrain while the Schrödinger one gives the temporal evolution for Ψ. Then, "to evolve" the system means 1) to choose $\Psi(\mathbf{x},t)$, 2) to solve the Poisson equation for $V(\mathbf{x},t)$, 3) to use such potential in order to find $\Psi(\mathbf{x},t+\Delta t)$ from the Schrödinger equation and repeat from 1) until convergence for physical quantities is lost.

The Code. The code used to evolve the SP system is constructed by using finite differences. Details about the specific methods and boundary conditions implemented will appear in a future work. As we actually evolve the discretized SP system instead of the continuous one, it is introduced an intrinsic numerical perturbation. It is not possible to avoid this perturbation but it is possible to maintain it under control. A concrete example, that is taken as a test for the code, will clarify this important point. I have seen in the preceding section that SP system has stable spherical solutions Ψ_0 with time independent density ρ. However, when we evolve one of such configurations, for instance $\hat{\Psi}_0$ for which $\hat{\Psi}_0(\mathbf{0},0) = 1$, their density becomes time dependent and its central density $\hat{\rho}_c(\mathbf{0},t)$ is not one. The most that we can expect is that our numerical solution converges to the value of the stationary solution and that the numerical perturbation behaves as a linear perturbation. That means that the value of the central density must converge to one and that its time behavior must be harmonic with a quasinormal mode of oscillation as predicted by the linear perturbation theory. Both conditions are satisfied by the code.

Axisymmetric Initial Configuration. The method followed to verify that non-spherical initial configurations evolve towards a stable stationary spherical configuration is as follows: 1) Given the numerical evolution of a non-spherical initial profile $\Psi(\mathbf{x},t)$ it is obtained, via a Fourier Transform of a physical quantity, e.g. the central density ρ_c^{Ψ}, the frequency γ of the fundamental mode of the system. It is assumed that γ is the quasinormal frequency of some stable configuration Ψ_0 which $\Psi(\mathbf{x},t)$ is evolving to. 2) With this γ and by using the rescaling properties (see above), it is straightforward to find the rescaling parameter λ which relates Ψ_0 to $\hat{\Psi}_0$. It is then possible via the rescaling properties to known all the non hat physical quantities for Ψ_0 3) I prove that the system is really evolving towards Ψ_0 if numerical values of physical quantities for $\Psi(\mathbf{x},t)$ converge, for different resolutions, to those of Ψ_0 4) Finally, I verify that the fate of Ψ is spherical using the ellipticity's temporal evolution, this quantity is defined as the spatial integration of the difference between the density in the z axis and the density in the x axis. I also verify that the virialization condition is satisfied for each evolved initial configuration. In figure 1, I present the results for the initial data given by $\Psi(\mathbf{x},0) = \Sigma_{l=0}^{3} A_l exp[-r^2/\sigma_l^2] Y_l^0$, where Y_l^0 are axisymmetric spherical harmonics, the coefficients A_l take values $A_0 = 9.0, A_1 = A_2 = A_3 = 1.0$ and $\sigma_0 = \sigma_1 = \sigma_2 = \sigma_3 = 1.5$. For this configuration I found that the frequency of its quasinormal mode is $\gamma = 0.128$, which implies a rescaling parameter $\lambda = 1.669$ then the central density and the mass of the state Ψ_0^{λ}, which Ψ is evolving to, are $\rho_c^{\lambda} = 7.75$ and $M^{\lambda} = 3.441$ respectively. The pair of those values are marked with a cross in figure 1, where the evolution of the mass M_{Ψ} versus the central density ρ_c^{Ψ} of the state Ψ for two grid resolutions is shown, the stars are the values that the system is approaching to in each case. Those values are

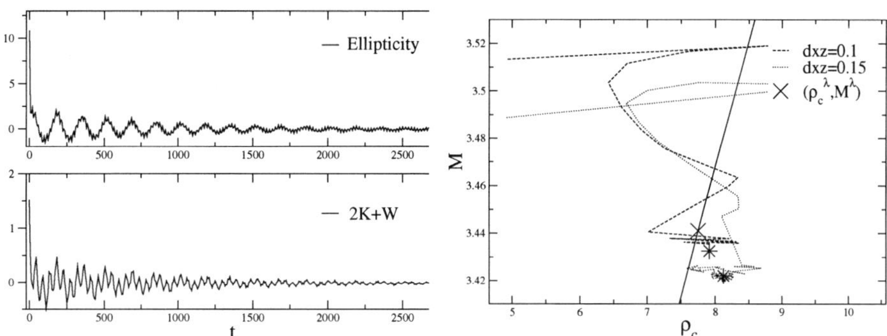

FIGURE 1. Evolution of different quantities for the initial configuration specified in the text. Left: As ellipticity goes to zero the configuration becomes spherical. *K* and *W* are the kinetic and potential energy respectively. The virialization condition is reached by the system. Right: Evolution of the central density and mass M. See text for details.

actually converging to the respective values for Ψ_0^λ with the Cauchy type convergence according to the expected convergence type for the numerical methods employed in the evolution.

CONCLUSIONS

In this paper it was shown that axisymmetric initial data for the SP system evolves towards spherical stables configurations. In this process the system relaxes through the emission of scalar field particles and reaches a virialized configuration.

Acknowledgments The author would like to thank F. Siddhartha Guzmán for sharing his numerical codes. This work was partially supported by a CONACyT grant. The runs were carried out in the Ek-bek cluster of LASUMA at Cinvestav.

REFERENCES

1. M. Alcubierre, F. S. Guzmán, T. Matos, D. Núñez, L. A. Ureña, and P. Wiederhold, *Class. Quantum Grav.* **19**, 5017 (2002).
2. F. S. Guzmán, and L. A. Ureña López, *Phys. Rev. D* **68**, 024023 (2003).
3. A. Arbey, J. Lesgourges, and P. Salatti, *Phys. Rev. D* **64**, 123528 (20001).
4. E. Seidel, and W.-M. Suen, *Phys. Rev. Letters* **66**, 1659 (1991).
5. J. H. Luscombe, A. M. Bouchard, and M. Luban, *Phys. Rev. B* **46**, 10262–10268 (1992).
6. G. Reinisch, J. Pacheo, and P. Valiron, *Phys. Rev. A* **63**, 042505 (2001).
7. D. Dell, S. Giovanazzi, G. Kurizki, and V. Akulin, *Phys. Rev. Letters* **84**, 5687–5690 (2000).
8. F. S. Guzmán, and L. A. Ureña López, *Phys. Rev. D* **69**, 124033 (2004).

Gravitational Perturbations in General Relativity

Claudia Moreno Gonzalez[1]

Departamento de Matemáticas,
Centro Universitario de Ciencias Exactas e Ingeniería, Universidad de Guadalajara,
Blvd. M. García Barragan No. 1421, S.R., Guadalajara, Jalisco, Mexico.

Abstract. In this article I will provide an introduction to the phenomena of gravitational waves. I begin talking about Einstein's theory of gravity where gravitational waves are introduced as an effect of the propagation of perturbations of space-time. The Einstein equations are of nonlinear character but using perturbation theory, they become linear. Gravitational waves are a new way of looking at the Universe.

INTRODUCTION

Linear differential equations have been used for describing a lot of phenomena related with perturbation transportation of different kind of information in physical, chemical and biological systems. In physics (General Relativity), perturbation theory had been very successful in Gravitational Waves (GW). GW are probably one of the most popular predictions of Einstein's General Relativity theory [1]. The scientific community in the world is working in the detection of gravitational radiation. The detection of GW imply experimental [2], phenomenological [3] and theoretical studies [4]; the reward is the prospect of new tests of fundamental physics.

The gravitational waves are oscillating propagations of gravitational fields, generated in the space-time. These waves are signals produced by violent events such as: 1) collision of black holes, 2) explosion of supernova nuclei or 3) radiation created at the birth of the Universe or 4) extra dimensions in string theory.

Theoretically, the phenomenon of gravitational waves is well understood. Current researchers are mainly trying to observe this radiation and to understand what these waves can tell us about our universe.

In this contribution, I want to give a rough overview over how gravitational waves arise from Einstein's theory, discussing some ideas about the role of gravitational waves in astrophysics.

[1] claudia.moreno@cucei.udg.mx

CP885, *Advanced Summer School in Physics 2006, Frontiers in Contemporary Physics—EAV06,*
edited by O. Miranda, M. Carbajal, L. M. Montaño, O. Rosas-Ortiz, and S. A. Tomás Velázquez

WAVE SOLUTION OF EINSTEIN'S EQUATIONS: LINEARIZED GRAVITY

Gravitational waves are analogous to light waves. Light waves are emitted by charged particles in movement while gravitational waves are emitted by masses in movement. Waves are mostly thought of as a linear phenomenon. Electromagnetic waves are indeed of linear nature and arise without any approximations (perturbation). Mechanical oscillations, however, are linear only in the approximation of small displacements. A pendulum for example is only a harmonic oscillator as long as the approximation $\sin(\theta) \equiv \theta$ holds. For a mass on a spring we only end up with harmonic as long as the restoring force is so small that the leading term is linear. We face the same kind of problem, when we think about gravitational radiation. The most inconvenient feature of Einstein's equations is that they are nonlinear. From the knowledge about wave phenomena in other fields it is clear that the easiest way to approach oscillations in the gravitational field will be to consider the limit of weak gravitational field, resulting in a linearized theory of gravity. There are several exact wave solutions in General Relativity, as well as numerical solutions. Linearization is nevertheless a very useful concept. Also from a physical standpoint this approximation is justifiable, since far away from massive objects the curvature of our universe is very small. Precisely, the weak field limit means that we consider space-time to be flat up to a small perturbation, such that terms of second order in the perturbation are negligible. In mathematical terms, the perturbation in the space-time metric $g_{\mu\nu}$ is expressed as,

$$g_{\mu\nu} \rightarrow g_{\mu\nu} + h_{\mu\nu} + O(h_{\mu\nu}^2); \tag{1}$$

here $h_{\mu\nu}$ is the first order metric perturbation. If $h_{\mu\nu} \ll 1$ we can drop terms of order $O(h_{\mu\nu}^2)$ in all equations and get consequently our linearized version of General Relativity. In this linearized version of gravity the indices of all terms which are linear in $h_{\mu\nu}$ can obviously be raised and lowered using the Minkowski metric, since corrections would be higher order terms. This suggest that the perturbation $h_{\mu\nu}$ can be understood as a separate field propagating in flat space similar to the electromagnetic field.

The Christoffel symbols and the Riemann tensor are given by [4, 5],

$$\begin{aligned} \delta g^{\mu\nu} &= -h^{\mu\nu}, \\ \delta\Gamma^{\mu}_{\nu\lambda} &= \frac{1}{2}\left(\nabla_\nu h^\mu_\lambda + \nabla_\lambda h^\mu_\nu - \nabla^\mu h_{\nu\lambda}\right), \\ \delta R_{\mu\nu\rho\sigma} &= h_{\lambda[\mu}R^\lambda_{\nu]\rho\sigma} + \nabla_\rho\nabla_{[\mu}h_{\nu]\sigma} - \nabla_\sigma\nabla_{[\mu}h_{\nu]\rho}. \end{aligned} \tag{2}$$

The perturbations for the Einstein tensor in vacuum, by the last expressions, are given by

$$\begin{aligned} \delta\left(R_{\mu\nu} - \frac{1}{2}g_{\mu\nu}R_{\rho\sigma}g^{\rho\sigma}\right) &= -\nabla^\rho\nabla_{(\mu}h_{\nu)\rho} + \frac{1}{2}\nabla_\rho\nabla^\rho h_{\mu\nu} + \frac{1}{2}\nabla_\mu\nabla_\nu h \\ &\quad + \frac{1}{2}g_{\mu\nu}\left(\nabla^\rho\nabla^\sigma h_{\rho\sigma} - \nabla_\rho\nabla^\rho h + R_{\rho\sigma}h^{\rho\sigma}\right) - \frac{1}{2}Rh_{\mu\nu}; \end{aligned} \tag{3}$$

where the trace h is $h \equiv h^{\mu}_{\mu}$.

On the other hand, supposing that the background field $g_{\mu\nu}$, satisfies the Einstein equations in vacuum,

$$G_{\mu\nu} \equiv R_{\mu\nu} - \frac{1}{2} R g_{\mu\nu} = 0, \tag{4}$$

from equations (3) and (4) we obtain

$$-\nabla^{\rho}\nabla_{(\mu}h_{\nu)\rho} + \frac{1}{2}\nabla_{\rho}\nabla^{\rho}h_{\mu\nu} + \frac{1}{2}\nabla_{\mu}\nabla_{\nu}h + \frac{1}{2}g_{\mu\nu}\left(\nabla^{\rho}\nabla^{\sigma}h_{\rho\sigma} - \nabla_{\rho}\nabla^{\rho}h\right) = 0. \tag{5}$$

For the Einstein equations with sources

$$\nabla^{\rho}\nabla_{(\mu}h_{\nu)\rho} - \frac{1}{2}\nabla_{\rho}\nabla^{\rho}h_{\mu\nu} - \frac{1}{2}\nabla_{\mu}\nabla_{\nu}h - \frac{1}{2}g_{\mu\nu}\left(\nabla^{\rho}\nabla^{\sigma}h_{\rho\sigma} - \nabla_{\rho}\nabla^{\rho}h\right) \equiv 8\pi T_{\mu\nu}. \tag{6}$$

The tensor $T_{\mu\nu}$, serves to complement equation (5) and, it physically represents a source for the field of gravitation.

SOURCES OF GRAVITATIONAL WAVES

Let us have a look at the linearized Einstein tensor again:

$$G_{\mu\nu} = \nabla^{\rho}\nabla_{(\mu}h_{\nu)\rho} + \frac{1}{2}\nabla_{\rho}\nabla^{\rho}h_{\mu\nu} + \frac{1}{2}\nabla_{\mu}\nabla_{\nu}h - \frac{1}{2}g_{\mu\nu}\left(\nabla^{\rho}\nabla^{\sigma}h_{\rho\sigma} - \nabla_{\rho}\nabla^{\rho}h\right). \tag{7}$$

To obtain a weak gravitational waves we choose a gauge that eliminates most of the inconvenient terms of the linearized Einstein equation. The perturbation $h_{\mu\nu}$ is so tiny, that is determined only up to a gauge transformation,

$$h_{\mu\nu} \rightarrow h_{\mu\nu} - 2\nabla_{(\mu}\varepsilon_{\nu)}. \tag{8}$$

Choosing a transverse traceless gauge and after some mathematical manipulations, we have

$$\Box h_{\mu\nu} = 0. \tag{9}$$

As you can see, this is a linear, homogeneous wave equation, which has a solution

$$h_{\mu\nu} = C_{\mu\nu}\exp(i\kappa_{\lambda}x^{\lambda}). \tag{10}$$

where $C_{\mu\nu}$ is a symmetric, constant, traceless tensor. It is a more difficult task to solve the linearized Einstein equations with a non-zero energy-momentum tensor. The analysis of how gravitational waves act on our spacetime once they propagate through vacuum is an easier task.

The first step is to find an appropriate gauge that makes calculations easy for linearized equations in non vacuum. The transverse traceless gauge reveals the basic properties of gravitational waves, such as polarizations and transversity. For calculation purposes, however, this is not the best.

The transverse gauge would eliminate terms containing $\nabla_\rho h^{\mu\nu}$ which would leave us with three terms. You might wonder why the terms containing the trace of $h_{\mu\nu}$ would not vanish in the transverse traceless gauge. Remember, however, that the fact that $h_{\mu\nu}$ was traceless ($h = 0$) was not directly enforced by the gauge. It was rather "by accident" that in the transverse gauge the traceless part of Einstein's equations happened to vanish. So we are still looking for a better gauge, i.e. a gauge that eliminates more terms in the Einstein tensor.

After staring for a while at equation (7) the reader might find that makes sense to rewrite the Einstein tensor in the following way:

$$G_{\mu\nu} \;=:\; \frac{1}{2}(\partial_\sigma\partial_\nu\overline{h}^\sigma_\mu + \partial_\sigma\partial_\nu\overline{h}^\sigma_\nu - \Box\eta_{\mu\nu}\partial_\rho\partial_\lambda\overline{h}^{\rho\lambda})$$

Here we define

$$\overline{h}_{\mu\nu} := h_{\mu\nu} - \frac{1}{2}h\eta_{\mu\nu}$$

$h_{\mu\nu}$ is often called the trace reversed perturbation, since $h^\mu_\nu = -h^\mu_\nu$.

After some mathematical manipulations, Einstein equations $G_{\mu\nu} = -16\pi GT_{\mu\nu}$ become

$$-2G_{\mu\nu} = \Box\overline{h}_{\mu\nu} = -16\pi GT_{\mu\nu}$$

The best way to solve this equation for an arbitrary matter distribution is to solve for a point source and integrate the solution for the point source over the volume of the extended source. This leaves us with the equation:

$$\overline{h}_{\mu\nu}(x^\sigma) = -16\pi G\int G(x^\sigma - y^\sigma)T_{\mu\nu}(y^\sigma)d^4y$$

where $G(x^\sigma - y^\sigma)$ is the Green function that solves

$$\Box G(x^\sigma - y^\sigma) = \delta(x^\sigma - y^\sigma),$$

o we will just write down the solution.

$$\overline{h}_{\mu\nu}(x^\sigma) = 4G\int \frac{1}{|\overrightarrow{x} - \overrightarrow{y}|}T_{\mu\nu}(t - |\overrightarrow{x} - \overrightarrow{y}|, \overrightarrow{y})d^3y$$

This equation tells us that the gravitational perturbation at any point in space and time is equal to the sum of the contributions of all point sources on the past light cone. Using Fourier transforms in the last equation we can find

$$\overline{h}_{ij}(x^\sigma) = \frac{2G}{r}\frac{d^2}{dt^2}\int y^i y^j T^{00}(t - |\overrightarrow{x} - \overrightarrow{y}|, \overrightarrow{y})d^3y.$$

where the quadrupole tensor is defined to be

$$I_{ij} = \int y^i y^j T^{00}(y^\sigma) d^3y.$$

Consequently, the only radiating sources are those with a non vanishing, time-dependent quadrupole moment. It is thus easy to exclude some matter distributions from the list of possible radiating sources just according to symmetry considerations [7, 8].

CONCLUSIONS

Using a mathematical analysis of the linearized Einstein equations we saw that, unlike the case of electromagnetic wave sources where the lowest contribution is a dipole term, for gravitational radiation it is required at least a quadrupole term. If General Relativity is a proper description of Nature, there must be gravitational radiation in our Universe. According to symmetry considerations, the radiating matter distributions could correspond to a couple of masses oscillating along a spring, two colliding black holes, two pulsars orbiting around each other or the radiation generated in the first minutes of the Universe, among others. There is not an experimental evidence of gravitational waves. However, by using perturbation theory we can investigate some quantitative properties of the radiation. For instance, we are able to answer the following questions: how much power is radiated, how large is the amplitude and what frequencies do we expect.

ACKNOWLEDGEMENTS

I want to thank to the Organizers of the School and gratefully acknowledge the fruitful discussions with Ricardo García. This work was supported by project CONACYT-UdeG 49924/2005.

REFERENCES

1. Bernard F. Schutzs, *A First Course in General Relativity*, Cambridge: Cambridge University Press, (1985).
2. Kip, Thorne, *Gravitacional Waves*, http://xxx.lanl.gov/PScache/gr-qc/pdf/9506/9506086.pdf, (1995).
3. Luis, Lehner, *Class. Quantum Gravity*, Vol. 18, (2001) R25.
4. Charles, Misner, Kip, Thorne, John, Wheeler, *Gravitation*, New York: W.H. Freeman and Company, (1973).
5. Carroll, Sean. *Spacetime and Geometry*, San Francisco: Addison Wesley, (2004).
6. Barton, Zwiebach, *A First Course in String Theory*, Cambridge: Cambridge University Press, (2004).
7. Claudia Moreno and Darío Núñez *Gravitational perturbations of the Kerr black hole due to arbitrary forces*, Int'l Journal of Modern Physics D, Vol. **11**, No. 8 (2002).
8. Ciufolini, I., Gorini, V., Moschella, U., Fre, P. *Gravitational Waves (Studies in High Energy Physics, Cosmology and Gravitation)*. U.K.: IOP Publishing Ltd., (2001).

Wormholes, Star-trek: Reality and Science Fiction

Tonatiuh Matos

Departamento de Física, Centro de Investigación y de Estudios Avanzados del IPN, A.P. 14-740, 07000 México D.F., Mexico.

Abstract. Wormholes are a speculative derivation of the Einstein's equations since the pioneer works of Albert Einstein and Nathan Rosen in 1935. The cosmology recently opened a door for the possibility of the existence of wormholes, because 73% of the matter of the Universe is made of dark energy, which could collapse to a star with an inner wormhole. If this could be possible, then the star-trek would be a reality.

INTRODUCTION

One of the most important factors for evolution of people is doubtless the communication. European nations evolve so strong because of the existence of the Mediterranean, an easily sailable sea which permits the communication between people and many regions of the Earth. This is a part of the explanation why humanity evolves so fast; the communication between people is now very efficient. Could you imagine the development of our specie if we could establish contact with beings from other planet, or from other region of the galaxy or from other galaxy? The consequences of this contact would be the beginning of a new era of the human beings. This is the reason why for many people is so important to study in some way the traveling trough starts and to visit our Universe. The main problem for this travel was given in the Einstein's work of special relativity, the maximal speed a space shift can reach is the speed of light. With this speed, in our frame, any space shift will need hundreds or thousands of years to reach any interesting region. And this is supposing that we are able to reach this speed. This travels are then impossible. The idea of the wormholes is that this "topological" objects are able to communicate to regions of the space-time, even if they are very far away. In principle, a space shift could travel trough this objects around the Universe without violating any physic law in a reasonable time. This is the reason why this speculations are so interesting for the physicists, even when we know that this objects are still science fiction, it is worth to keep looking for the possibility of their existence.

As we mention in the abstract of this work, the wormhole (WH) solutions of the Einstein equations were started by Einstein himself, and Nathan Rosen [1]. In this pioneer work they were interested in giving a field representation of particles. The idea was further developed by Ellis, [2] and others, where they tried to model "bridges" between two regions of the space-time. The idea consists in the following. First we

CP885, *Advanced Summer School in Physics 2006, Frontiers in Contemporary Physics—EAV06*, edited by O. Miranda, M. Carbajal, L. M. Montaño, O. Rosas-Ortiz, and S. A. Tomás Velázquez

write the Schwarzschild metric

$$ds^2 = -(1-\frac{2M}{r})dt^2 + \frac{dr^2}{1-\frac{2M}{r}} + r^2 d\Omega^2 \tag{1}$$

where M is the mass of the black hole and $d\Omega^2 = d\theta^2 + \sin^2(\theta)d\varphi$. Observe that the radial coordinate r is such that $0 < r < \infty$. If we perform the coordinate transformation $l^2 = r - 2M$, we arrive at

$$ds^2 = -\frac{1}{1+\frac{2M}{l^2}}dt^2 + 4(l^2+2M)dl^2 + (l^2+2M)^2 d\Omega^2 \tag{2}$$

where now the coordinate l runs from $-\infty < l < \infty$. In these coordinates, the factor in front of $d\Omega^2$ is never zero. Let us analyze this metric. First, we make a photo of the black hole, i.e., we make $t = const$. This metric has spherical symmetry which implies that the black hole looks exactly the same for any θ. We can watch the black hole from the equator, this means we make $\theta = \pi/2$. Under this conditions, the metric can be rewritten as

$$ds^2 = 4(l^2+2M)dl^2 + (l^2+2M)^2 d\varphi^2 \tag{3}$$

Now, we can compare this metric with the three dimensional cylindrically symmetric metric

$$ds^2 = d\rho^2 + dz^2 + \rho^2\, d\varphi^2 = \left(1+(z_{,\rho})^2\right)d\rho^2 + \rho^2\, d\varphi^2 \tag{4}$$

where we have assumed that $z = z(\rho)$. We can compare (3) and (4), making $\rho = l^2 + 2M$ and $\left(1+(z_{,\rho})^2\right) = (l^2+2M)/l^2 = \rho/(\rho-2M)$. Integrating z we obtain $z = \sqrt{8M\,|\rho - 2M|}$. If we go around the φ coordinate, we obtain the figure Fig1. This is an Einstein bridge communicating two spaces.

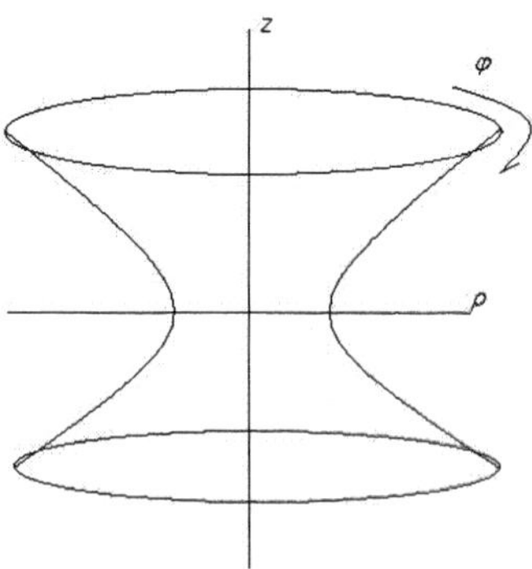

FIGURE 1. The plot of the function $z(\rho)$ for the Schwarzschild metric. The coordinate φ goes around the figure.

This can be interpreted as follows. Observe that the radius of the sphere around the singularity of the black hole never closes. This is so because the area of this sphere is $A(l) = 4\pi(l^2 + 2M)^2$, but the minimum of this area is $A(0) = 4\pi(2M)^2$, which is never zero. There, in the minimum area region, the radius of the sphere is $\rho_{min} = 2M$. However, this metric has a serious problem. The gravitational potential of this metric is given by $e^{2u} = 1 - M/r$, where e^{2u} is the factor in front of the $-dt^2$ term, and u might be interpreted in these coordinates as the "Newtonian potential". Close to the horizon, this interpretation fails, but close to the horizon the gravitational forces can be enormous. A deeper analysis of these forces requires the study of the geodesic deviations of this metric. Nevertheless, if an astronaut would go trough the black hole, we would expect that the mass of the black hole should not be so big, in order that the tidal forces of the black hole do not destroy the astronaut. Suppose the black hole has a mass similar to the Earth's one. Then the mass parameter is $M = 8.8$mm (in units where $G = c = 1$) and the radius is too small for an astronaut to go trough. If the radius is of order of kilometers, then the tidal forces will destroy the astronaut. The first work which has attracted a lot of attention for the study of the wormholes was presented in a series of papers by Morris and Thorne [3]. In these papers, they propose the idea of considering such solutions as actual connections between two separated regions of the Universe. Perhaps, this is the first time these solutions evolved from a sort of science-fiction scenario to a solid scientific topic, even at the point of considering the actual possibility of their existence. The idea of a wormhole is just to create a short way across the Universe, like a worm through an apple. It turned out that the solution proposed by Ellis [2] actually could be interpreted as the identification, or union, of two different regions, no matter how far apart they were or even if those regions were in the same space-time. You could still identify two regions. Ellis obtained a solution which allows one to go from one region to another by means of this identification.

In the eighties, Carl Sagan wrote the novel *Contact*, where the astronomer Ellie Arroway traveled through a wormhole to visit other civilizations. One of the beings of the other civilization adopted the form of her father in order to make easy the contact between them. Nevertheless, Sagan wrote a science-fiction novel and he wanted to be sure that there was not a violation of any physical law. Some years after the publication of *Contact*, Sagan asked to physicist how it could be possible to travel across the Universe without violating Lorentz invariance or any other physical law. Kip Thorne and his students gave an answer to the question. They found the following metric for a wormhole

$$ds^2 = -dt^2 + \frac{dr^2}{1 - \frac{b^2}{r^2}} + r^2 d\Omega^2 \tag{5}$$

As in the black hole case, the coordinate transformation $r^2 = l^2 + b^2$ produces

$$ds^2 = -dt^2 + dl^2 + (l^2 + b^2) d\Omega^2 \tag{6}$$

This metric has the feature that the tidal forces are small and the radius of the throat is b, which can be as big as you want. This is then a wormhole for transportation. Nevertheless, they found that such solutions require an "exotic" type of matter able to violate the energy conditions which we usually impose on any typical kind of matter, (see [4] for a detailed review of this subject). The solutions exist but they need to be generated by matter which apparently does not exist. Actually one needs something very peculiar to warp space-time or to make holes in it. The energy momentum tensor implies that the matter density of this wormhole is negative. The source of this wormhole is a scalar field, such that $\phi \sim \arctan(l/b)$, but with negative kinetic part. This feature was a serious drawback for the possibility of their existence in nature, so WH's remained in the realm of fiction.

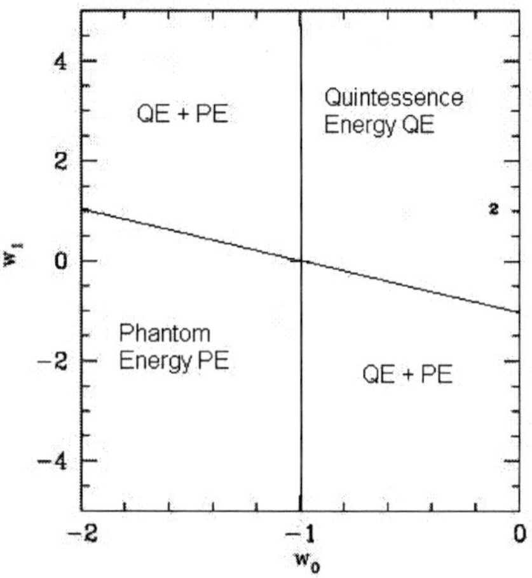

FIGURE 2. Plot of the different regions for the dark energy state equation in order to fit the several observational constraints.

Nevertheless, the situation changed radically in the last years. In 1998 it was shown that the Universe is in a stage of accelerated expansion. This implies the existence of an unknown sort of matter, a kind of anti-gravitational or repulsive matter. The biggest surprise is that this "new substance" should dominate the matter contents of the Universe: it represents more than 73% of the whole matter of the Universe and it is called *dark energy*. This new type of matter forms the overwhelming majority of matter in the Universe and seems to be found everywhere [5]. More recently, some works have also discussed the plausibility of energy-condition violations at the classical level of the dark energy (see for example [6]). This study consists in the following. If the dark energy of the Universe is a perfect fluid, it should be a state equation like $p = \omega\rho$. The most popular supposition for the factor ω is that it is lineal in the scale factor, which means

$$\omega = A_0 + A_1 a = \omega_0 + \omega_1 \frac{z}{1+z} \tag{7}$$

where $a = 1/(1+z)$ is the scale factor of the Universe and gives the rate of how much the Universe is expanding, z is the redshift of the object we are observing. This ansatz is very popular because it summarizes the different hypothesis of the dark energy with simple values. For example, if the dark energy is the cosmological constant, the values for ω are $\omega_0 = -1$, $\omega_1 = 0$, etc. The most interesting hypothesis on the nature of the dark energy is that it is a scalar field or a Phantom field, that is, a scalar field for which the kinetic term has opposite sign. In Figure 2 we see the different values for the constants of the different hypothesis of the dark energy.

After the discovery of the dark energy, there is an agreement in the scientific community that matter which violates some of the energy conditions may exist. Thus, the idea that the WH's can be rejected because of the type of the required matter is not as tenable as it was once. In Figure 3 it is shown the results from the observations on Ia supernova (SNIa) and gamma ray burst (GRB) as standard candles. We see that the observations let a great space for the phantom energy, which can cause the existence of WH's.

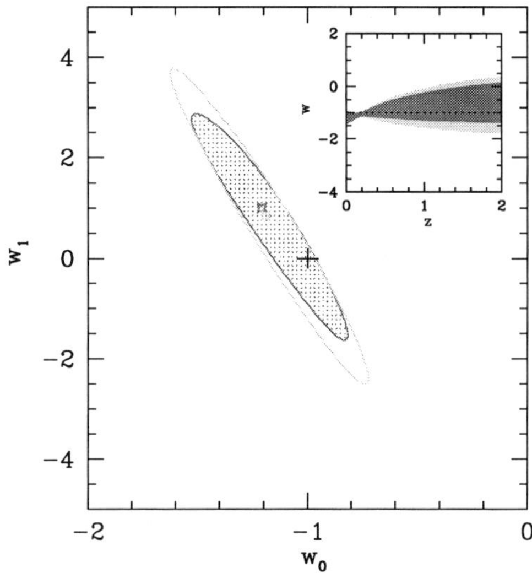

FIGURE 3. Plot of the results for the different observations, taking into account the SNIa and the GRB together. This figure is courtesy of the authors of reference [7]

A mayor problem faced by WH solutions is their stability. One expect that WH should be traversable, that is, an astronaut should be able to go from one side of the throat to the other in a finite time as measured by the astronaut and by another observer far away from him, and without feeling large tidal forces. The stability problem of the "bridges" has been studied since the $60's$ by Penrose [8] in connection with the stability of Cauchy horizons. However, the stability of the throat of a WH was just recently studied numerically by Shinkay and Hayward [9]. They shown that the WH, as proposed by Thorne [3], when perturbed by a scalar field with a stress-energy tensor with the usual sign, may collapse to a black hole and the throat should be closed. In the same way, when the perturbation is due to a scalar field of the same type as that generating the WH, the throat grows exponentially, showing that the solution is highly unstable. It is in this sense that the possibility of the existence of traversable wormholes is again limited. To solve this problem, it is intuitively clear that a rotating solution would have a higher possibility of being stable. We expect that stationary axial symmetric solutions, more general than the one proposed by Thorne, could be stable. Then, the next step is to construct a rotating scalar field wormhole [12]. The idea is to use solution generation techniques developed in the late $80's$ for the Einstein's equations (see [11]), for which it is possible, in the chiral formulation [13], to derive the Kerr solution starting from the Schwarzschild one. In [12] these techniques were applied on the WH proposed by Ellis and Thorne. The final result is the following ansatz for the line element:

$$\begin{aligned} ds^2 &= -f\,(dt + a\cos\theta\, d\varphi)^2 \\ &+ \frac{1}{f}\left[dl^2 + \left(l^2 - 2\,l\,l_1 + {l_0}^2\right)\left(d\theta^2 + \sin^2\theta\, d\varphi^2\right)\right], \end{aligned} \tag{8}$$

where $f = f(l)$ is an unknown function to be determined by the field equations, l_0, l_1 are constant parameters with units of distance such that $l_0^2 > l_1^2$, $l_0 \neq 0$, and a is the rotational parameter, i.e., the angular momentum per unit mass (we use the speed of light $c = 1$). In these coordinates the distance l covers the complete manifold, going from minus to plus infinity. Notice that, modulo f, there is already the throat or bridge feature of the WH's in such a line element, as the coefficient of the angular variables is again never zero.

The formation process of a WH is still an open question. We suppose that some scalar field fluctuation collapses in such a way that it forms a rotating scalar field configuration. This configuration has three regions; the interior, where the rotation is non-zero and two exterior regions, one on each side of the throat, where the rotation stops. The inner boundaries of this configuration are defined where the rotation vanishes. The interior field is the source of the WH. In [12] it has been conjectured that the rotation will keep the throat from being unstable. We follow reference [12] for the construction of the solutions and its analysis.

The Einstein equations with an stress-energy tensor describing an opposite sign massless scalar field, ϕ (see [14]), are given by

$$R_{\mu\nu} = -\frac{8\pi G}{c^4}\,\phi_{,\mu}\,\phi_{,\nu}, \tag{9}$$

with $R_{\mu\nu}$ the Ricci tensor. Using the ansatz (8), the components ^t_t and the sum of ^t_t and ^r_r in equation (9) reduce to

$$\left(\left(l^2 - 2l_1\,l + {l_0}^2\right)\frac{f'}{f}\right)' + \frac{a^2 f^2}{l^2 - 2l_1\,l + {l_0}^2} = 0, \tag{10}$$

$$\left(\frac{f'}{f}\right)^2 + \frac{4\left({l_0}^2 - l_1^2\right) + a^2 f^2}{\left(l^2 - 2l_1\,l + {l_0}^2\right)^2} - \frac{16\pi G}{c^4}\,\phi'^2 = 0, \tag{11}$$

where prime stands for the derivative with respect to l. Recall that the Klein Gordon equation is a consequence of the Einstein's ones, so we only need to solve (10) and (11). Even though this is a solution for the complete space-time, we can see that it as describing the internal parts of a rotating wormhole, from the throat to some distance on each mouth, and match it with an external, asymptotically flat solution, such as Kerr or the static wormhole presented above. In what follows we present this last matching for the radial part of the solution, showing that it can be a smooth one. Then, the matched function for the rotating WH solution reads

$$f=\begin{cases}\exp(\lambda-\frac{\pi}{2}) & \text{if } \; l>l_+\\ 2\,\frac{\phi_0\sqrt{D\left(l_0{}^2-l_1{}^2\right)}\,e^{\phi_0\left(\lambda-\frac{\pi}{2}\right)}}{a^2+De^{2\phi_0\left(\lambda-\frac{\pi}{2}\right)}} & \text{if } \; l_-\leq l\leq l_+\\ \phi_1\exp(\lambda+\frac{\pi}{2}) & \text{if } \; l<l_-\end{cases} \tag{12}$$

where l_- and l_+ respectively are the matching points on the left and right hand side (see Fig.4). It can be seen that the interior solution matches with the rhs exterior once provided that the parameter D becomes:

$$D=2\,\phi_0\sqrt{\left(l_0{}^2-l_1{}^2\right)\left(\left(l_0{}^2-l_1{}^2\right)\phi_0{}^2-\frac{a^2}{E^2}\right)}+2\,\left(l_0{}^2-l_1{}^2\right)\phi_0{}^2-\frac{a^2}{E^2} \tag{13}$$

where the constant E is determined by the radio l_+ where the two solutions match, it is given by:

$$E=\exp\left[\phi_0\left(\arctan\left(\frac{l_+-l_1}{\sqrt{l_0^2-l_1^2}}\right)-\frac{\pi}{2}\right)\right]. \tag{14}$$

In order to have a real solution everywhere we impose the constraint $4M^2=\left(l_0{}^2-l_1{}^2\right)\phi_0{}^2>\frac{a^2}{E^2}$. On the other hand, the matching of the interior solution is smooth with the lhs exterior one, if the constant ϕ_1 is chosen such that

$$\phi_1=2\,\frac{\phi_0\sqrt{D(l_0{}^2-l_1{}^2)}\,e^{2\,\phi_0\,\lambda_-}}{a^2+De^{2\,\phi_0\left(\lambda_--\frac{\pi}{2}\right)}}, \tag{15}$$

where λ_- is evaluated at some $l=l_-$ on the lhs. In Fig. 4 we see the plot of f. Observe that the matching on the rhs could be smooth if l_+ is sufficiently large and the rotation parameter a is sufficiently small. On the contrary, for small l_+ or/and big rotation parameter the matching on the rhs is continuous but not necessarily smooth. On the lhs and for the scalar field the matching is always smooth. Following the procedure above described to obtain the shape of the throat, we obtain a figure similar to Fig. 1 as for the non-rotating case.

Let us write the solution in terms of the mass parameter M. We start with the constant D, it reads

$$D=M^2\left(4\sqrt{4-\frac{J^2}{E^2}}+8-\frac{J^2}{E^2}\right)=M^2\,d^2 \tag{16}$$

where $J=a/M$. Observe that d does not depend on the mass parameter $-M=mG/c^2$, where m is the total mass of the scalar field star. Furthermore, if we want a traversable

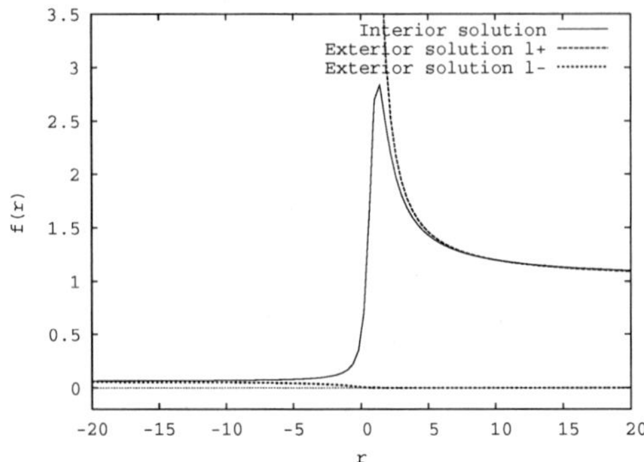

FIGURE 4. Plot of the function f for the exterior and interior solutions, observe that for these values the matching of the solutions on the rhs is smooth, while the matching is always smooth on the lhs. The values of the parameters are $l_1 = 0.5, l_+ = 10, \phi_0 = 2, l_0 = 1, a = 0.6$ and we plot for radius from -20 to 20.

wormhole, we expect a gravitational field similar to the Earth's one, this means $m \sim$ Earth's mass. For this value the mass parameter is $-M \sim 0.01$ meters. But we want that a spaceship can go trough the throat of the wormhole, we can suppose that $l_0 \sim 10$ meters, then, $l_1 \sim 10$ meters as well, depending on the value of ϕ_0. Thus, the interior solution reads

$$f_{int} = 4\frac{-M\sqrt{D}e^{\phi_0(\lambda-\frac{\pi}{2})}}{a^2+De^{2\phi_0(\lambda-\frac{\pi}{2})}} = 4\frac{d\,e^{\phi_0(\lambda-\frac{\pi}{2})}}{J^2+d^2e^{2\phi_0(\lambda-\frac{\pi}{2})}} \tag{17}$$

where the function λ is given by

$$\lambda = \arctan\left(-\phi_0\frac{l}{2M}\right) \tag{18}$$

Observe that $l_{Sch} = -2M$ is the Schwarzschild radius. The scalar field star which produce the wormhole could be thousands of meters long. Compare with the mass parameter M of orders of millimeters, the matching could be considered as if it were at infinity. In that case the parameter $E = 1$. Thus the solution has only 3 parameters, M, J and ϕ_0. In this unites $|J| \leq 2$. Then the parameter d is bounded to $2 \leq d \leq 4$. Nevertheless, the matching condition on the left hand side (in l_-) does not depend on the mass parameter M, it is given by

$$\phi_1 = 4\frac{d\,e^{2\phi_0\lambda_-}}{J^2+d^2e^{2\phi_0(\lambda_- -\frac{\pi}{2})}}. \tag{19}$$

where $\lambda_- = \lambda(l_-)$. We can take l_- such that $\lambda_- \simeq -\pi/2$. Thus, on the left hand side the metric at minus infinity can be written as

$$ds^2_{\text{far}} = -\phi_1 dt_+^2 + \frac{1}{\phi_1}\left(dl_+^2 + l_+^2 d\Omega^2\right) = -dt_-^2 + dl_-^2 + l_-^2 d\Omega^2, \tag{20}$$

where now $\Delta t_- = \sqrt{\phi_1}\Delta t_+$ and $\Delta l_- = \frac{1}{\sqrt{\phi_1}}\Delta l_+$.

The constant ϕ_1 can be very big. Amazingly the major values for ϕ_1 are obtained for small rotation, $J << 1$. It can be seen that ϕ_1 has a maximum value when ϕ_0 is

$$\phi_{0max} = -\frac{1}{2\pi}\ln\left(\frac{J^2E^2}{4\sqrt{4E^2-J^2}\,E+8E^2-J^2}\right) \tag{21}$$

Thus, an observer on the left hand side space (on l_-) will feel that the time goes fast for small changes in Δt_+ and will measure small changes of space for big changes on Δl_+. It is left to find a phantom field star with small rotation and scalar charge given by

$$\sqrt{\frac{8\pi G}{c^4}}q_\phi = 2M\sqrt{\frac{2}{\phi_0^2}+\frac{1}{2}} \tag{22}$$

with a scalar charge given by $\phi_0 = \phi_{0max}$. For example, if the phantom scalar star has a rotation parameter like $J = 10^{-10}$, then $\phi_1 \sim 1.4\times 10^5$ when the scalar charge is $q_\phi \sim 0.3M\ m_{Plank}$. For a star with an Earth's mass, this charge is equivalent to $q_\phi \sim 0.003\ m_{Plank}$ per meter. These results are very similar to those propose by Carl Sagan in his science fiction story. A spaceship can travel trough the throat in a small time across very, very big distances. Suppose that the astronaut on the spaceship travels 18 hours = 6.48×10^4 seconds long, like in the Carl Sagan's Ellie Arroway did. An observer on the rhs space-time will measure $\Delta t_+ = \frac{1}{\sqrt{\phi_1}}\Delta t_- = 205$ seconds, this means 3.4 minutes. On the other side, if Arroway travels only some meters, on the Earth people will measure some kilometers. Furthermore, an observer in our side of the space-time will measure a speed

$$\frac{\Delta l_+}{\Delta t_+} = \phi_1\frac{\Delta l_-}{\Delta t_-} \tag{23}$$

where $\Delta l_-/\Delta t_-$ is the speed of the Arroway proper spaceship. For the example we are studying here, this means that we see a speed $\sim 10^5$ times bigger than the spaceship proper speed. For example, if the spaceship can flight ~ 100 km per second, as a normal earthlings spaceship can do, the local observer will detect a speed $\sim 10^7$ km per second, this is ~ 100 times the speed of light. This velocity is enough for traveling to other stars.

Even when the wormholes still belong to the science fiction arena, we can say that the existence of the dark energy and the Casimir effect have open the door for the possibility of traveling across the Universe.

ACKNOWLEDGMENTS

The author wish to thank Dario Nuñez for many helpful discussions and jokes. This work was partly supported by CONACyT México, under grants 32138-E, 42748 and 45713-F.

REFERENCES

1. A. Einstein, and N. Rosen, Phys. Rev. **48**, 73, (1935).
2. H. G. Ellis, J. Math. Phys., **14**, 395, (1973).
3. M. S. Morris, K. S. Thorne, Am. J. of Physics, **56**, N. 5, 365, (1988).
4. M. Visser, *Lorentzian wormholes: form Einstein to Hawking*, I. E. P. Press, Woodsbury, N. Y. 1995.
5. F. S. N. Lobo, Phys.Rev. D71 (2005) 084011, e-print: gr-qc/0502099.
6. T. Roman, e-print: gr-qc/0409090.
7. C. Firmani, V. Avila–Reese, G. Ghisellini[1] and G. Ghirlanda, e-print: astro-ph/0605430.
8. R. Penrose, Battele Rencontres, ed. by B. S. de Witt and J. A. Wheeler, Benjamin, New York, (1968).
9. H. Shinkay and S. A. Hayward, Phys. Rev. **D** , (2002), e-print: gr-qc/0205041.
10. E. Teo, e-print: gr-qc/9803098, P. K. F. Kuhfitting, e-print: gr-qc/0401023, e-print: gr-qc/0401028, e-print: gr-qc/0401048.
11. Stephani, H., Kramer, D., MacCallum, M., Hoenselaers, C., Herlt, E., *Exact solutions to Einstein's field equations*, 2nd, ed., Cambridge, U. P., U. K., (2003).
12. Tonatiuh Matos and Dario Nuñez. Rotating Scalar Field Wormhole. *Class. Quant. Grav.*, **23**, (2006), 4485-4495. Available at: *gr-qc/0508117*.
13. T. Matos, Gen.Rel.Grav.19:481-492,1987. T. Matos, Phys.Rev.D38:3008-3010,1988. T. Matos, Annalen Phys.46:462-472,1989. R. Becerril, T. Matos, Phys.Rev.D41:1895-1896,1990. T. Matos, G. Rodriguez, In *Cocoyoc 1990, Proceedings, Relativity and gravitation: Classical and quantum* 456-462. R.B. Becerril, T.C. Matos, In *Cocoyoc 1990, Proceedings, Relativity and gravitation: Classical and quantum* 504-509. T. Matos, G. Rodriguez, R. Becerril, J.Math.Phys.33:3521-3535,1992. Tonatiuh Matos, J.Math.Phys.35:1302-1321,1994. [GR-QC 9401009]. Tonatiuh Matos, CIEA-GR-94-06 (n.d.) 8p. [HEP-TH 9405082]. Tonatiuh Matos, Dario Nunez, Hernando Quevedo, Phys.Rev.D51:310-313,1995. [GR-QC 9510042]. Tonatiuh Matos, Cesar Mora, Class.Quant.Grav.14:2331-2340,1997. [HEP-TH 9610013]. Tonatiuh Matos, Cesar Mora, Class.Quant.Grav.14:2331-2340,1997. [HEP-TH 9610013]. Tonatiuh Matos, Phys.Lett.A249:271-274,1998. [GR-QC 9810034]. T. Matos, U. Nucamendi, P. Wiederhold, J.Math.Phys.40:2500-2513,1999. T. Matos, M. Rios, Gen.Rel.Grav.31:693-700,1999. Tonatiuh Matos, Dario Nunez, Gabino Estevez, Maribel Rios, Gen.Rel.Grav.32:1499-1525,2000. [GR-QC 0001039]. Tonatiuh Matos, Dario Nunez, Maribel Rios, Class.Quant.Grav.17:3917-3934,2000. [GR-QC 0008068].
14. Jian-gang Hao, Xin-zhou Li.Phys.Rev. D67 (2003) 107303. L.P. Chimento, Ruth Lazkoz. Phys.Rev.Lett. 91 (2003) 211301. German Izquierdo, Diego Pavon. astro-ph/0505601. S. V. Sushkov.Phys.Rev. D71 (2005) 043520.
15. D. Kramer, H. Stephani, M. MacCallum and E. Herlt. Exact Solutions of Einstein Field Equations. VEB Deutscher Verlag der Wissenschaften, Berlin, 1980.

Nonlinear Electrodynamics and Wormhole Type Solutions for Einstein Field Equations

R. García-Salcedo * and Aarón V. B. Arellano †

*Cinvestav-Monterrey.
Av. Cerro de las Mitras 2565. Col. Obispado. Monterrey, N.L. Mexico 64060.
† Facultad de Ciencias, UAEMéx.
El Cerrillo Piedras Blancas. C.P. 50200, Toluca, Mexico.

Abstract. In this contribution we review the wormhole-type solutions for Einstein field equations. In addition, we show that it is not possible to construct these solutions with nonlinear electromagnetic fields in static spherically symmetric case.

INTRODUCTION

A specific model of nonlinear electrodynamics (NLED) was proposed by Born and Infeld in 1934 [1] founded on a principle of finiteness, namely, that a satisfactory theory should avoid physical quantities becoming infinite. The Born-Infeld model was inspired mainly to remedy the fact that the standard picture of a point particle possesses an infinite self-energy, and it consisted of placing an upper limit on the electric field strength, as velocity in special relativity and considering a finite electron radius. Later, Plebanski showed that the Born-Infeld theory satisfy physically acceptable requirements and presented other examples of nonlinear electrodynamic Lagrangians [2]. Furthermore, a recent revival of nonlinear electrodynamics has been verified, mainly in Dp-branes and supersymmetric extensions, and non-Abelian generalizations [3]. There are applications of NLED in general relativity such as cosmological models, where NLED produces an the inflationary epoch and the late accelerated expansion of the universe [4]. In a recent paper NLED is the ingredient which allows to avoid singularities in some spacetimes [5].

Indeed, it is interesting to notice that there are regular exact solutions in general relativity (GR) were black holes couple with nonlinear electrodynamics [6]. NLED used in such models satisfy the weak energy condition (WEC) and the Maxwell electrodynamics are reproduced in the weak field limit. Also it was shown that GR coupled with NLED gives magnetic black holes and monopoles [7], as well as regular electrically charged structures that have a de Sitter center [8]. The stability of these regular solutions has also been explored [9].

A wormhole is a solution for the Einstein's gravitational field equations that represents a hypothetical tunnel through spacetime. The detailed calculations describing a wormhole are complex and still under debate among physicists because they require a knowledge of general relativity (GR), however, with the techniques of standard calculus [10], a student could be able to reproduce the images of wormholes based on geomet-

CP885, *Advanced Summer School in Physics 2006, Frontiers in Contemporary Physics—EAV06,*
edited by O. Miranda, M. Carbajal, L. M. Montaño, O. Rosas-Ortiz, and S. A. Tomás Velázquez

ric principles [11]. The wormholes have become popular since the Morris and Thorne's paper [12].

In this contribution, we shall be interested in showing that there are no static or spherically symmetric wormhole solutions coupled to NLED with this type of solutions. This property was probed by Bronnikov [7, 13], and confirmed by A. Berrocal and F. Lobo [14]. Next, we shall analyze the (3+1)-dimensional stationary and axisymmetric case coupled to nonlinear electrodynamics.

WORMHOLE SOLUTION

It is interesting to mention that Einstein himself was interested in wormholes, to give a field representation of particles [15]. The idea was further developed by Ellis [16], where, instead of particles, wormholes were modeled as "bridges" between two regions of spacetime. The idea of considering such solutions as actual connections between two separated regions of the Universe, has attracted a lot of attention since the seminal work of Morris and Thorne [12].

In this contribution, we study the static and spherically symmetric spacetime metric of such wormholes which is given by [12]

$$ds^2 = -e^{\Phi(r)}dt^2 + \frac{dr^2}{1-\frac{b(r)}{r}} + r^2(d\theta^2 + \sin^2\theta d\phi^2), \tag{1}$$

where $\Phi(r)$ is the shift function and $b(r)$ the shape function. This solution must everywhere obey the Einstein field equations of general relativity. The radial coordinate has a range that increases from a minimum value r_0, corresponding to the wormhole throat, up to infinity.

On the other hand, to traverse a wormhole, it is necessary to demand the absence of event horizons that are identified as the surface with $e^{2\Phi(r)}$ tending to zero, thus, $\Phi(r)$ must be finite everywhere. A fundamental property of wormhole physics is the flaring out condition, which is deduced from the mathematical conditions of embedding and is given by

$$(b - b'r)/b^2 > 0. \tag{2}$$

Notice that at the throat, i.e. $b(r_0) = r_0$, the flaring out condition reduces to $b'(r_0) < 1$. The condition $(1 - b/r) > 0$ is also imposed.

WORMHOLE SOLUTION COUPLED WITH NLED

We use the following units through this work, $G = 1 = c$. The action of $(3+1)-$dimensional general relativity coupled to nonlinear electrodynamics is given by

$$S = \int \sqrt{-g}\left[\frac{R}{16\pi} + L(F)\right] d^4x, \tag{3}$$

where R is the Ricci scalar and $L(F)$ is the gauge-invariant electromagnetic Lagrangian that depends on a single invariant F [2], defined by $F \equiv \frac{1}{4}F^{\mu\nu}F_{\mu\nu}$. We shall not consider the case where L depends on the invariant $G \equiv \frac{1}{4}F_{\mu\nu}{}^{*}F^{\mu\nu}$.

Varying the action with respect to the gravitational field provides the Einstein field equations $G_{\mu\nu} = 8\pi T_{\mu\nu}$, where the stress-energy tensor is given by

$$T_{\mu\nu} = g_{\mu\nu}L(F) - F_{\mu\alpha}F_{\nu}{}^{\alpha}L_F\,. \tag{4}$$

Taking into account the symmetries of the geometry, the only non-zero compatible terms for the electromagnetic tensor are $F_{tr} = E(r)$ and $F_{\theta\phi} = B(r)$. The spacetime metric representing a spherically symmetric and static $(3+1)-$dimensional wormhole takes the form of Eq.(1).

One may prove the non existence of (3+1)-dimensional static and spherically symmetric traversable wormholes in nonlinear electrodynamics through an analysis of the null energy condition (NEC) violation. Recall the fundamental condition in wormhole physics is the violation of the NEC , which is defined as $T_{\hat{\mu}\hat{\nu}}k^{\hat{\mu}}k^{\hat{\nu}} \geq 0$, where $k^{\hat{\mu}}$ is any null vector. Considering the orthonormal reference frame with $k^{\hat{\mu}} = (1,\pm 1,0,0)$, it turns out that solving Eq. (4) for the metric of Ec. (1) gives the following result

$$T_{\hat{\mu}\hat{\nu}}k^{\hat{\mu}}k^{\hat{\nu}} = \frac{1}{8\pi}\left[\frac{b'r-b}{r^3} + \left(1-\frac{b}{r}\right)\frac{\Phi'}{r}\right]. \tag{5}$$

It is a simple matter to prove that, for this geometry, the NEC is violated at the throat when $r = r_0$, i.e., $T_{\hat{\mu}\hat{\nu}}k^{\hat{\mu}}k^{\hat{\nu}} < 0$. The relevant components for the stress-energy tensor are then

$$T_{\hat{t}\hat{t}} = -L - e^{-2\Phi}\left(1-\frac{b}{r}\right)E^2 L_F\,, \tag{6}$$

$$T_{\hat{r}\hat{r}} = L + e^{-2\Phi}\left(1-\frac{b}{r}\right)E^2 L_F\,. \tag{7}$$

We consider $|e^{-2\Phi}(1-b/r)E^2 L_F| < \infty$, as $r \to r_0$, to ensure that the stress-energy tensor components are regular. From the Einstein's field equations, we verify the following condition

$$\Phi' = -\frac{b'r-b}{2r(r-b)}\,, \tag{8}$$

which may be integrated to yield the solution $e^{2\Phi} = (1-b/r)$, rendering a non-traversable wormhole solution, as it possesses an event horizon at the throat, $r = r_0$.

Note that the NEC, for the stress-energy tensor defined by (4), is identically zero for arbitrary r, i.e., $T_{\hat{\mu}\hat{\nu}}k^{\hat{\mu}}k^{\hat{\nu}} = 0$. In particular this implies that the flaring-out condition at the throat is not satisfied, showing, therefore, the non-existence of $(3+1)-$dimensional static and spherically symmetric traversable wormholes coupled to nonlinear electrodynamics. The analysis outlined in this Section is consistent with that of Refs. [7, 13], where it was pointed out that nonlinear electrodynamics, with any Lagrangian of the form $L(F)$, coupled to general relativity cannot support static and spherically symmetric $(3+1)-$dimensional traversable wormholes.

It is interesting to explore the possibility of evolving wormhole spacetimes conformally related to the respective static geometries, within the context of nonlinear electrodynamics [17]. At this moment, we are exploring the possibility to include in this analysis the second invariant G in the electromagnetic Lagrangian.

ACKNOWLEDGMENTS

RGS gratefully acknowledges the Organizers of the School and the fruitful discussions with Nora Breton and Claudia Moreno.

REFERENCES

1. M. Born and L. Infeld, Nature **132**, 970 (1933). M. Born and L. Infeld, Proc. Roy. Soc. London A **144**, 425 (1934).
2. J. F. Plebański, *Lectures on Nonlinear Electrodynamics*, Copenhagen (1968).
3. N. Seiberg and E. Witten, JHEP **9909**, 032 (1999) [arXiv:hep-th/9908142].
4. M. Novello, S. E. P. Bergliaffa and J. Salim, Phys. Rev. D 69 127301 (2004) [arXiv:astro-ph/0312093]; R. Garcia-Salcedo and N. Breton, Int. J. Mod. Phys. A **15**, 4341-4354 (2000) [arXiv:gr-qc/0004017]; R. Garcia-Salcedo and N. Breton, Class. Quant. Grav. **20**, 5425 (2003) [arXiv:hep-th/0212130]; D. N. Vollick, Gen. Rel. Grav. **35**, 1511 (2003) [arXiv:hep-th/0102187].
5. R. Garcia-Salcedo and N. Breton, Class. Quant. Grav. **22**, 4783 (2005) [arXiv:gr-qc/0410142]
6. E. Ayón-Beato and A. García, Phys. Rev. Lett. **80**, 5056 (1998); M. Cataldo and A. García, Gen. Rel. Grav. **31**, 629 (1999) [arXiv:gr-gc/9911084].
7. K. A. Bronnikov, Phys. Rev. D **63**, 044005 (2001) [arXiv:gr-qc/ 0006014].
8. I. Dymnikova, Class. Quant. Grav. **21**, 4417 (2004) [arXiv:gr-qc/0407072].
9. N. Breton, Phys. Rev. D **72**, 044015 (2005) [arXiv:hep-th/0502217].
10. James Stewart, *Calculus*, 5th ed., Brooks/Cole, Belmont, CA, (2003); Deborah Hughes-Hallett, et al., *Calculus*, 3rd ed., Wiley, New York, (2002).
11. Frank Y. Wang (2005)[arXiv:physics/0505108].
12. M. S. Morris and K. S. Thorne, Am. J. Phys. **56**, 395 (1988).
13. K. A. Bronnikov and S. V. Grinyok, Grav. Cosmol.**11**, 75 (2006) [arXiv:gr-qc/0509062].
14. Aarón V. B. Arellano and Francisco S. N. Lobo [arXiv:gr-qc/0604095] submitted to CQG (2006).
15. A. Einstein and N. Rosen, Phys. Rev. **48**, 73 (1935).
16. H. G. Ellis, J. Math. Phys., **14**, 395 (1973).
17. Aarón V. B. Arellano and Francisco S. N. Lobo, in preparation (2006).

Supertubes and Superconducting Membranes

Rubén Cordero and Zelin Miguel-Pilar

Departamento de Física, Escuela Superior de Física y Matemáticas del IPN,
Edificio 9, Unidad Profesional "Adolfo López Mateos", Zacatenco, 07738 México D.F., Mexico.
e-mail: cordero@esfm.ipn.mx, zelin@esfm.ipn.mx.

Abstract. We show the equivalence between configurations that arise from string theory of type IIA, called supertubes, and superconducting membranes at the bosonic level. We find equilibrium and oscillating configurations for a tubular membrane carrying a current along its axis.

INTRODUCTION

M theory can be thought as an extension of string theory to a theory of membranes (branes). It was shown in [1] that a cylindrical, or tubular, D2-brane in a Minkowski vacuum spacetime can be supported against collapse by the angular momentum produced by crossed electric and magnetic Born-Infeld (BI) fields. This D2-brane cylindrical configuration is called supertube.

In the cosmological context, some field theories predict the existence of topological defects like strings or domain walls possibly carrying permanent current. These extended objects are called superconducting membranes. This current is responsible for the existence of stable string configurations called vortons [2]. In the domain walls case is viable to obtain stable configurations due to the presence of currents [3].

In this work we are going to focus on the dynamical similarities between supertubes and superconducting membranes. Besides we will obtain a D2-brane stable configuration with an arbitrary cross section carrying a current along its axis.

SUPERTUBES

We consider a Dp-brane Σ in a N-dimensional background spacetime M, with a metric tensor $g_{\mu\nu}$ $(\mu,\nu = 0,\ldots,N-1)$. The Dp-brane worldvolume is an oriented time-like manifold m of dimension $p+1$. We denote the embedding functions X^μ of the Dp-brane as $x^\mu = X^\mu(\xi^a)$, where x^μ are the local coordinates in the background spacetime and ξ^a $(a,b = 0,1,...,p)$ are the worldvolume coordinates. The induced metric is given by $\gamma_{ab} = g_{\mu\nu}X^\mu_{,a}X^\nu_{,b}$, where $X^\mu_{,a}$ denotes partial derivative of X^μ with respect to ξ^a [4]. The action describing the dynamics of the worldvolume of the Dp-brane is the Dirac-Born-Infeld action

$$S = \int_m d^{p+1}\xi \sqrt{-det(\gamma_{ab}+F_{ab})}, \tag{1}$$

CP885, *Advanced Summer School in Physics 2006, Frontiers in Contemporary Physics—EAV06,* edited by O. Miranda, M. Carbajal, L. M. Montaño, O. Rosas-Ortiz, and S. A. Tomás Velázquez

where $F_{ab} = 2\partial_{[a}A_{b]}$ is the electromagnetic field strength on the worldvolume. The equations of motion are [5]

$$\widetilde{T}^{ab}K^i_{ab} = 0, \tag{2}$$

$$\nabla_a J^{ab} = 0, \tag{3}$$

where $K^i_{ab} = -n^i_\mu \nabla^\mu_a X_b$ is the extrinsic curvature on m, $\widetilde{T}^{ab} = \frac{\sqrt{-M}}{\sqrt{-\gamma}}(M^{-1})^{(ab)}$ is the stress-energy tensor and $J^{ab} = -\frac{\sqrt{-M}}{\sqrt{-\gamma}}(M^{-1})^{[ab]}$ is the covariant bicurrent on the worldvolume. Here $M_{ab} \equiv \gamma_{ab} + F_{ab}$, $(M^{-1})^{ab}$ denotes its inverse, $M = \det(M_{ab})$, $(M^{-1})^{(ab)}$ and $(M^{-1})^{[ab]}$ denotes the symmetric and antisymmetric parts of M^{-1} respectively, $\gamma = \det(\gamma_{ab})$ and n^i_μ are the orthonormal vectors to the tangent vectors $X^\mu_{,a}$. Equation (2) is the generalization of Newton's second law where $\widetilde{T}^{ab}$ represents the mass and K^i_{ab} represents the acceleration [6]. The conservation of the stress-energy tensor is a consequence of (3). Now we focus on the D2-brane case. The stress-energy tensor is $\widetilde{T}^{ab} = \frac{1}{\sqrt{1+F^2}}\left[(1+F^2)\gamma^{ab} + F^{ac}F_c{}^b\right]$, the bicurrent is $J^{ab} = \frac{F^{ab}}{\sqrt{1+F^2}}$, $M = \gamma(1+F^2)$ where $F^2 \equiv \frac{1}{2}F^{ab}F_{ab}$. The former expressions are the building blocks for our theory.

SUPERCONDUCTING MEMBRANES

In this part we consider a p-dimensional relativistic superconducting membrane Σ in a N-dimensional background spacetime M. The effective action for the dynamics of a superconducting membrane is a function of a scalar field ϕ,

$$S = \int_m d^{p+1}\xi\sqrt{-\gamma}\mathscr{L}(\omega), \tag{4}$$

where $\mathscr{L} = \mathscr{L}(\omega(\gamma_{ab}, \phi_{,a}))$ is the Lagrangian density, ω is given by $\omega = \gamma^{ab}\nabla_a\phi\nabla_b\phi$. The equations of motion for the superconducting membrane are [7]

$$T^{ab}K^i_{ab} = 0, \tag{5}$$

$$\nabla_a J^a = \frac{1}{\sqrt{-\gamma}}\partial_a(\sqrt{-\gamma}J^a) = 0, \tag{6}$$

where the stress-energy tensor is $T^{ab} = \mathscr{L}\gamma^{ab} - 2\frac{d\mathscr{L}}{d\omega}\gamma^{ac}\gamma^{bd}\phi_c\phi_d$, $J^a = 2\frac{d\mathscr{L}}{d\omega}\gamma^{ab}\nabla_b\phi$ is the electromagnetic current. The conservation of the stress-energy tensor is a consequence of equation (6).

EQUIVALENCE BETWEEN SUPERTUBES AND SUPERCONDUCTING MEMBRANES

Taking into account the conservation of the current (6) for the case of a D2-brane we can propose the following expression

$$\sqrt{-\gamma}J^a = \varepsilon^{abc}\partial_b A_c \tag{7}$$

where ε^{abc} is the Levi-Civita symbol and A_c is vector field living on the worldvolume. Now we are going to show the equivalence between the action (4) and the following dual action

$$S = \int_m d^{p+1}\xi \sqrt{-\gamma}\widetilde{\mathscr{L}}(\widetilde{\omega}), \tag{8}$$

here we define $\widetilde{\omega} \equiv -J^a J_a = F^2$, i.e. we can consider the former action as a function of A_c. The consistency of the equation of motion for the field A_c, $\nabla_a\left(\frac{\partial \widetilde{\mathscr{L}}}{\partial A_{a,b}}\right) = 0$ with (6) imposes the condition $4\frac{d\widetilde{\mathscr{L}}(\widetilde{\omega})}{d\widetilde{\omega}}\frac{d\mathscr{L}(\omega)}{d\omega} = 1$. From the definition of $\widetilde{\omega}$ we get the relation $4\left(\frac{d\mathscr{L}(\omega)}{d\omega}\right)^2 \omega = -\widetilde{\omega}$. Combining the last two expression we arrive at

$$\widetilde{\mathscr{L}}(\widetilde{\omega}) = \mathscr{L}(\omega) - 2\omega\frac{d\mathscr{L}(\omega)}{d\omega}, \tag{9}$$

which is a relation between the Lagrangian density and its dual. Finally, from the last expressions, it can be shown that $\widetilde{T}^{ab} = T^{ab}$, and this fact produces the equivalence between (2) and (5). Taking the Nielsen Lagrangian density [8] for superconducting membranes (equivalent to a Dirac-Nambu-Goto membrane in an extended spacetime) $\mathscr{L}(\omega) = \sqrt{k_1 + k_2\omega}$, where k_1, k_2 are constants, and by means of the equation (9) it is possible to find the expression for the dual Lagrangian $\widetilde{\mathscr{L}}(\widetilde{\omega}) = \sqrt{\frac{k_1}{k_2}}\sqrt{k_2 + \widetilde{\omega}} = \sqrt{\frac{k_1}{k_2}}\sqrt{k_2 + F^2}$. If we put $k_1 = k_2 = 1$, we have the equivalence between the Nielsen action and Dirac-Born-Infeld action for supertubes.

TUBULAR D2-BRANES

Now we consider a D2-brane with tubular geometry, in a Minkowski space-time background, with the parametrization $X^\mu = (\tau, \mathbf{x})$, where $\mathbf{x} = \mathbf{x}_\perp(\tau,\theta) + z\hat{\mathbf{k}}$, and $\mathbf{x}_\perp$ is on the transverse subspace of the brane and the axis of the tube is along $\hat{\mathbf{k}}$. In order to satisfy equation (3), we can write the field strength in terms of the scalar function ϕ, by means of $F^{ab} = \frac{\sqrt{1+F^2}}{\sqrt{-\gamma}}\varepsilon^{abc}\partial_c\phi$. We consider a superficial charge density on the brane and two currents: one current around its transversal section and the other constant current along its axis. They are given by $\partial_\tau\phi = \dot{\phi}, \partial_\theta\phi = \phi'$ and $\partial_z\phi = c$ where c is a constant.

Due to our parametrization, equation (2) has non z dependence and it reduces to a two dimensional problem on the crossed section of the brane $\partial_i(\Gamma^{ij}\partial_j x^\mu) = 0$, where

$$\Gamma_{ij} = \frac{1}{\sqrt{-\gamma(1-\gamma^{cd}\partial_c\phi\partial_d\phi)}}\begin{pmatrix} -\gamma_{11}(1+c^2) - \phi'^2 & \gamma_{01}(1+c^2) + \phi'\dot{\phi} \\ \gamma_{01}(1+c^2) + \phi'\dot{\phi} & -\gamma_{00}(1+c^2) - \dot{\phi}^2 \end{pmatrix}. \tag{10}$$

If we impose the condition $\Gamma_{ij} = \eta_{ij}$, where η_{ij} is the flat Minkowski metric, we have a wave equation and the most general solution presents the form $\mathbf{x} = \frac{1}{2}(\mathbf{a}(\theta - \tau) + \mathbf{b}(\theta + \tau)) + z\hat{\mathbf{k}}$. Using this solution, the constraints give the expression for the scalar field: $\phi(\tau,\theta) = \frac{1}{2}(f(\theta - \tau) + g(\theta + \tau)) + cz$. The gauge fixing produces the constraints equations $||\mathbf{a}'||^2 + \left(\frac{g'}{\sqrt{1+c^2}}\right)^2 = 1$ and $||\mathbf{b}'||^2 + \left(\frac{f'}{\sqrt{1+c^2}}\right)^2 = 1$, here f' and g' denotes derivatives with respect to its correspondent arguments. When $g'^2 = 1$ or $f'^2 = 1$ we obtain constant values of $||\mathbf{a}||$ or $||\mathbf{b}||$ respectively. In this case, the transverse position of the brane is unchanged in time, and it can take an arbitrary form. If we want closed tubular solutions, the periodicity conditions on the $||\mathbf{a}||$, $||\mathbf{b}||$, f and g functions can be imposed. Then, we propose oscillating periodic solutions of the form

$$\mathbf{a} = A(\sin(\theta-\tau)\hat{\mathbf{i}} + \cos(\theta-\tau)\hat{\mathbf{j}}), \quad \mathbf{b} = B(\sin(\theta+\tau)\hat{\mathbf{i}} + \cos(\theta+\tau)\hat{\mathbf{j}}), \tag{11}$$

$$\phi = \frac{\sqrt{1+c^2}}{2}(\sqrt{1-A^2}(\theta-\tau) + \sqrt{1-B^2}(\theta+\tau)) + cz, \tag{12}$$

where A and B are constants with $0 < A, B < 1$. If we take $A = 1$ for simplicity and $B = 0$, the modulus $||\mathbf{a}||$ is a constant: this is an oscillating supertube configuration. This solution generalized the result presented in [9].

CONCLUSIONS

We showed the equivalence between the equations of motion for Dp-branes derived from a Dirac-Born-Infeld action and the corresponding equations for superconducting membranes derived from an effective action. We obtain an equilibrium and an oscillating tubular brane configuration carrying a current along its axis.

ACKNOWLEDGMENTS

The authors received financial support from CGPI grant 20061047. ZM acknowledges partial support from PIFI and a Cinvestav-IPN grant for the Advanced Summer School 2006. RC thanks COFAA, SNI and CONACyT through grant CO1-41639, for financial support.

REFERENCES

1. D. Mateos and P. K. Townsend, *Phys. Rev. Lett.* **87**, 011602 (2001).
2. A. Vilenkin and E. P. S. Shellard, *Cosmic Strings and Other Topological Defects*, Cambridge University Press, Cambridge, (1994).
3. Z. Miguel, *P-branas, Bachelor's Thesis*, Instituto Politécnico Nacional, México, (2004).
4. R. Capovilla and J. Guven, *Phys. Rev. D* **51**, 6736 (1995).
5. R. Cordero and E. Rojas, *Gen. Rel. Grav.* **37**, 659 (2005).
6. B. Carter, "Brane dynamics for treatment of cosmic strings and vortons" in *Proc. 2nd Mexican School on Gravitation and Mathematical Physics, Tlaxcala, 1996*, edited by A. Garcia, C. Lammerzahl, A. Macias, D. Nuñez, Science Network Publishing, Konstanz, 1997. hep-th/9705172.
7. R. Cordero and E. Rojas, *Phys. Lett B* **470**, 45 (1999).
8. N. Nielsen, *Nucl. Phys. B* **167**, 249 (1980).
9. J. J. Blanco-Pillado, *JHEP* **0508**, 015 (2005).

II. PARTICLES AND FIELDS

Neutrinos: In and Out of the Standard Model

Stephen Parke

Theoretical Physics Department,
Fermi National Accelerator Laboratory,
P.O. Box 500, Batavia, IL 60510, USA
email: parke@fnal.gov

Abstract. The particle physics Standard Model has been tremendously successful in predicting the outcome of a large number of experiments. In this model Neutrinos are massless. Yet recent evidence points to the fact that neutrinos are massive particles with tiny masses compared to the other particles in the Standard Model. These tiny masses allow the neutrinos to change flavor and oscillate. In this series of Lectures, I will review the properties of Neutrinos In the Standard Model and then discuss the physics of Neutrinos Beyond the Standard Model. Topics to be covered include Neutrino Flavor Transformations and Oscillations, Majorana versus Dirac Neutrino Masses, the Seesaw Mechanism and Leptogenesis.

In the Standard Model the neutrinos, $(\nu_e,\ \nu_\mu,\ \nu_\tau)$, are massless and interact diagonally in flavor,

$$\begin{array}{llll} W^+ & \rightarrow e^+ + \nu_e & Z & \rightarrow \nu_e + \bar{\nu}_e \\ W^+ & \rightarrow \mu^+ + \nu_\mu & Z & \rightarrow \nu_\mu + \bar{\nu}_\mu \\ W^+ & \rightarrow \tau^+ + \nu_\tau & Z & \rightarrow \nu_\tau + \bar{\nu}_\tau . \end{array} \quad (1)$$

Since they travel at the speed of light, their character cannot change from production to detection. Therefore, in flavor terms, massless neutrinos are relatively uninteresting compared to quarks.

NEUTRINO OSCILLATIONS IN VACUUM:

If neutrinos have mass, then time passes for them and they can change character since they are not traveling at the speed of light. Typically, the neutrino states that interact with the W and Z bosons are not necessarily the states that propagate simply in time but they are related by a unitary matrix,

$$\begin{pmatrix} \nu_\mu \\ \nu_\tau \end{pmatrix} = \begin{pmatrix} \cos\theta & \sin\theta \\ -\sin\theta & \cos\theta \end{pmatrix} \begin{pmatrix} \nu_1 \\ \nu_2 \end{pmatrix} \quad (2)$$

where $(\nu_\mu,\ \nu_\tau)$ are the flavor states, eg $W^+ \rightarrow \mu^+ + \nu_\mu$ and $(\nu_1,\ \nu_2)$ are the mass eigenstates. The angle θ is the mixing angle to be determined experimentally and eventually explained by the theory of fermion masses. The mass eigenstates propagate in time as $|\nu_j\rangle \rightarrow e^{-ip_j \cdot x}|\nu_j\rangle$ with $p_j^2 = m_j^2$. (The greek (latin) letters $\alpha, \beta \ldots$ $(i, j \ldots)$ refer to flavor (mass) eigenstates.)

CP885, *Advanced Summer School in Physics 2006, Frontiers in Contemporary Physics—EAV06,* edited by O. Miranda, M. Carbajal, L. M. Montaño, O. Rosas-Ortiz, and S. A. Tomás Velázquez

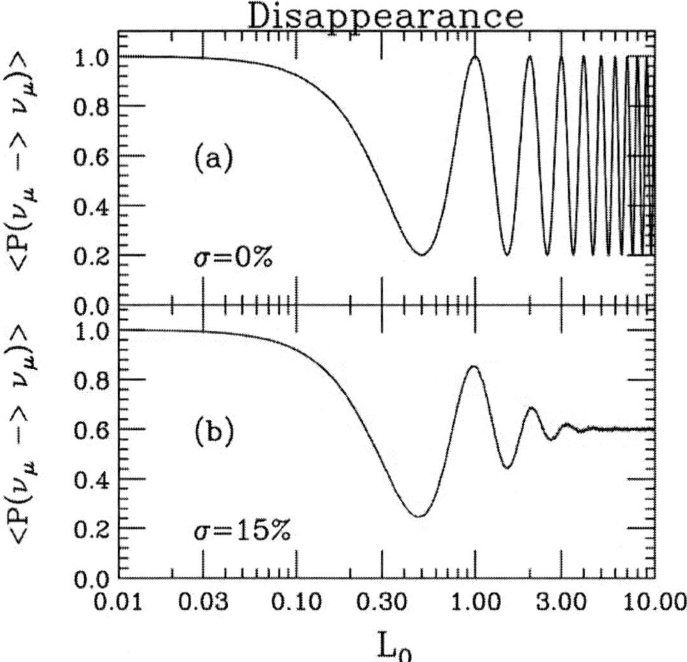

FIGURE 1. The survival probability for a muon neutrino versus distance traveled in units of the oscillation length, $4\pi E/\delta m^2$: (a) for fixed neutrino energy, (b) using a Gaussian energy spread equal to 15% of the mean energy of the neutrino. Notice that even for this narrow band beam the oscillations have disappeared after three oscillations!

Thus, the life of a neutrino can be represented as follows (at the amplitude level):

$$\text{At Production:} \quad |\nu_\mu\rangle = \cos\theta|\nu_1\rangle + \sin\theta|\nu_2\rangle$$

$$\text{During Propagation:} \quad |\nu_1\rangle \to e^{-ip_1\cdot x}|\nu_1\rangle \text{ and } |\nu_2\rangle \to e^{-ip_2\cdot x}|\nu_2\rangle$$

$$\text{At Detection:} \quad \begin{cases} |\nu_1\rangle = \cos\theta|\nu_\mu\rangle - \sin\theta|\nu_\tau\rangle \\ |\nu_2\rangle = \sin\theta|\nu_\mu\rangle + \cos\theta|\nu_\tau\rangle \end{cases}$$

Thus, the transition probability for a neutrino to change flavor is

$$P(\nu_\mu \to \nu_\tau) = |\cos\theta(e^{-ip_1\cdot x})(-\sin\theta) + \sin\theta(e^{-ip_2\cdot x})\cos\theta|^2. \tag{3}$$

Using the same E formulation, we have that $p_j = \sqrt{E^2 - m_j^2} \approx E - \frac{m_j^2}{2E}$ and therefore

$$P(\nu_\mu \to \nu_\tau) = \sin^2\theta\cos^2\theta|e^{-im_2^2L/2E} - e^{-im_1^2L/2E}|^2 = \sin^2 2\theta\sin^2\Delta \tag{4}$$

where $\Delta \equiv \delta m^2 L/4E$ is the kinematic phase, with $\delta m^2 = m_2^2 - m_1^2$. The disappearance probability is given by

$$P(\nu_\mu \to \nu_\mu) = 1 - P(\nu_\mu \to \nu_\tau) = 1 - \sin^2 2\theta \sin^2 \Delta. \quad (5)$$

If we put the $\hbar$'s and c's into the appearance probability we find

$$P(\nu_\mu \to \nu_\tau) = \sin^2 2\theta \sin^2 \left(\frac{\delta m^2 c^4 L}{4\hbar c E} \right). \quad (6)$$

In the semi-classical limit, $\hbar \to 0$, the oscillation length goes to zero and the oscillations are averaged out. This is the same limit as letting δm^2 become large. This is precisely what happens in the quark sector. In Fig. 1 we have shown the oscillation probability for both fixed energy and a Gaussian spread of 15% of the mean neutrino energy. Notice that oscillations are observable only for a limited range of distance. At small distance the simple flavor description is a good one. But at very large distance using the probability description with mass eigenstates works well since the oscillations are averaged out. The neutrino mass eigenstates are effectively incoherent. Thus, in terms of probabilities[1]

At Production:	the fraction of $\lvert\nu_\mu\rangle$ that is $\lvert\nu_1\rangle$ is $\cos^2\theta$
	the fraction of $\lvert\nu_\mu\rangle$ that is $\lvert\nu_2\rangle$ is $\sin^2\theta$
During Propagation:	flavor fractions in $\lvert\nu_1\rangle$ and $\lvert\nu_2\rangle$ remain unchanged
At Detection:	the fraction of $\lvert\nu_1\rangle$ that is $\lvert\nu_\mu\rangle$ is $\cos^2\theta$
	the fraction of $\lvert\nu_1\rangle$ that is $\lvert\nu_\tau\rangle$ is $\sin^2\theta$
	the fraction of $\lvert\nu_2\rangle$ that is $\lvert\nu_\mu\rangle$ is $\sin^2\theta$
	the fraction of $\lvert\nu_2\rangle$ that is $\lvert\nu_\tau\rangle$ is $\cos^2\theta$

Thus, in the ν_μ beam, the fraction of ν_1 is $f_1 = \cos^2\theta$ and ν_2 is $f_2 = \sin^2\theta$, independently of the neutrino energy, and the survival probability is

$$\begin{aligned} P(\nu_\mu \to \nu_\mu) &= f_1 \cos^2\theta + f_2 \sin^2\theta \\ &= \cos^4\theta + \sin^4\theta = 1 - \sin^2 2\theta \left\langle \sin^2\Delta \right\rangle, \end{aligned} \quad (7)$$

since $\left\langle \sin^2\Delta \right\rangle = 1/2$. Notice that the full treatment given earlier is really only useful for distances around (1/5 to 5 times, say) the oscillation length, $L_0 = 4\pi E/\delta m^2$. At small distance, the oscillations haven't built up enough to be significant, whereas as at the large distance the oscillations are average out.

---

[1] ν_μ is the neutrino produced in association with μ^+.

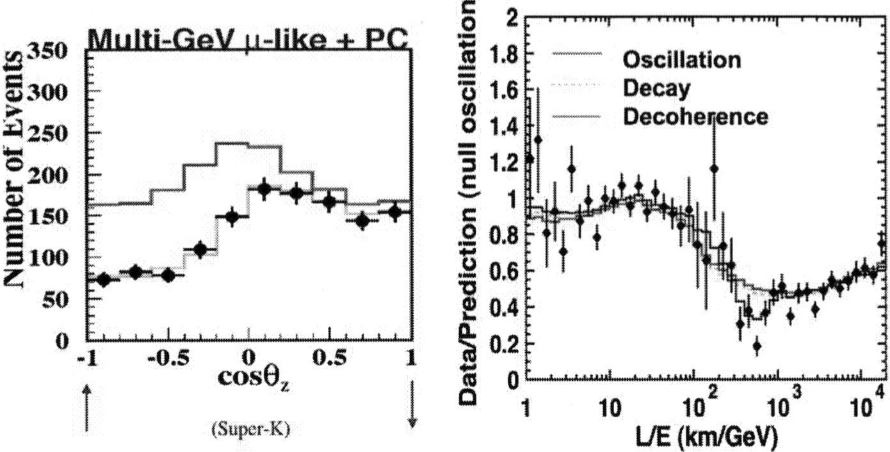

FIGURE 2. SuperKamiokande's evidence for neutrino oscillations both in the zenith angle and L/E plots.

EVIDENCE FOR NEUTRINO OSCILLATIONS:

Atmospheric and Accelerator Neutrinos

SuperKamiokande(SK) has very compelling evidence for ν_μ disappearance in their atmospheric neutrino studies, see [1]. In Fig. 2 the zenith angle dependence of the multi-GeV ν_μ sample is shown together with their L/E plot. This data fits very well the simple two component neutrino hypothesis with

$$\delta m^2_{atm} = 2 - 3 \times 10^{-3} eV^2 \qquad \text{and} \qquad \sin^2\theta_{atm} = 0.50 \pm 0.15 \tag{8}$$

This corresponds to a L/E for oscillations of 500 km /GeV and nearly maximal mixing. No evidence for the involvement of the ν_e is observed so the assumption is that $\nu_\mu \rightarrow \nu_\tau$.

Two beams of ν_μ neutrinos have been sent to two detectors located at large distance: K2K experiment, [2], is from KEK to SK with a baseline of 250 km and the MINOS experiment, [3], from Fermilab to the Soudan mine with a baseline of 735 km. Both experiments see evidence for ν_μ disappearance which is summarized in Fig. 3

Reactor and Solar Neutrinos:

The KamLAND reactor experiment, [5], sees evidence for neutrino oscillations and not only at a different L/E than the atmospheric and accelerator experiments but also this oscillation involves the ν_e. These flavor transitions have also been seen in solar neutrino

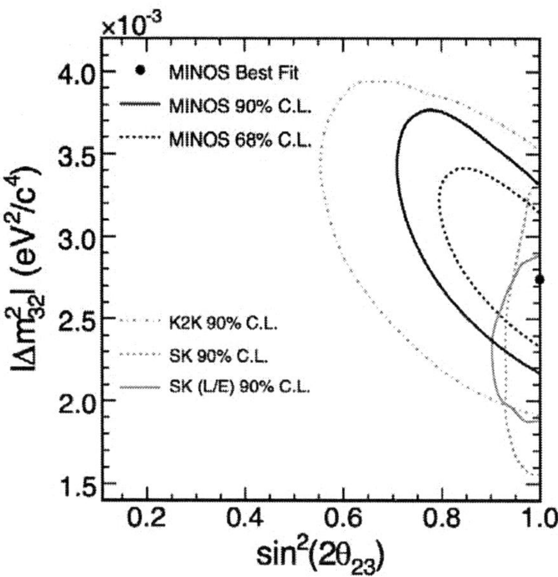

FIGURE 3. The allowed regions in the δm^2_{atm} v $\sin^2\theta_{atm}$ plane for MINOS data as well as for K2K data and two of the SK analyses. MINOS's best fit point is at $\sin^2\theta_{atm} = 1$ and $\delta m^2_{atm} = 2.7 \times 10^{-3}$ eV^2.

experiments. The best fit values for $\delta m^2_\odot$ and $\sin^2\theta_\odot$ are

$$\delta m^2_\odot = 8.0 \pm 0.4 \times 10^{-5} eV^2 \qquad \text{and} \quad \sin^2\theta_\odot = 0.31 \pm 0.03. \tag{9}$$

Thus, the L/E for this oscillation is 15 km/MeV which is 30 times larger than the atmospheric scale and the mixing angle, though large, is not maximal.

Fig. 4 shows the disappearance probability for the $\bar{\nu}_e$ from many reactor experiments as well as the flavor content of the ^{8}Boron solar neutrino flux measured by SNO, [6], and SK, [7]. The reactor result can be understood in terms of vacuum neutrino oscillations and the fit to the disappearance probability, Eq. [4], suitably averaged over E and L, provides a good fit.

Solar neutrinos are somewhat more complicated because of the matter effects that the neutrinos experience from the production region until they exit the sun, at least for the ^{8}Boron neutrinos. The pp and ^{7}Be neutrinos are little effected by the matter and undergo quasi-vacuum oscillations whereas the ^{8}Boron neutrinos exit the sun mainly as a ν_2 mass eigenstate because of matter effects and therefore do not undergo oscillations. This difference is primarily due to the difference in the energy of the neutrinos: pp (^{7}Be) have a mean energy of 0.2 MeV (0.9 MeV) whereas ^{8}B have a mean energy of 10 MeV and the matter effect is proportional to energy of the neutrino.

The kinematic phase for solar neutrinos is

$$\Delta_\odot = \frac{\delta m^2_\odot L}{4E} = 10^{7\pm1}. \tag{10}$$

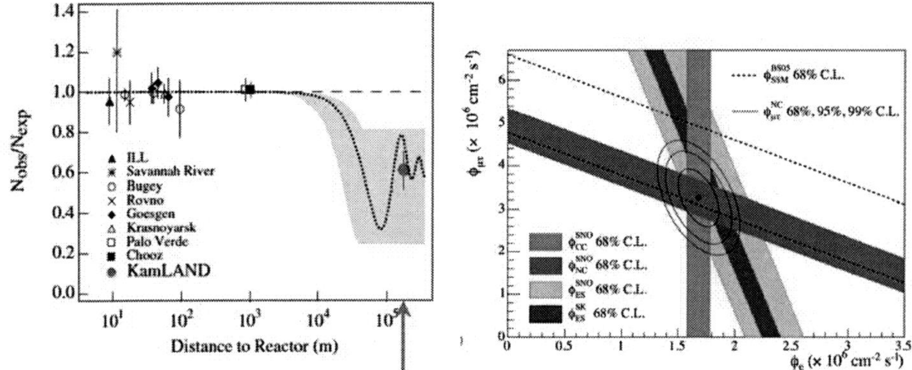

FIGURE 4. The disappearance of $\bar{\nu}_e$ observed by reactor experiments as a function of distance from the reactor. The flavor content of the ^{8}Boron solar neutrinos for the various reactions for SNO and SK. CC: $\nu_e + d \rightarrow e^- + p + p$, NC: $\nu_x + d \rightarrow \nu_x + p + n$ and ES: $\nu_\alpha + e^- \rightarrow \nu_\alpha + e^-$.

Therefore, the solar neutrinos are “effectively incoherent” when they reach the earth. Hence the ν_e survival probability is given by[2]

$$\begin{aligned} \langle P_{ee} \rangle &= f_1 \cos^2\theta_\odot + f_2 \sin^2\theta_\odot \\ \text{where} \quad f_1 + f_2 &= 1 \quad \text{and} \quad \cos^2\theta_\odot + \sin^2\theta_\odot = 1. \end{aligned} \tag{11}$$

Now the pp and ^{7}Be solar neutrinos behave essentially as in vacuum and therefore $f_1 \approx \cos^2\theta_\odot = 0.69$ and $f_2 \approx \sin^2\theta_\odot = 0.31$ whereas the mass eigenstate fraction for the ^{8}B are substantially different, see Fig. 5.

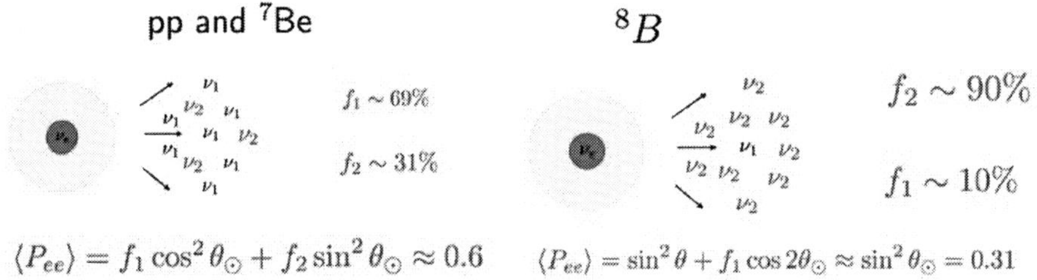

FIGURE 5. The sun produces ν_e in the core but once they exit the sun thinking about them in the mass eigenstate basis is useful. The fraction of ν_1 and ν_2 is energy dependent above $\sim$ 1 MeV and has a dramatic effect on the ^{8}Boron solar neutrinos, as first observed by Davis.

---

[2] Given the relationship between the quantities in this expression there are many equivalent ways to write the same expression.

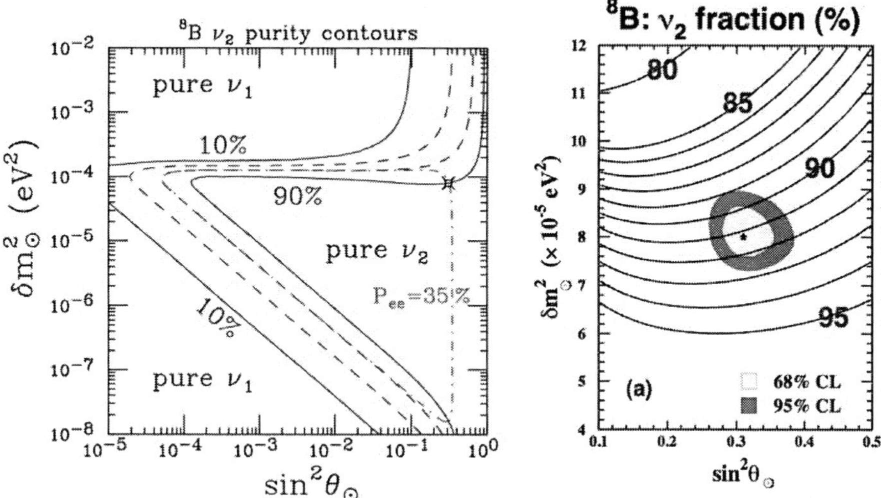

FIGURE 6. The ν_2 fraction (%) in the $\delta m^2_\odot$ versus $\sin^2\theta_\odot$ plane. (a) The solid and dashed (blue) lines are the 90, 65, 35 and 10% iso-contours of the fraction of the solar ^{8}B neutrinos that are ν_2's. The current best fit value, indicated by the open circle with the cross, is close to the 90% contour. The iso-contour for an electron neutrino survival probability, $\langle P_{ee}\rangle$, equal to 35% is the dot-dashed (red) "triangle" formed by the 65% ν_2 purity contour for small $\sin^2\theta_\odot$ and a vertical line in the pure ν_2 region at $\sin^2\theta_\odot = 0.35$. Except at the top and bottom right hand corners of this triangle the ν_2 purity is either 65% or 100%. (b) Focuses in on the current allowed region. The 68 and 95% CL are shown by the shaded areas with the best fit values indicated by the star using the combined fit of KamLAND and solar neutrino data given in [6] .

In a two neutrino analysis, the *day-time* CC/NC of SNO, which is equivalent to the day-time average ν_e survival probability, $\langle P_{ee}\rangle$, is given by

$$\left.\frac{\mathrm{CC}}{\mathrm{NC}}\right|_{\mathrm{day}} = \langle P_{ee}\rangle = f_1\cos^2\theta_\odot + f_2\sin^2\theta_\odot, \tag{12}$$

where f_1 and $f_2 = 1 - f_1$ are understood to be the ν_1 and ν_2 fractions, respectively, averaged over the ^{8}B neutrino energy spectrum weighted with the charged current cross section. Therefore, the ν_1 fraction (or how much f_2 differs from 100%) is given by

$$f_1 = \frac{\left(\left.\frac{\mathrm{CC}}{\mathrm{NC}}\right|_{\mathrm{day}} - \sin^2\theta_\odot\right)}{\cos 2\theta_\odot} = \frac{(0.347 - 0.311)}{0.378} \approx 10 \pm\ ??\ \%, \tag{13}$$

where the central values of the recent SNO analysis, [6], have been used. Due to the correlations in the uncertainties between the CC/NC ratio and $\sin^2\theta_\odot$ we are unable to estimate the uncertainty on f_1 from their analysis. Note, that if the fraction of ν_2 were 100%, then $\frac{\mathrm{CC}}{\mathrm{NC}} = \sin^2\theta_\odot$.

Using the analytical analysis of the Mikheyev-Smirnov-Wolfenstein (MSW) effect given in Ref. [8], the mass eigenstate fractions are given by

$$f_2 = 1 - f_1 = \langle \sin^2\theta^N_\odot + P_x\cos 2\theta^N_\odot\rangle_{^8\mathrm{B}}, \tag{14}$$

Life of a Boron-8 Solar Neutrino:

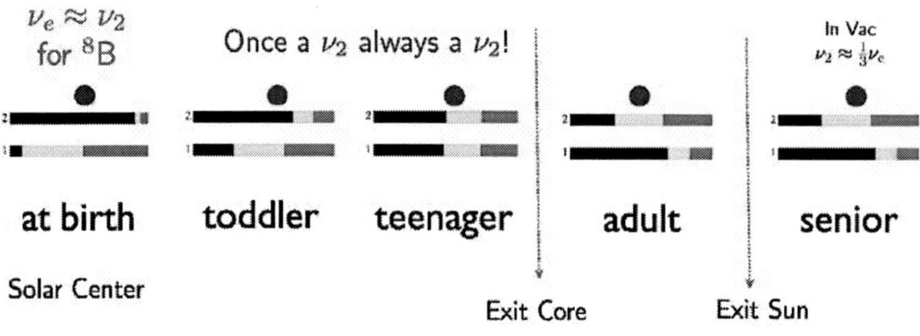

ν_e ■ ν_μ ■ ν_τ ■

FIGURE 7. Life of a ^{8}Boron solar neutrino from its birth at the center of the sun to its "death" in a detector at the earth. Notice how the flavor content of the the ν_2 mass eigenstate evolves as the neutrino travels through the solar core.

where $\theta_\odot^N$ is the mixing angle defined at the ν_e production point and P_x is the probability of the neutrino to jump from one mass eigenstate to the other during the Mikheyev-Smirnov resonance crossing. The average $\langle \cdots \rangle_{^8\mathrm{B}}$ is over the electron density of the ^{8}B ν_e production region in the center of the Sun predicted by the Standard Solar Model and the energy spectrum of ^{8}B neutrinos weighted with SNO's charged current cross section. Fig. 6 shows the iso-contours of this averaged ν_2 fraction using a threshold of 5.5 MeV on the kinetic energy of the recoil electrons, this figure is taken from Ref. [9]. Thus, the ^{8}B energy weighted average fraction of ν_2's observed by SNO is

$$f_2 = 91 \pm 2\% \quad \text{at the 95\% CL.} \tag{15}$$

Hence, the ^{8}B solar neutrinos are the purest mass eigenstate neutrino beam known so far and SK famous picture of the sun taken with neutrinos is more than 80% ν_2!!!

NU STANDARD MODEL:

The Neutrino Standard Model has emerged as follows[3]:

- 3 light (m_i <1 eV) Majorana Neutrinos: $\Rightarrow$ only 2 δm^2

[3] If MiniBooNE confirms the LSND result then this section will require major revision.

$|\delta m^2_{atm}| \sim 2.5 \times 10^{-3}\ \text{eV}^2$ and $\delta m^2_{solar} \sim +8.0 \times 10^{-5}\ \text{eV}^2$

- Only three Active flavors (no steriles): $e,\ \mu,\ \tau$
- Unitary Mixing Matrix: 3 angles ($\theta_{12},\ \theta_{23},\ \theta_{13}$), 1 Dirac phase ($\delta$), 2 Majorana phases ($\alpha, \beta$)

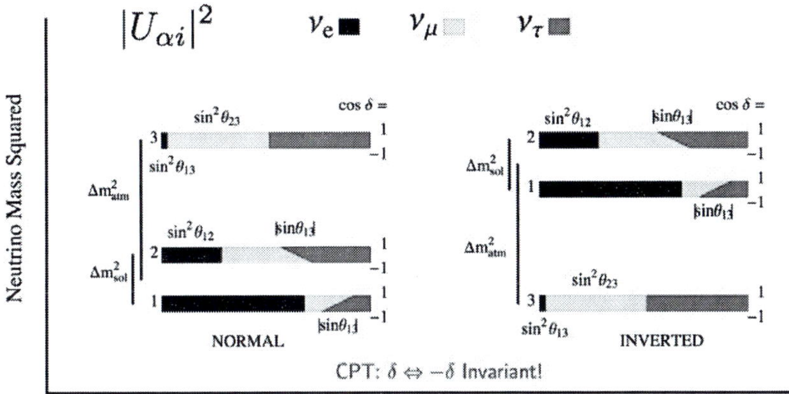

FIGURE 8. Flavor content of the three neutrino mass eigenstates showing the dependence on the cosine of the CP violating phase, δ. If CPT is conserved, the flavor content must be the same for neutrinos and anti-neutrinos. This figure was adapted from Ref. [10].

where the MNS mixing matrix relating flavor to mass eigenstates, $|\nu_\alpha\rangle = U_{\alpha i}|\nu_i\rangle$ is given by

$$U_{\alpha i} =$$

$$\begin{pmatrix} 1 & & \\ & c_{23} & s_{23} \\ & -s_{23} & c_{23} \end{pmatrix} \begin{pmatrix} c_{13} & & s_{13}e^{-i\delta} \\ & 1 & \\ -s_{13}e^{i\delta} & & c_{13} \end{pmatrix} \begin{pmatrix} c_{12} & s_{12} & \\ -s_{12} & c_{12} & \\ & & 1 \end{pmatrix} \begin{pmatrix} 1 & & \\ & e^{i\alpha} & \\ & & e^{i\beta} \end{pmatrix} \quad (16)$$

where $s_{ij} = \sin\theta_{ij}$ and $c_{ij} = \cos\theta_{ij}$. The (23) sector is identified with the atmospheric δm^2_{atm} and the (12) sector is identified with the solar $\delta m^2_\odot$. The (13) sector is responsible for the ν_e flavor transitions at the atmospheric scale so far unobserved, see [11]. Therefore,

$$\begin{aligned} \sin^2\theta_{12} &= 0.31 \pm 0.03 \\ \sin^2\theta_{23} &= 0.50 \pm 0.15 \\ \sin^2\theta_{13} &< 0.04 \end{aligned}$$

and the mass splittings[4] are

$$|\delta m^2_{32}| = 2.7 \pm 0.4 \times 10^{-3}\text{eV}^2 \qquad \text{and} \qquad \delta m^2_{21} = +8.0 \pm 0.4 \times 10^{-5}\text{eV}^2.$$

The mass of the lightest neutrino is unknown but the heaviest one must be lighter than about 1 eV. These mixing angles and mass splittings are summarized in Fig. 8 which also shows the dependence of the flavor fractions on the CP violating Dirac phase, δ. The Majorana phases are unobservable in oscillations since oscillations depend on $U^*_{\alpha i}U_{\beta i}$ but they have observable CP conserving effects in neutrinoless double beta decay.

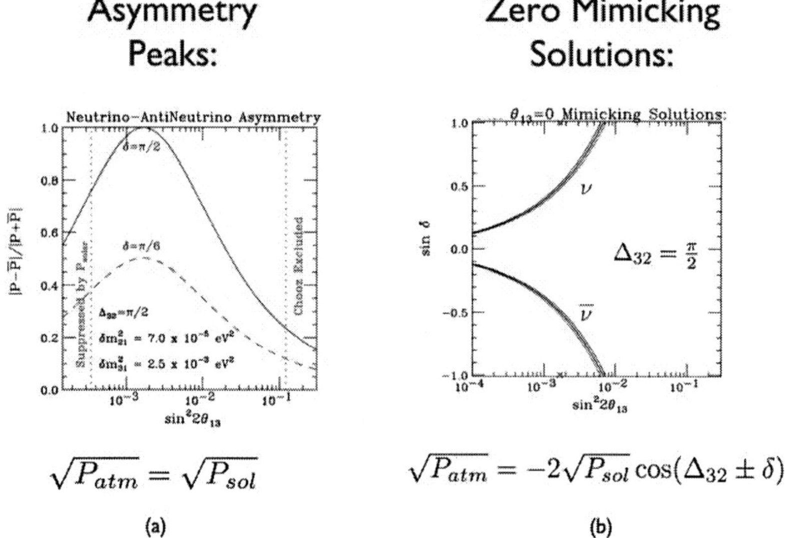

FIGURE 9. (a) The neutrino-antineutrino asymmetry as function of $\sin^2 2\theta_{13}$ at the first vacuum oscillation maximum. The asymmetry peaks when $\sin^2 2\theta_{13} = 0.002$. (b) The zero mimicking solutions at the first vacuum oscillation maximum. Along these lines there is no evidence of non-zero θ_{13}.

---

[4] The δm^2 MINOS actually measures is

$$\frac{(|U_{\mu 2}|^2|\delta m^2_{32}| + |U_{\mu 1}|^2|\delta m^2_{31}|)}{(|U_{\mu 2}|^2 + |U_{\mu 1}|^2)}.$$

Genuine Three Flavor Effects: $\nu_\mu \rightarrow \nu_e$

The most likely genuine three flavor effects to be first observed are $\nu_\mu \rightarrow \nu_e$ and/or its CP and T conjugate processes. That is, in one of following transitions

$$\begin{array}{ccccc} & & \text{CP} & & \\ & \nu_\mu \rightarrow \nu_e & \Longleftrightarrow & \bar{\nu}_\mu \rightarrow \bar{\nu}_e & \\ \text{T} & \Updownarrow & & \Updownarrow & \text{T} \\ & \nu_e \rightarrow \nu_\mu & \Longleftrightarrow & \bar{\nu}_e \rightarrow \bar{\nu}_\mu & \\ & & \text{CP} & & \end{array}$$

Processes across the diagonal are related by CPT. The first row will be explored in very powerful conventional beams, Superbeams, whereas the second row could be explored in Nu-Factories or Beta Beams.

In vacuum, the probability for $\nu_\mu \rightarrow \nu_e$ is derived like so, [12],

$$\begin{aligned} P(\nu_\mu \rightarrow \nu_e) &= \mid U^*_{\mu 1} e^{-im_1^2 L/2E} U_{e1} + U^*_{\mu 2} e^{-im_2^2 L/2E} U_{e2} + U^*_{\mu 3} e^{-im_3^2 L/2E} U_{e3} \mid^2 \\ &= |2U^*_{\mu 3} U_{e3} \sin\Delta_{31} e^{-i\Delta_{32}} + 2U^*_{\mu 2} U_{e2} \sin\Delta_{21}|^2 \\ &\approx |\sqrt{P_{atm}} e^{-i(\Delta_{32}+\delta)} + \sqrt{P_{sol}}|^2 \end{aligned} \quad (17)$$

where $\sqrt{P_{atm}} = \sin\theta_{23} \sin 2\theta_{13} \sin\Delta_{31}$ and $\sqrt{P_{sol}} \approx \cos\theta_{23} \sin 2\theta_{12} \sin\Delta_{21}$. For anti-neutrinos δ must be replaced with $-\delta$ and the interference term changes

$$2\sqrt{P_{atm}}\sqrt{P_{sol}} \cos(\Delta_{32}+\delta) \;\Rightarrow\; 2\sqrt{P_{atm}}\sqrt{P_{sol}} \cos(\Delta_{32}-\delta).$$

This allows for the possibility that CP violation maybe able to be observed in the neutrino sector since it allows for $P(\nu_\mu \rightarrow \nu_e) \neq P(\bar{\nu}_\mu \rightarrow \bar{\nu}_e)$.

In matter, $\sqrt{P_{atm}}$ and $\sqrt{P_{sol}}$ are modified as follows

$$\begin{aligned} \sqrt{P_{atm}} &\Rightarrow \sin\theta_{23} \sin 2\theta_{13} \frac{\sin(\Delta_{31} \mp aL)}{(\Delta_{31} \mp aL)} \Delta_{31} \\ \sqrt{P_{sol}} &\Rightarrow \cos\theta_{23} \sin 2\theta_{12} \frac{\sin(aL)}{(aL)} \Delta_{21} \end{aligned} \quad (18)$$

where $a = \pm G_F N_e/\sqrt{2} \approx (4000\,km)^{-1}$ and the sign is positive for neutrinos and negative for anti-neutrinos. This change follows since in both the (31) and (21) sectors the product $\{\delta m^2 \sin 2\theta\}$ is approximately independent of matter effects. In Fig. 10 the bi-probability plots are shown for both T2K, [13], and NOνA, [14] . It is possible that these two experiments will determine the mass ordering (normal or inverted hierarchy, see Fig. reffig: pmns-sq), and observe CP violation in the neutrino sector.

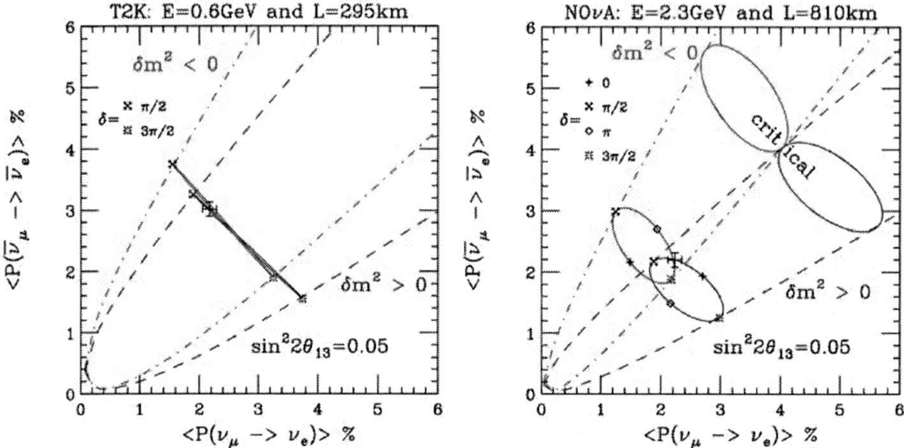

FIGURE 10. The bi-probability plots for both T2K and NOνA. The matter effects and hence the separation between the hierarchies is 3 times large for T2K than NOνA primarily due to the fact NOνA has three times the baseline as T2K. See [15] to understand how to use these plots to untangle CP violation and the mass hierarchy.

NEUTRINO MASS

Absolute Neutrino Mass

Tritium beta decay, neutrinoless double beta decay and cosmology all have the potential to provide us information on the absolute scale of neutrino mass. The Katrin tritium beta decay experiment, [16], has sensitivity down to 200 meV for the "mass" of ν_e defined as

$$m_{\nu_e} = |U_{e1}|^2 m_1 + |U_{e2}|^2 m_2 + |U_{e3}|^2 m_3. \tag{19}$$

Neutrinoless double beta decay, see [17] for review, measures the following combination of neutrino mass,

$$m_{\beta\beta} = |\sum m_i U_{ei}^2| = |m_1 c_{13}^2 c_{12}^2 + m_2 c_{13}^2 s_{12}^2 e^{2i\alpha} + m_3 s_{13}^2 e^{2i\beta}|, \tag{20}$$

assuming the neutrinos are Majorana. It maybe possible to eventually reach below 10meV for $m_{\beta\beta}$ in double beta decay.

Cosmology measures the sum of the neutrino masses,

$$m_{cosmo} = \sum_i m_i. \tag{21}$$

If $\sum m_i \approx 50\ eV$ the universe's critical density would be saturated. The current limit, [18], is a few % of this number, $\sim$1eV. Given the systematic uncertainties inherent in cosmology, a convincing limit of less than 100 meV seems difficult.

Fig. 11 shows the allowed values for these masses for both the normal and inverted hierarchy.

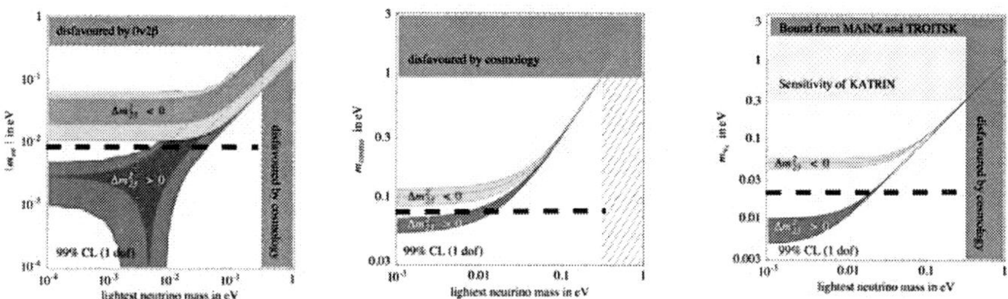

FIGURE 11. The "mass" measured in double β-decay, in cosmology and Tritium β-decay versus the mass of the lightest neutrino. Below the dashed lines, only the normal hierarchy is allowed. This figure was adapted from hep-ph/0503246 [19].

Majorana v Dirac

Fermion mass is a coupling of left handed to right handed states. Consider a massive fermion at rest, then one can consider this state as a linear combination of a massless right handed particle and a massless left handed particle as shown in Fig. 12. For a particle with an electric charge, like an electron, the left handed particle must have the same charge as the right handed particle. This is a Dirac mass. For a neutral particle, like a sterile neutrino, there is another possibility, the left handed particle could be coupled to the right handed anti-particle, this is the Majorana mass.

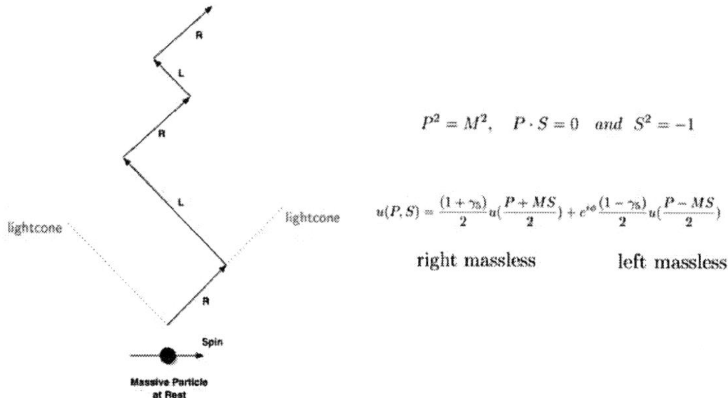

FIGURE 12. The left diagram shows how a massive particle at rest can be considered as a linear combination of two massless particles, one right handed and one left handed. The equation at the right shows the decomposition of a massive Dirac spinor into two massless spinors with different momenta, one right handed and the other left handed (from the Appendix of [20]).

Therefore for a neutral particle there is the possibility of having both Dirac and Majorana masses, as

$$\begin{array}{lcccl} \text{Left Chiral} & \nu_L & \Longleftrightarrow & \bar{\nu}_R & \\ & \Updownarrow & & \Updownarrow & \text{Dirac Mass} \\ \text{Right Chiral} & \nu_R & \Longleftrightarrow & \bar{\nu}_L & \\ & & \begin{array}{c}\text{Majorana}\\ \text{Mass}\end{array} & & \end{array}$$

For the neutrino, the left chiral field couples to $SU(2) \times U(1)$ therefore a Majorana mass term is forbidden by gauge symmetry. However, the right chiral field carries no quantum numbers. Therefore, the Majorana mass term is unprotected by any symmetry and it is expected to be very large. The Dirac mass terms are expected to be of the order of the charge lepton or quark masses. Thus, the mass matrix for the neutrinos is as in Fig. 13.

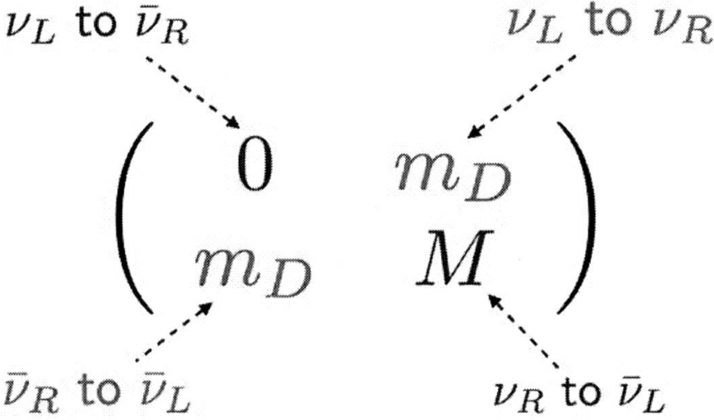

FIGURE 13. The neutrino mass matrix with the various right to left couplings. m_D is the Dirac mass terms while 0 and M are Majorana masses for the charged and uncharged (under $SU(2) \times U(1)$) chiral components.

After diagonalizing the neutrino mass matrix, one is left with two Majorana neutrinos, one heavy Majorana neutrino with mass ~M and one light Majorana neutrino with mass m_D^2/M. This is the famous seesaw mechanism, [21]. The light neutrino is the one observed in current experiments whereas the heavy neutrino is responsible for leptogenesis at very high energy scales since its decays are CP violating and depend on the Majorana phases in the MNS matrix, Eq. 16.

Majorana neutrinos not only allow for neutrinoless double beta decay but also for the possibility that the a muon neutrino, say, produces a positive charged muon, violating lepton number. However, this process would be suppressed by $(m_\nu/E)^2$ which is tiny, 10^{-20}, and, therefore is unobservable.

SUMMARY

Neutrino Mass $\Leftrightarrow$ Flavor Change

Open questions:

- Majorana v Dirac
- Light Steriles ???
- Mass Hierarchy $m_3 > m_2 > m_1$ OR $m_2 > m_1 > m_3$
  (labeling such that $|U_{e3}|^2 < |U_{e2}|^2 < |U_{e1}|^2$)
- fraction of ν_e in ν_3 ($< 4\%$) (value of $\sin^2\theta_{13}$)
- Is CP violated ? ($\sin\delta \neq 0$)
- Mass of Heaviest Neutrino
- Mass of Lightest Neutrino
- New Interactions, Surprises !!!

ACKNOWLEDGMENTS

I would like to thank all of the students and all of the organizers of the Cinvestav Advanced Summer School for giving me the opportunity to present these lectures on neutrinos. I especially thank Chairman Omar Miranda for his wonderful hospitality. The support by Academia Mexicana de Ciencias (AMC-Mexico) and The United States-Mexico Foundation for Science (FUMEC) is acknowledged. Fermilab is operated by URA under DOE contract DE-AC02-76CH03000.

REFERENCES

1. Y. Ashie *et al.* [Super-Kamiokande Collaboration], Phys. Rev. D **71**, 112005 (2005).
2. E. Aliu *et al.* [K2K Collaboration], Phys. Rev. Lett. **94**, 081802 (2005);
   M. H. Ahn *et al.* [K2K Collaboration], Phys. Rev. Lett. **90**, 041801 (2003).
3. E. Ables *et al.* [MINOS Collaboration], FERMILAB-PROPOSAL-0875.
4. [MINOS Collaboration], arXiv:hep-ex/0607088.
5. K. Eguchi *et al.* [KamLAND Collaboration], Phys. Rev. Lett. **90**, 021802 (2003). T. Araki *et al.* [KamLAND Collaboration], Phys. Rev. Lett. **94**, 081801 (2005).
6. B. Aharmim *et al.* [SNO Collaboration], arXiv:nucl-ex/0502021.
7. S. Fukuda *et al.* [Super-Kamiokande Collaboration], Phys. Lett. B **539**, 179 (2002).
8. S. J. Parke, Phys. Rev. Lett. **57**, 1275 (1986); S. J. Parke and T. P. Walker, Phys. Rev. Lett. **57**, 2322 (1986); S. J. Parke, "RESONANT NEUTRINO OSCILLATIONS," FERMILAB-CONF-86-131-T, *Proceedings of the Topical Mtg. of 14th SLAC Summer Inst. on Particle Physics, Stanford, CA, Aug 5-8, 1986*
9. H. Nunokawa, S. Parke and R. Z. Funchal, "What fraction of B-8 Solar neutrinos arrive at the Earth as a ν_2 mass eigenstate", Fermilab-Pub-05-049, arXiv:hep-ph/0601198.
10. O. Mena and S. J. Parke, Phys. Rev. D **69**, 117301 (2004) [arXiv:hep-ph/0312131].
11. M. Apollonio *et al.* [CHOOZ Collaboration], Phys. Lett. B **466**, 415 (1999).
12. A. Cervera *et al.*, Nucl. Phys. B **579**, 17 (2000) [Erratum-*ibid.* B **593**, 731 (2001)].
13. Y. Hayato *et al.*, Letter of Intent, available at `http://neutrino.kek.jp/jhfnu/`
14. D. S. Ayres *et al.* [NOvA Collaboration], hep-ex/0503053. FERMILAB-PROPOSAL-0929, March 21, 2005. Revised NOνA Proposal available at `http://www-nova.fnal.gov/NOvA_Proposal/Revised_NOvA_Proposal.html`

15. O. Mena and S. J. Parke, "Untangling CP violation and the mass hierarchy in long baseline experiments," Phys. Rev. D **70**, 093011 (2004) [arXiv:hep-ph/0408070].
16. A. Osipowicz *et al.* [KATRIN Collaboration], "KATRIN: A next generation tritium beta decay experiment with sub-eV sensitivity for the electron neutrino mass," arXiv:hep-ex/0109033.
17. S. R. Elliott and P. Vogel, Ann. Rev. Nucl. Part. Sci. **52**, 115 (2002) [arXiv:hep-ph/0202264].
18. D. N. Spergel *et al.*, "Wilkinson Microwave Anisotropy Probe (WMAP) three year results: Implications for cosmology," arXiv:astro-ph/0603449.
19. A. Strumia and F. Vissani, Nucl. Phys. B **726**, 294 (2005) [arXiv:hep-ph/0503246].
20. G. Mahlon and S. J. Parke, "Angular Correlations in Top Quark Pair Production and Decay at Hadron Phys. Rev. D **53**, 4886 (1996) [arXiv:hep-ph/9512264].
21. M. Gell-Mann, P. Ramond and R. Slansky, in *Supergravity*, edited by P. van Nieuwenhuizen and D. Freedman, (North-Holland, 1979), p. 315;
T. Yanagida, in *Proceedings of the Workshop on the Unified Theory and the Baryon Number in the Universe*, edited by O. Sawada and A. Sugamoto (KEK Report No. 79-18, Tsukuba, 1979), p. 95;
R. N. Mohapatra and G. Senjanovic, Phys. Rev. Lett. **44**, 912 (1980).
M. Fukugita and T. Yanagida, Phys. Lett. B **174**, 45 (1986).

Top-Quark Flavor Changing Neutral Couplings in Extensions of the Standard Model

Miguel A. Pérez

Departamento de Física, Cinvestav, Apdo. Postal 14-740, 07000 México D.F., Mexico

Abstract. We survey the predictions made by various extensions of the standard model of electroweak interactions on the decay modes of the top quark which involve the flavor-changing neutral couplings tcV and tcH, with $V = \gamma, g, Z$.

INTRODUCTION

Processes that are forbidden or highly suppressed in the standard model (SM) of electroweak interactions constitute a natural playground to test any virtual effects induced by new physics. This has been the case in rare Z decays [1], in the trilinear couplings VVV [2] and WHV [3], with $V = \gamma, Z$ and also in the decay model involving flavor-changing neutral couplings (FCNC) of the Higgs boson [4], leptons [5] and the top quark [6]. In particular, it has been pointed out that the searches of FCNC effects in the LHC/ILC accelerators may constitute one of the best ways to look for physics beyond the SM [6].

Although the top quark was discovered eleven years ago [7], many of his properties are still poorly known [8]. The top-quark mass has been measured at the 1% precision level only recently: $m_Z = 172.5 \pm 2.3 GeV$ [9]. With a decay width of about 1.5 GeV and a lifetime of only $10^{-25} sec$, it decays so quickly that it does not have time to hadronize. It decays almost entirely into a bW pair with a branching ratio close to unit. The other two decay modes expected in the SM $t \to sW, dW$ have branching ratios of order 10^{-6}, and have not been detected yet.

In the present paper we survey the FCNC decay modes of the top quark $t \to cV, cH, cVV$. As soon as it was confirmed that these decay modes are highly suppressed in the SM, with branching ratios of order $10^{-11} - 10^{-14}$ [10], it was also realized that some of these decay modes can be enhanced by several orders of magnitude in scenarios beyond the SM [6], and some of them within the LHC reach. With a LHC luminosity of $100 fb^{-1}$, 80 million of $t\bar{t}$ pairs per year will make possible to reach FCNC branching ratios of order 10^{-5}[11]. The interest in FCNC top-quark physics is thus expected to increase in the near future.

SM COUPLINGS

The SM couplings predicted by the SM for the top quark to the electroweak gauge bosons W, Z, γ, g have not yet been measured directly. The $V - A$ nature of the tbW vertex is being tested at the Tevatron via measurements of the W-helicity with results consistent

CP885, *Advanced Summer School in Physics 2006, Frontiers in Contemporary Physics—EAV06,*
edited by O. Miranda, M. Carbajal, L. M. Montaño, O. Rosas-Ortiz, and S. A. Tomás Velázquez

TABLE 1. Summary of the predictions that are potentially visible at the LHC and ILC for BR-FCNC top-quark decay modes. References to specific results are included in the text. The column for the effective Lagrangian approach (ELA) includes the respective bounds obtained for these decay modes from low-energy precision measurements.

Decay	THDM II	THDM III	MSSM	R/MSSM
$t \to c\gamma$			10^{-6}	10^{-5}
$t \to cZ$		10^{-6}	10^{-6}	10^{-4}
$t \to cg$	10^{-5}	10^{-4}	10^{-4}	10^{-3}
$t \to cH$	8×10^{-5}	10^{-3}	10^{-4}	10^{-5}
$t \to cW^+W^-$		10^{-3}		
$t \to cZZ$		10^{-3}		
$t \to c\gamma\gamma$		10^{-4}		
$t \to c\gamma Z$		10^{-4}		
$t \to cgg$		10^{-4}	10^{-5}	
$t \to cHg$		10^{-5}		

	TC2	L-R	LR-MSSM	Extra q	ELA
$t \to c\gamma$			10^{-6}		2×10^{-3}
$t \to cZ$	10^{-5}		10^{-4}	10^{-4}	1.6×10^{-2}
$t \to cg$	10^{-5}		10^{-5}	10^{-6}	3.4×10^{-2}
$t \to cH$		10^{-4}		4×10^{-5}	$0.1 - 3 \times 10^{-3}$
$t \to cW^+W^-$	10^{-3}				

with the *SM* expectations [9]. The ttZ vector and axial vector couplings are tightly but indirectly constrained by the LEP data [12]. The right-handed tbW coupling, predicted to be zero at tree level in the *SM*, is severely bounded to be about 10^{-3} by the observed $b \to s\gamma$ rate [13]. On the other hand, with the expected luminosity in the *LHC* and *ILC*, it will be possible to measure the ttV couplings at the few percent level [13].

On the other hand, and update of the original calculations [10] for the FCNC top-quark decays give the following results for their branching ratios [11]: $BR(t \to c\gamma, cZ) \sim 10^{-11}$, $BR(t \to cg) \sim 10^{-12}$, $BR(t \to cH) \sim 10^{-15}$.

NEW PHYSICS IN TOP-QUARK FCNC

The FCNC decays of the top quark have been found to be very sensitive to physics beyond the SM. Some extensions of the SM predict spectacular enhancements on these FCNC branching ratios, and some of them are definitely within the reach of LHC. In table 1 we summarize the predictions for the branching ratios that are potentially visible at the LHC [6] in the following models: two Higgs doublet models of type II and III (THDM II, III); minimal standard supersymmetric model (MSSM) and its version with R-parity violation (R-MSSM); top-assisted technicolor models (TC2); left-right symmetric models (LR) and its supersymmetric version (LR-SUSY); and finally models with extra $Q = 2/3$ singlet quarks.

According to the results shown in table 1, searches for FCNC top-quark effects at LHC may constitute one of the best ways to look for physics beyond the SM. The experimental feasibility of detecting such effects at LHC may constitute one of the best ways to look for physics beyond the SM. In particular, the experimental feasibility of detecting such effects in top-quark production and decays seems to be better than the situation expected in FCNC effects of the Z gauge boson [1], the Higgs boson [4], or even in the case of the expected CP-violating effects in top quark decays [15].

CONSTRAINTS FROM LOOP OBSERVABLES

The possibility of extracting bounds on the effective tcV and tcH vertices form loop observables have been studied in different low-energy processes [6].Even though these bounds can not be considered model-independent constraints, they may be regarded as order of magnitude estimates which could be used in the search of new physics effects in the following generation of colliders.

The measurement of the inclusive branching ratio for the $FCNC$ process $b \to s\gamma$ has been used to put constraints on the $tc\gamma$ and tcg couplings:$BR(t \to c\gamma) < 2.2 \times 10^{-3}$, $BR(t \to cg) < 3.4 \times 10^{-2}$ [16]. The tcZ and tcH couplings have also been constrained by using the electroweak precision observables measured at LEP: $BR(t \to cZ) < 1.6 \times 10^{-2}$, $BR(t \to cH) < 0.9 - 29 \times 10^{-4}$ for $116\ GeV < m_H < 170\ GeV$ [17]. In particular, the limit on $BR(t \to cZ)$ is similar to the bound recently reported by the DELPHI collaboration [18]. In table 1 we have included these bounds, which were obtained within the framework of the effective Lagrangian approach (ELA).

ACKNOWLEDGMENTS

The present work was supported by CONACyT (Mexico).

REFERENCES

1. M.A. Pérez, G. Tavares-Velasco, J.J. Toscano, Int. J. Mod. Phys. A19, 159 (2004); M.A. Pérez, M. Soriano, Phys. Rev. D46, 125 (1992).
2. F. Larios, M.A. Pérez, G. Tavares-Velasco, J.J. Toscano, Phys. Rev. D63, 113014 (2001).
3. R. Martínez, M.A. Pérez, Nucl. Phys. B347, 105 (1990); J. Hernández-Sánchez, M.A. Pérez, G. Tavares-Velasco, J.J. Toscano, Phys. Rev. D69, 095008 (2004); A. Arhrib et al., hep-ph/0607182.
4. J.L. Díaz-Cruz, J.J. Toscano, Phys. Rev. D66, 116005 (2002).
5. F. Larios, R. Martínez, M.A. Pérez, Phys. Lett. B345, 259 (1995).
6. F. Larios, R. Martínez, M.A. Pérez, Int. J. Mod. Phys. A21, 3473(2006).
7. F. Abe et al. (CDF Collab.) Phys. Rev. Lett. 74, 2626 (1995); S. Abachi et al., (DO Collab.) *ibid* 2623 (1995).
8. D. Chakraborty, J. Konigsberg and D.L. Rainwater, Ann. Rev. Nucl. Part. Sci. 53, 301 (2003).
9. *http : //www.cdf.fnal.gov/physics/new/top/top.html; htttp : //www.do.fnal.gov/Run2Physics /top/top − public_web_pages/top_public.html.*
10. J.L. Díaz-Cruz, R. Martínez, M.A. Pérez, A. Rosado, Phys. Rev. D41, 891 (1990); G. Eilam, J.L. Hewett, A. Suri, Phys. Rev. D44, 1473 (1991); [Erratum: ibid 59, 039901 (1999)].
11. A. Aguilar-Saavedra, Acta Phys. Polonica B35, 2695 (2004).
12. S. Eidelman et al. [Particle Data Group], Phys. Lett. B592, 1 (2004).
13. F. Larios, M.A. Pérez, C.P. Yuan, Phys. Lett. B457, 334 (1999).
14. U. Baur et al., hep-ph/0606264.

15. D. Atwood et al., Phys. Rep. 347, 1 (2001).
16. R. Martínez, M.A. Pérez, J.J. Toscano, Phys. Lett B340, 91 (1994); T. Han et al., Phys. Rev. D55, 7241 (1996).
17. F. Larios, R. Martínez, M.A. Pérez, Phys. Rev. D72, 057504 (2005); T. Han, R.D. Peccei, X. Zhang, Nucl. Phys. B454 527 (1995).
18. I. Abdalla et al. (DELPHI Collab.), Phys. Lett. B590, 21 (2004).

A Monochromatic Neutrino Beam to Obtain U(e3) and the CP Phase

C. Espinoza*, J. Bernabeu*, J. Burguet-Castell* and M. Lindroos†

*Universitat de València and IFIC, E-46100 Burjassot, València, Spain
†AB-division, CERN, Geneva, Switzerland

Abstract. The goal for future neutrino facilities is the determination of the $[U_{e3}]$ mixing and CP violation in neutrino oscillations. This will require precision experiments with a very intense neutrino source. The future experiments such as T2K, NOVA and Double CHOOZ will measure the $[U_{e3}]$ mixing. In order to explore CP violation, we present [1] a novel method to create a monochromatic neutrino beam based on the recent discovery of nuclei that decay fast through electron capture in a superallowed Gamow-Teller transition. The boost of such radioactive ions will generate an intense monochromatic directional neutrino beam when decaying at high energy in a storage ring with long straight sections. We show that the capacity of such a facility to discover new physics is impressive, so that the principle of energy dependence in the oscillation probability of the suppressed $\nu_e \to \nu_\mu$ channel is operational to separate out the two parameters of the mixing θ_{13} and the CP-violating phase δ.

INTRODUCTION

The observation of *CP* violation needs an experiment in which the emergence of another neutrino flavor is detected rather than the deficiency of the original flavor of the neutrinos. At the same time, the interference needed to generate CP-violating observables can be enhanced if both the atmospheric and solar components have a similar magnitude. This happens in the suppressed $\nu_e \to \nu_\mu$ transition. The appearance probability $P(\nu_e \to \nu_\mu)$ as a function of the distance between source and detector (L) is given by [2]

$$\begin{aligned} P(\nu_e \to \nu_\mu) &\simeq s_{23}^2 \sin^2 2\theta_{13} \sin^2\left(\frac{\Delta m_{13}^2 L}{4E}\right) + c_{23}^2 \sin^2 2\theta_{12} \sin^2\left(\frac{\Delta m_{12}^2 L}{4E}\right) \\ &+ \tilde{J} \cos\left(\delta - \frac{\Delta m_{13}^2 L}{4E}\right) \frac{\Delta m_{12}^2 L}{4E} \sin\left(\frac{\Delta m_{13}^2 L}{4E}\right), \end{aligned} \quad (1)$$

where $\tilde{J} \equiv c_{13} \sin 2\theta_{12} \sin 2\theta_{23} \sin 2\theta_{13}$. The three terms of Eq. (1) correspond, respectively, to contributions from the atmospheric and solar sectors and their interference. The four measured parameters $(\Delta m_{12}^2, \theta_{12})$ and $(\Delta m_{23}^2, \theta_{23})$ have been fixed throughout this paper to their mean values [3].

Neutrino oscillation phenomena are energy dependent (see Fig.1) for a fixed distance between source and detector, and the observation of this energy dependence would disentangle the two important parameters: whereas $|U_{e3}|$ gives the strength of the appearance probability, the *CP* phase acts as a phase-shift in the interference pattern. These properties suggest the consideration of a facility able to study the detailed energy depen-

CP885, *Advanced Summer School in Physics 2006, Frontiers in Contemporary Physics—EAV06,*
edited by O. Miranda, M. Carbajal, L. M. Montaño, O. Rosas-Ortiz, and S. A. Tomás Velázquez

dence by means of fine tuning of a boosted monochromatic neutrino beam. As shown below, in an electron capture facility the neutrino energy is dictated by the chosen boost of the ion source and the neutrino beam luminosity is concentrated at a single known energy which may be chosen at will for the values in which the sensitivity for the (θ_{13}, δ) parameters is higher. This is in contrast to beams with a continuous spectrum, where the intensity is shared between sensitive and non sensitive regions. Furthermore, the definite energy would help in the control of both the systematics and the detector background. In the beams with a continuous spectrum, the neutrino energy has to be reconstructed in the detector. In water-Cerenkov detectors, this reconstruction is made from supposed quasielastic events by measuring both the energy and direction of the charged lepton. This procedure suffers from non-quasielastic background, from kinematic deviations due to the nuclear Fermi momentum and from dynamical suppression due to exclusion effects [4]. In our case, such an energy reconstruction is not necessary.

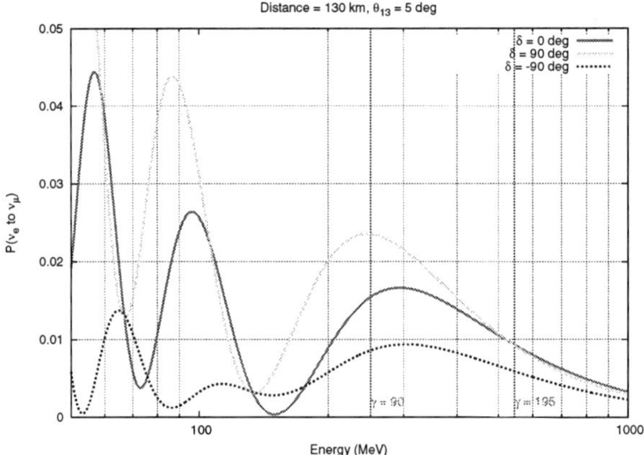

FIGURE 1. $P(\nu_e \rightarrow \nu_\mu)$ as a function of the LAB energy E, with fixed L and connecting mixing. The three curves refer to different values of the CP violating phase δ. The two vertical lines are the energies of our simulation study.

NEUTRINOS FROM ELECTRON CAPTURE

Electron Capture is the process in which an atomic electron is captured by a proton of the nucleus leading to a nuclear state of the same mass number A, replacing the proton by a neutron, and a neutrino. Its probability amplitude is proportional to the atomic wavefunction at the origin, so that it becomes competitive with the nuclear β^+ decay at high Z. Kinematically, it is a two body decay, so that the neutrino energy is well defined and given by the difference between the initial and final atomic masses (Q_{EC}) minus the excitation energy of the final nuclear state. In general, the high proton number Z nuclear beta-plus decay (β^+) and electron-capture (EC) transitions are very "forbidden", i.e., disfavored, because the energetic window open Q_β / Q_{EC} does not contain the important Gamow-Teller strength excitation seen in (p,n) reactions. There are a few cases, however,

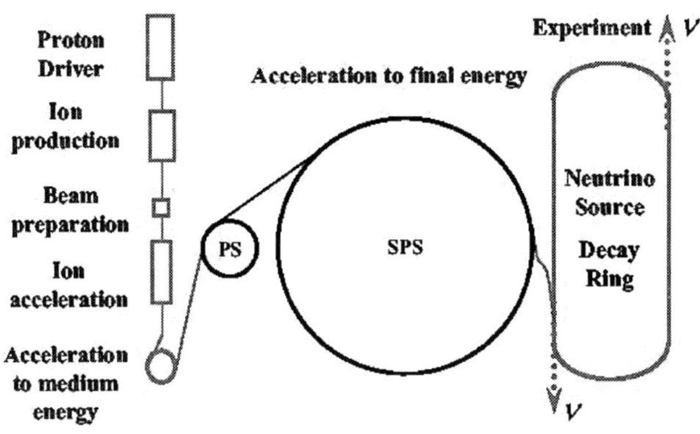

FIGURE 2. A proposal for a CERN EC neutrino beam facility.

where the Gamow-Teller resonance can be populated. The properties of a few examples [5] of interest for neutrino beam studies are given in Table 1.

A proposal for an accelerator facility with an EC neutrino beam is shown in Fig.2. It is based on the most attractive features of the beta beam concept [6]: the integration of the CERN accelerator complex and the synergy between particle physics and nuclear physics communities. However, the electron-capture facility [1] will require a different approach to acceleration and storage of the ion beam compared to the standard beta-beam [7], as the ions cannot be fully stripped. Partly charged ions have a short vacuum life-time [8] due to a large cross-section for stripping through collisions with rest gas molecules in the accelerators. The isotopes discussed here have a half-life comparable to, or smaller than, the typical vacuum half-life of partly charged ions in an accelerator with very good vacuum. The fact that the total half-life is not dominated by vacuum losses will permit an important fraction of the stored ions sufficient time to decay through electron-capture before being lost out of the storage ring through stripping. The ions mentioned before turn out to be very good candidates.

TABLE 1. Four fast decays in the rare-earth region above ^{146}Gd leading to the giant Gamow-Teller resonance. Energies are given in keV. The first column gives the life-time, the second the branching ratio of the decay to neutrinos, the third the relative branching between electron capture and β^+, the fourth is the position of the giant GT resonance, the fifth its width, the sixth the total energy available in the decay, the seventh is the neutrino energy $E_\nu = Q_{EC} - E_{GR}$ and the eighth its uncertainty.

Decay	$T_{1/2}$	BR_ν	EC/β^+	E_{GR}	Γ_{GR}	Q_{EC}	E_ν	ΔE_ν
$^{148}Dy \rightarrow ^{148}Tb^*$	$3.1m$	1	96/4	620	≈ 0	2682	2062	≈ 0
$^{150}Dy \rightarrow ^{150}Tb^*$	$7.2m$	0.64	100/0	397	≈ 0	1794	1397	≈ 0
$^{152}Tm2^- \rightarrow ^{152}Er^*$	$8.0s$	1	45/55	4300	520	8700	4400	520
$^{150}Ho2^- \rightarrow ^{150}Dy^*$	$72s$	1	77/33	4400	400	7400	3000	400

NEUTRINO FLUX

A neutrino of energy E_0 that emerges from radioactive decay in an accelerator will be boosted in energy. The measured energy distribution of the ion at the moment of decay can be expressed as $E = E_0/[\gamma(1-\beta\cos\theta)]$ where the angle (θ) is the deviation between the actual neutrino detection and the ideal detector position in the prolongation of one of the long straight sections of the Decay Ring of Fig. 2. The neutrinos are concentrated inside a narrow cone around the forward direction. The energy distribution of the Neutrino Flux arriving to the detector in absence of neutrino oscillations is given by the Master Formula

$$\frac{d^2N_\nu}{dSdE} = \frac{1}{\Gamma}\frac{d^2\Gamma_\nu}{dSdE}N_{ions} \simeq \frac{\Gamma_\nu}{\Gamma}\frac{N_{ions}}{\pi L^2}\gamma^2\delta(E-2\gamma E_0), \tag{2}$$

with a dilation factor $\gamma >> 1$. It is remarkable that the result is given only in terms of the branching ratio and the neutrino energy and it is independent of nuclear models. In Eq. (2), N_{ions} is the total number of ions decaying to neutrinos. In the forward direction the neutrino energy is fixed by the boost $E = 2\gamma E_0$, with the entire neutrino flux concentrated at this energy. As a result, such a facility will measure the neutrino oscillation parameters by changing the γ's of the decay ring (energy dependent measurement) and there is no need of energy reconstruction in the detector, as emphasized before.

THE VIRTUES OF TWO ENERGIES

We have made a simulation study in order to reach conclusions about the measurability of the unknown oscillation parameters. The ion type chosen is ^{150}Dy (see Table 1), with neutrino energy at rest given by 1.4 MeV due to a unique nuclear transition from 100% electron capture in going to neutrinos. Some 64% of the decay will happen as electron-capture, the rest goes through alpha decay. We have used a neutrino source of 10^{18} ^{150}Dy decaying ions/year, during a total running time of 10 years. The statistics is shared between the two runs at different γ's. We have taken two energies (see Fig. 1), defined by $\gamma_{max} = 195$ as the maximum energy possible at CERN with the present accelerator complex, and a minimum, $\gamma_{min} = 90$, in order to avoid atmospheric neutrino background in the detector below a certain energy. The detector has an active mass of 440 $kton$ and is located at a distance $L = 130$ km which equals the distance from CERN to the underground laboratory LSM in Frejus. Statistics from both appearance and disappearance events is accumulated. Although the survival probability does not contain any information on the CP phase, its measurement helps in the cut of the allowed parameter region.

The improvement over the standard beta-beam reach is due to the concentration of the intensity at appropriate energies (see Fig.1): whereas $\gamma = 195$ leads to an energy above the oscillation peak with almost no dependence of the $\delta-$phase, the value $\gamma = 90$, leading to energies between the peak and the node, is highly sensitive to the phase of the interference. These two energies are thus complementary to fix the values of (θ_{13}, δ) as we can see in Fig. 3. In practice what actually happens is that the shape of the strip for the allowed values is different for the two energies. This fact implies that the intersection region is squeezed when we look at the results of the combined energies. As a consequence, there is a better separation of the parameters (θ_{13} and δ)), even with lower statistics.

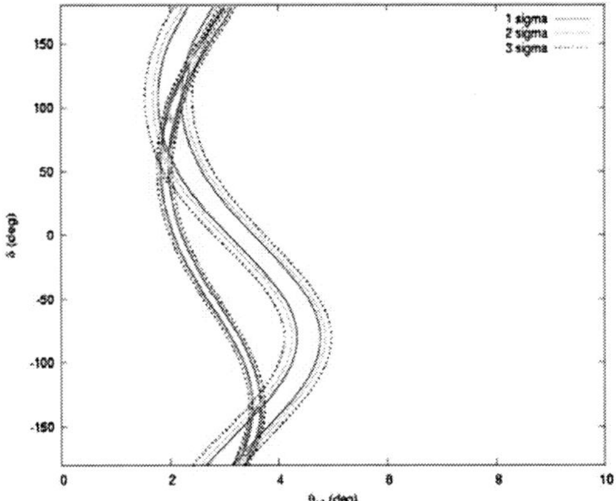

FIGURE 3. The virtues of two energies.

The Physics Reach is represented by means of the plot in the parameters (θ_{13}, δ) as given in Fig.4, with the expected results shown as confidence level lines for the assumed values $(8°, 0°)$, $(5°, 90°)$, $(2°, 0°)$ and $(1°, -90°)$. The exclusion plot for the $\theta_{13} \neq 0$ sensitivity is particularly impressive [9]. It is significant even at $1°$, for all possible values of δ.

To achieve a precise value of δ, it has been shown [9] that a combination of electron-capture neutrino beams with $\beta^-(^6He)$ antineutrino beams is very promising. An alternative could be to move to an smaller ratio E/L, where the sensitivity to δ is higher, by going to the longer baseline of 650 Km of Canfranc. This is being explored now.

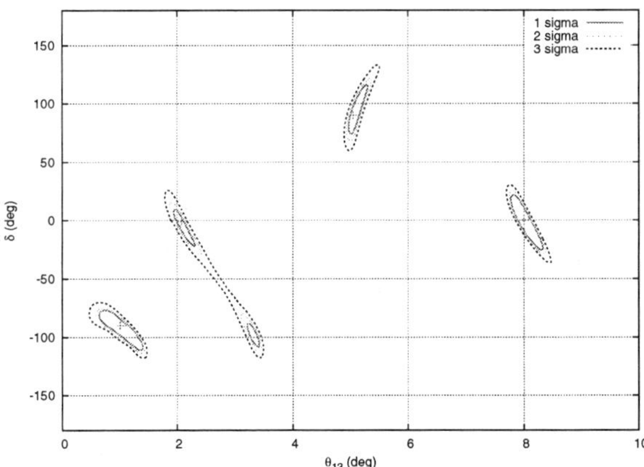

FIGURE 4. Physics Reach for the presently unknown (θ_{13}, δ) parameters, using two definite energies in the electron-capture facility discussed in this paper.

CONCLUSIONS AND PROSPECTS

The main conclusion is that **the principle of an energy dependent measurement is working and a window is open to the discovery of CP violation in neutrino oscillations**, in spite of running at two energies only. The opportunity is better for higher values of the mixing angle θ_{13}, the angle linked to the mixing matrix element $|U_{e3}|$ and for small mixing one would need to enter into the interference region of the neutrino oscillation by going to higher distance between source and detectors. To prove that the phase shift induced by δ in our EC design is due to a genuine CP-violating effect, one could combine in the facility the running with EC ^{150}Dy neutrinos and with β^- (^6He) antineutrinos.

A detailed study of production cross-sections, target and ion source designs, ion cooling and accumulation schemes, possible vacuum improvements and stacking schemes is required in order to reach a definite answer on the achievable flux. The experience gained at present on the acceleration and storage of partly charged ions and the calculations for the decay ring yield less than 5% of stripping losses per minute. A Dy atom with only one electron left would still yield more than 40% of the yield of the neutral Dy atom.

With an isotope having an EC half-life of 1 minute and a source rate of 10^{13} ions per second, a rate of 10^{18} ν's per year along one of the straight sections of the storage ring could be achieved [10]. A new effort is on its way to revisit the rare-earth region on the nuclear cart and measure the EC properties of possible candidates. The discovery of isotopes with half-lives of less than 1 minute decaying mainly through electron-capture to a single Gamow-Teller resonance in super allowed transitions, certainly would push the concept of a monochromatic neutrino beam facility. The Physics Reach that we have explored here is impressive and demands such a study. Most important is a realistic simulation to determine the full capability of this facility.

ACKNOWLEDGMENTS

This work has been supported by the Grant FPA2005-01678. C. E. is indebted to HELEN Program for financial help and to Cinvestav-Mexico for its hospitality.

REFERENCES

1. J. Bernabeu, et al; JHEP **0512**, 014 (2005), [hep-ph/0505054].
2. A. Cervera *et al.*, Nucl. Phys. B **579** (2000) 17; Erratum-ibid B **593** (2001) 731.
3. M. C. Gonzalez-Garcia, Global Analysis of Neutrino, 2004, [hep-ph/0410030].
4. J. Bernabeu, Nucl. Phys. B **49** (1972) 186.
5. E. Nacher, Beta decay studies in the $N \sim Z$ and the rare-earth regions using Total Absorption Spectroscopy techniques, Ph. D. Thesis, Univ. Valencia (2004).
6. P. Zucchelli, Phys. Lett. B **532** (2002) 166.
7. B. Autin *et al.*, Proc. Nufact 02, London, UK, 2002, J. Phys. G: Nucl. Part. Phys. **29** (2003) 1785.
8. B. Franzke, Proc. 1981 Particle Accelerator Conference, IEEE Trans. Nucl. Sci. **NS-28** (1981) 2116.
9. J. Bernabeu et al., Nucl. Phys. Proc. Suppl. **155** (2006) 222, [hep-ph/0510278].
10. M. Lindroos, et al; PoS **HEP2005**, 365 (2006).

Bounds of Non-Standard Neutrino Interactions Using a Model Independent Method

J. Barranco

Departamento de Física, Centro de Investigación y de Estudios Avanzados del IPN, Apdo. Postal 14-740 07000 México D.F., Mexico

Abstract. A neutrino magnetic moment bound is estimated using a model independent analysis from solar neutrinos data. The same method is applied to determine the neutrino electron scattering coupling constants. After obtaining the coupling constants values, we obtain constrains to the mass of an extra heavy neutral gauge boson and to a generic non-standard neutrino interaction.

INTRODUCTION

Neutrino physics has started a new process: the precision measurement of neutrino properties. Since experiments have shown that oscillatory behavior of neutrino offers a satisfactory explanation of the Solar and Atmospheric Neutrino problem, then, the old possible solutions such as Resonant Spin Flavor Precession (RSFP), Non-standard flavor changing interactions (NSI-FC) and others are possible only as sub-leading effects. At present, the available neutrino data permits us to constrain the non-standard neutrino properties, although in general, this is a difficult task because the large number of free parameters. In order to estimate bounds in a more simple way we show that the method previously used in Ref. [1] is very useful when the new physics parameters affect only the $\nu - e$ scattering cross section. We apply this method in order to estimate bounds in neutrino magnetic moment (μ_ν) and universal or non-universal non-standard $\nu - e$ interactions. The last can be parametrized by an effective Lagrangian [2]. Such Lagrangians can be generated as the low energy limit of super-symmetric theories [3] or theories with extended symmetries [4].

MODEL INDEPENDENT PROBABILITY SCHEME

One way to extract averaged survival probabilities in a model independent method was done in Ref. [1]. It consists of dividing the ν spectrum in three parts, each one with a constant averaged probability: i) Low energy probability P_L, that accounts only for the neutrinos coming from *pp* ii) Intermediate energy probability P_I, accounting for ^{7}Be, *pep* and CNO neutrino energies and finally iii) High energy probability P_H, that lies in the ^{8}B neutrino energy region.

CP885, *Advanced Summer School in Physics 2006, Frontiers in Contemporary Physics—EAV06,* edited by O. Miranda, M. Carbajal, L. M. Montaño, O. Rosas-Ortiz, and S. A. Tomás Velázquez

TABLE 1. Solar neutrino data including the experimental uncertainties

Experiment	data/SSM	Experiment	data/SSM
^{37}Cl	0.310 ± 0.030	SNO^{Scatt}	0.413 ± 0.05
^{71}Ga	0.540 ± 0.039	SNO^{CC}	0.300 ± 0.01
Super-K	0.413 ± 0.015	SNO^{NC}	0.880 ± 0.06

With these assumptions we can write the theoretical rate as

$$\begin{aligned} R^{th} &= \frac{\int dE_\nu(\phi_{^8B}\lambda_{^8B}(E_\nu)+\phi_{hep}\lambda_{hep}(E_\nu))\sigma(E_\nu)}{\sum_i \int dE_\nu \phi_i \lambda_i(E_\nu)\sigma(E_\nu)} P_H \\ &+ \frac{\int dE_\nu(\phi_{^7Be}\lambda_{^7Be}(E_\nu)+\phi_{pep}\lambda_{pep}(E_\nu)+\phi_{^{15}O}\lambda_{^{15}O}(E_\nu)+\phi_{^{13}N}\lambda_{^{13}N}(E_\nu))\sigma(E_\nu)}{\sum_i \int dE_\nu \phi_i \lambda_i(E_\nu)\sigma(E_\nu)} P_I \\ &+ \frac{\int dE_\nu \phi_{pp}\lambda_{pp}(E_\nu)\sigma(E_\nu)}{\sum_i \int dE_\nu \phi_i \lambda_i(E_\nu)\sigma(E_\nu)} P_L \end{aligned} \quad (1)$$

Because $\nu - e$ scattering processes are sensitive to all flavors the theoretical rate is given by

$$R^{Scatt} = \frac{\int dE_\nu \phi_{^8B}\lambda_{^8B}\sigma^{\nu_e}}{\int dE_\nu \phi_{^8B}\lambda_{^8B}\sigma^{\nu_e}} P_H + \frac{\int dE_\nu \phi_{^8B}\lambda_{^8B}\sigma^{\nu_{\tau,\mu}}}{\int dE_\nu \phi_{^8B}\lambda_{^8B}\sigma^{\nu e}}(1-P_H) \quad (2)$$

Therefore, using the spectrum $\lambda_i(E_\nu)$ from the Bahcall's solar model [5] and the corresponding cross section for each experiment we can construct the following set of equation for the theoretical rates:

$$\begin{aligned} R^{th}_{Cl} &= 0.803 f_B P_H + 0.197 P_I \\ R^{th}_{Ga} &= 0.109 f_B P_H + 0.335 P_I + 0.556 P_L \\ R^{th}_{SK} &= f_B[P_H + 0.154(1-P_H)] \\ R^{Scatt}_{SNO} &= f_B[P_H + 0.154(1-P_H)] \\ R^{CC}_{SNO} &= f_B P_H \\ R^{NC}_{SNO} &= f_B \end{aligned} \quad (3)$$

We can now perform the χ^2 analysis taking as free parameters the three averaged probabilities P_L, P_I, P_H and allowing the ^{8}B's flux to be free (f_B). The analysis is done with the experimental values of the rates reported by each detector. Values are shown in Table 1.

The values for the averaged probabilities obtained from the χ^2 are the following:

$$P_L = 0.713 \pm 0.12\,, \quad P_I = 0.326 \pm 0.15\,, \quad P_H = 0.336 \pm 0.03 \quad (4)$$

which are in good agreement with the best LMA probability.

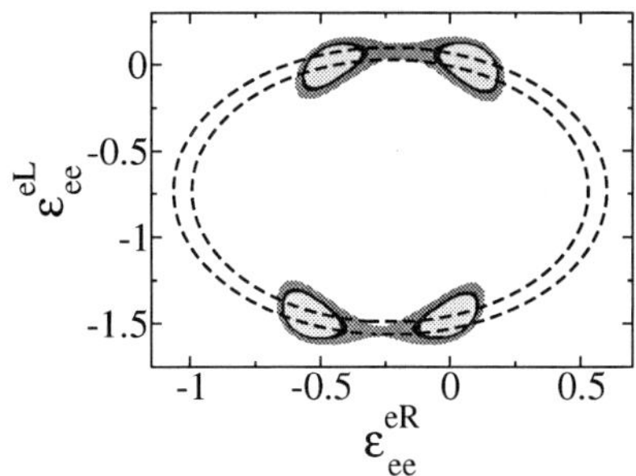

FIGURE 1. Allowed region for ε_{ee}^{eL} and ε_{ee}^{eR} obtained in this analysis and compared with the allowed regions obtained by means of reactor data [2].

BOUNDS ON THE NEUTRINO MAGNETIC MOMENT

Once we have seen that averaged probabilities agree with the LMA solution, we perform a similar analysis adding a non-zero neutrino magnetic moment. It is known that in this case, a diagram that includes a photon exchange should be added. The weak and electromagnetic cross sections do not interfere between them as far as neutrino mass can be neglected and hence the differential cross section of the electron- neutrino scattering due to magnetic moment effects is

$$\frac{d\sigma(E_\nu,T)}{dT} = \frac{\alpha^2\pi}{m_e^2\mu_B^2}\left(\frac{1}{T}-\frac{1}{E_\nu}\right)\mu_{eff}^2 \tag{5}$$

where μ_{eff} is the effective neutrino magnetic moment in Bohr magneton units. After including this cross section the theoretical rates due to scattering processes in SK and SNO change to

$$\begin{aligned} R_{SK}^{th} &= f_B[P_H + 0.154(1-P_H)] + 0.323 f_B \mu_{eff}^2 \\ R_{SNO}^{ES} &= f_B[P_H + 0.154(1-P_H)] + 0.323 f_B \mu_{eff}^2 \end{aligned} \tag{6}$$

By doing a χ^2 analysis with this new parameter (μ_{eff}) we obtain an upper bound to the neutrino magnetic moment of $\mu_{eff} \leq 2\times 10^{-11}\mu_B$ at 90 % C.L.

BOUNDS ON NSI FROM NEUTRINO ELECTRON SCATTERING

Models with a primordial E_6 gauge symmetry generate additional neutral heavy bosons Z' with a mass much bigger than the SM Z-boson ($M_{Z'} \gg M_Z$). Therefore, the neutral current couplings of the SM fermions are modified due to the mixing of the SM Z-boson with the additional Z'. Neglecting radiative corrections it is possible to write an effective Lagrangian that describes the low-energy neutral current phenomena:

$$-\mathscr{L}_{\text{NSI}}^{eff} = \varepsilon_{\alpha\beta}^{fP} 2\sqrt{2} G_F (\bar{\nu}_\alpha \gamma_\rho L \nu_\beta)(\bar{f}\gamma^\rho P f). \tag{7}$$

This Lagrangian must be added to the SM one.
In particular for the E_6 model the additional coupling constants in the $\nu - e$ scattering are [4] :

$$\begin{aligned}
\varepsilon_{ee}^{eL} &= 2\gamma \sin^2\theta_W \rho_{\nu e}^{NC} \left(\frac{3c_\beta}{2\sqrt{6}} + \frac{s_\beta}{3}\sqrt{\frac{5}{8}} \right)^2 \\
\varepsilon_{ee}^{eR} &= 2\gamma \sin^2\theta_W \rho_{\nu e}^{NC} \left(\frac{c_\beta}{2\sqrt{6}} - \frac{s_\beta}{3}\sqrt{\frac{5}{8}} \right) \left(\frac{3c_\beta}{\sqrt{24}} + \frac{s_\beta}{3}\sqrt{\frac{5}{8}} \right)
\end{aligned} \tag{8}$$

where $\gamma = (M_Z/M_{Z'})^2$ and $c_\beta = \cos\beta$, $s_\beta = \sin\beta$ are parameters that characterize particular realizations of the E_6 model. In particular, for the χ model $\cos\beta = 1$, for the η model $\cos\beta = 0$ and finally the ψ model $\cos\beta = \sqrt{3/8}$. These couplings are universal and modify the cross section as $\sigma^{Tot} = \sigma^{SM} + \delta\sigma(M_{Z'})$. After a χ^2 analysis for the χ model we obtain a lower bound $M_{Z'} \geq 150$ GeV at 90 % C.L. This bound is poor in comparison with the Tevatron result ($M_{Z'} > 595$ GeV, 95% C.L. [6]). Nevertheless, it is important to show that with a more accurate measurement of the rates in SK and SNO there will be better bounds by using neutrinos as messengers.
Finally, consider a different type of NSI that affects only ν_e coupling constants ε_{ee}^{eL} and ε_{ee}^{eR}, that is, a non-universal correction. In this case, only the electron type interaction will be modified $\sigma^{\nu_e} = \sigma_{SM}^{\nu_e} + \delta\sigma(\varepsilon_{ee}^{eL}, \varepsilon_{ee}^{eR})$. We can write the correction to the $\nu - e$ scattering cross section due to NSI as:

$$\delta\sigma(\varepsilon_{ee}^{eL}, \varepsilon_{ee}^{eR}) = \left(\frac{2\varepsilon_{ee}^{L} g_L + |\varepsilon_{ee}^{L}|^2}{g_L^2 + g_R^2/3} + \frac{2\varepsilon_{ee}^{R} g_R + |\varepsilon_{ee}^{R}|^2}{3g_L^2 + g_R^2} \right) \tag{9}$$

where g_L and g_R are the SM coupling constants.
From the corresponding χ^2 analysis we obtained that the allowed region for the NSI parameters is between the ellipses shown in Fig. 1. We can see that they are competitive with previous analysis from reactor neutrinos [2] also shown in the same figure.

Acknowledgments This work has been supported by CONACYT. Special thanks to M. Tortola, O. G. Miranda, and J. W. F. Valle for very productive discussions.

REFERENCES

1. V. Barger, D. Marfatia and K. Whisnant Phys. Lett. B **617** 78-86,2005; Phys.Lett.B**509**:19-29,2001. V. Berezinsky and M. Lissia, Phys. Lett. B **521**, 287 (2001) [arXiv:hep-ph/0108108].
2. J. Barranco, O. G. Miranda, C. A. Moura and J. W. F. Valle, Phys. Rev. D **73**, 113001 (2006) [arXiv:hep-ph/0512195].
3. V. D. Barger, R. J. N. Phillips and K. Whisnant, Phys. Rev. D **44**, 1629 (1991).
4. M. C. Gonzalez-Garcia and J. W. F. Valle, Nucl. Phys. B **345**, 312 (1990).
5. J. Bahcall's homepage http://www.sns.ias.edu/ jnb/
6. W.-M. Yao et al., J. Phys. G 33, 1 (2006)

The Renormalization Group Running of the Higgs Quartic Coupling: Unification vs. Phenomenology

J.H. Montes de Oca Y.*, S.R. Juárez W.* and P. Kielanowski†

*Departamento de Física, Esc. Sup. de Fís. y Mat., Instituto Politécnico Nacional, U.P. "Adolfo López Mateos" Edif. 9, C.P. 07738, México, D.F., Mexico
†Departamento de Física, Cinvestav, Av. IPN 2508, C.P. 07360, México, D.F., Mexico

Abstract. Within the framework of the standard model (SM) of elementary particles, we obtained numerical solutions for the running Higgs mass, considering the renormalization group equations at the one and two loop approximation. Through the triviality condition (TC) and stability condition (SC) on the Higgs quartic coupling λ_H the bounds on the Higgs running mass have been fixed. The numerical results are presented for two special cases. One considering an unification of the three gauge couplings at the energy $E_U = 10^{13}$ GeV and the other using the current experimental data for the gauge couplings.

INTRODUCTION

The gauge group of the Standard Model is $SU(3)_c \otimes SU(2) \otimes U(1)$. The sector that corresponds to the strong interaction is $SU(3)_c$ and the gauge group of the electroweak interaction is $SU(2) \otimes U(1)$. The gauge coupling parameters for $SU(3)_c$, $SU(2)$ and $U(1)_Y$ are g_3, g_2 and g_1, respectively. The values of these parameters g_i, $(i = 1,2,3)$ do depend on the renormalization point energy. The *running* with the energy of these parameters is determined by the Renormalization Group Equations (RGE). These gauge couplings provide a convenient means to describe changes in the behavior of the interactions under changes in the energy scale.

The Higgs mechanism is an important ingredient in the SM. The Higgs mechanism requires a scalar field ϕ (Higgs) and a potential $V(\phi)$ to generate the masses of the leptons, the quarks and the gauge bosons through a spontaneous symmetry breaking. This potential is: $V_\phi = -\mu^2\phi^\dagger\phi + \frac{1}{2}\lambda_H(\phi^\dagger\phi)^2$, where the Higgs field ϕ is a doublet with weak isospin $T = \frac{1}{2}$ and hypercharge $Y = 1$, Ref. [1]. The λ_H is called the Higgs quartic coupling. This potential has a non-vanishing minimum value from which we can determine the vacuum expectation value $vev = 246.22$ GeV (at the Z boson mass scale, $m_Z = 91.187$ GeV, Ref. [2]). After expanding the ϕ field around its minimum value and transforming to the unitary gauge, one obtains in the one loop case the following relation for the Higgs current mass, $m_H = v(t)\sqrt{\lambda_H(t)}$, Ref. [3]. Accordingly, the bounds for the Higgs mass are related to the bounds of Higgs quartic coupling. These bounds are obtained by means of two model consistency conditions. When the SM is renormalized, inevitably, an arbitrary energy scale μ called the renormalization scale is introduced and the SM parameters acquire a renormalization point energy dependence through the vari-

CP885, *Advanced Summer School in Physics 2006, Frontiers in Contemporary Physics—EAV06,* edited by O. Miranda, M. Carbajal, L. M. Montaño, O. Rosas-Ortiz, and S. A. Tomás Velázquez
 978-0-7354-0385-7/07/$23.00

able $t = \ln(\frac{E}{\mu})$. The renormalization group equations (RGE) arise from transformations among different energy scales. The RGE are ordinary differential equations for the SM parameters when infinitesimal changes in the scale μ are performed. Between the several schemes to extract the divergences, in this study we choose the modified minimal subtraction ($\overline{MS}$) scheme.

NON-SUPERSYMMETRIC UNIFICATION

The idea that all electroweak and strong interactions can be derived from a unified theory is a very interesting one and deserves to be explained in an elegant and simple manner. The extension of the idea of the unification of the interactions, which underlies the SM, is the origin of the Grand Unified Theories (GUTs). Nowadays, the big question of grand unification is how it is possible to have a universal gauge coupling in the theory and yet have three unequal gauge couplings at the weak scale with $g_1 < g_2 < g_3$. The simplest model $SU(5)$ that has emerged under this conceptual structure has been ruled out by the increased accuracy in the measurement of the Weinberg's angle and by the early bounds on the proton lifetime. Another attractive way to give an explanation to the unification is provided by the concept of supersymmetry (SUSY). As the SM has been so successful in predicting the outcome of a large number of experiments, an enormous amount of work has been performed in relation to phenomenological studies of the non-supersymmetric field theories without the assumption of grand unification or within the framework of a more fundamental scheme. In this context, and as a starting point to obtain and compare results, we consider an analysis of an hypothetical case in which an unification takes place in the SM regime. We assume the intersection of the three gauge coupling parameters at the energy E_U. The average values of the low energy experimental data for the gauge couplings are $(g_1, g_2, g_3) = (0.461, 0.652, 1.221)$ at the m_Z scale [2]. Once evolved to the top quark mass scale (our departure energy), $m_t = 172.5$ GeV, they become $(g_1, g_2, g_3) = (0.463, 0.648, 1.173)$, Ref. [4]. In this analysis, we consider the previous current experimental values (CEV) in one evaluation and compare results with the ones obtained in a second evaluation using new theoretical hypothetical values (THV) $(g_1, g_2, g_3^U) = (0.463, 0.648, 0.9393)$ at the m_t scale, that evolve in the two loop approximation in such a way that they intersect at an *unification* point, where $g_i(E_U) = 0.5454$ $(i = 1,2,3)$, $E_U \approx 10^{13}$ GeV, and $t_U = 24.783$.

The value of the new strong interaction gauge coupling at the m_Z scale, is $g_3^U(m_Z) = 0.9636$ which corresponds to an $\alpha_3^U(m_Z) = 0.0702$. This value is consistent with the one given by the non-supersymmetric unified theories which consider the low energy value $\alpha_3(m_Z) = 0.073 \pm 0.001$, cited in the particle listings in [2].

SOLUTION OF THE RGE

In the SM the RGE constitute a system of non-linear, coupled differential equations [1]

$$\frac{dx}{dt} = \frac{1}{16\pi^2}\beta_x^{(1)} + \frac{1}{(16\pi^2)^2}\beta_x^{(2)} + \cdots, \tag{1}$$

where $t = \ln(\frac{E}{\mu})$; $x = g_i$, $y_{u,d}$, v, λ_H, Z, with $y_{u(d)}$ the Yukawa coupling for up (down) sector, the Z is related with the anomalous dimension of the Higgs field and $\beta_x^{(i)} = \beta_x^{(i)}(g_i, y_{u,d}, v, \lambda_H, Z)$.

The elements of the RGE can be expanded in terms of the λ parameter (Cabbibo angle) [5] in order to deal with new approximated and simplified RGE. In the one loop approximation, terms of order λ^4 and higher are neglected. Under this assumptions, it is possible to find analytical solutions for the quartic coupling and the gauge coupling parameters [3]:

$$g_i^{(1)}(t) = \frac{g_i^0}{\sqrt{1 - \frac{2b_i(g_i^0)^2(t-t_0)}{(4\pi)^2}}}, \tag{2}$$

$$\lambda_H^{(1)}(t) = -\frac{(4\pi)^2}{12}\frac{W_1'(t) - \frac{12}{(4\pi)^2}\lambda_H^0 W_2'(t)}{W_1(t) - \frac{12}{(4\pi)^2}\lambda_H^0 W_2(t)}, \tag{3}$$

where $(b_1, b_2, b_3) = (\frac{41}{10}, -\frac{19}{6}, -7)$, $g_i^0 = g_i(t_0)$, $\lambda_H^0 = \lambda_H^{(1)}(t_0)$ with $t_0 = t(m_t) = 0$. The W_i are auxiliary functions used to transform the Riccati λ_H equation into a second order linear differential equation.

Due to its complexity, the more precise solutions for the two loops approximation RGE were obtained in a numerical way [6]. The evolution of the gauge coupling parameters is shown in Fig. 1.

There are two consistency conditions for the SM: the stability condition (ST), which implies that $\lambda_H(t) > 0$ and the triviality condition (TC), which implies that λ_H must have non singular values $\lambda_H(t) < \infty$.

In the one loop approximation, the allowed region for SM is obtained using the analytical solution for the $\lambda_H(t)$. The lower limit for the allowed region is determined

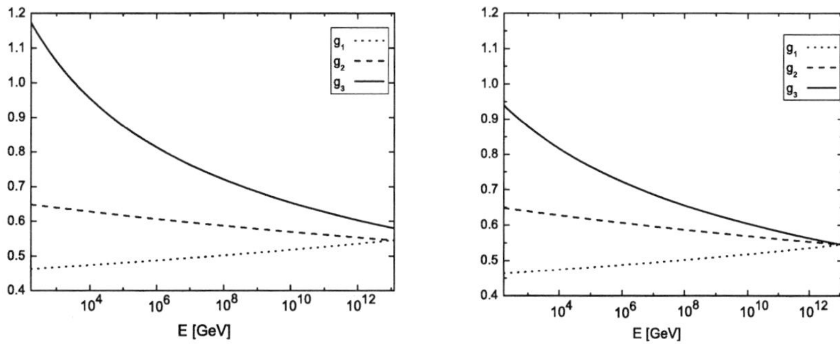

FIGURE 1. The gauge coupling running in the two loop approximation. On the left with the CEV vs. on the right under the unification assumption with the THV.

by the numerator of Eq. (3) through the ST and the upper limit is determined through the zero value of the denominator of Eq. (3) (TC); in both cases for $m_t \leq E \leq E_U$.

The stability condition is the same for the one and the two loop approximation. In the two loop case the λ_H does not become singular, it gets an asymptotic value $\lambda_H^{(2)as} \approx 24.28$, Ref. [7].

The Fig. 2 shows the $\lambda_H(E)$ evolution for the two boundaries of λ_H, and the interval of allowed values of $\lambda_H(m_t)$ is: $0.357 \leq \lambda_H(m_t) \leq 0.6351$ for the two loop RGE and CEV; $0.5119 \leq \lambda_H(m_t) \leq 0.6686$ for two loop RGE and THV. This results lead to the following bounds for the running Higgs mass: 145.9 GeV $\leq m_H \leq$ 194.6 GeV for the CEV case and 174.71 GeV $\leq m_H \leq$ 199.7 GeV for the THV case.

The range of energies at which the SM is consistent in terms of $\lambda_H(m_t)$ is shown in Fig. 3. The allowed zone is the one between the two curves.

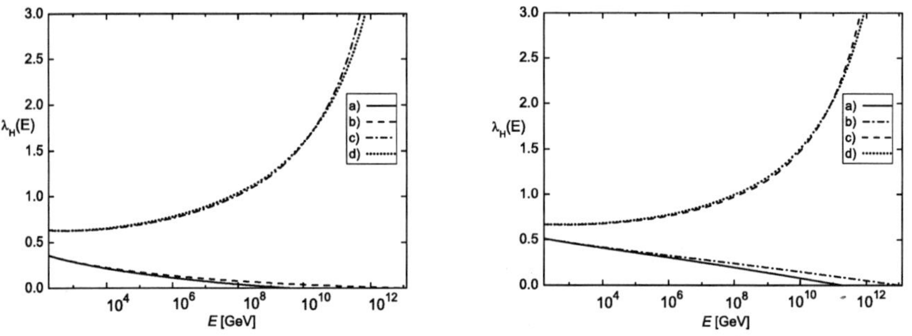

FIGURE 2. The energy dependence of the Higgs quartic coupling. a) one loop and SC, b) two loops and SC, c) one loop and TC, d) two loops and TC. On the left with CEV vs. on the right with THV.

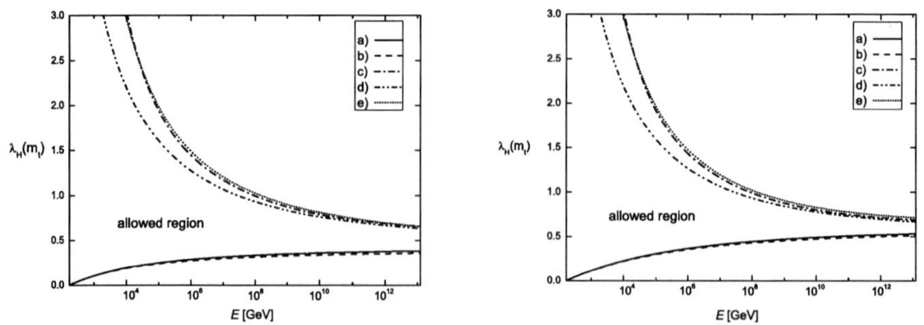

FIGURE 3. Bounds for the SM validity. Graphics obtained from the conditions: a) one loop (SC) with $\lambda_H(E) = 0$, b) two loops (SC) with $\lambda_H(E) = 0$, c) (TC) with λ_H in the pole, two loops for d) $\lambda_H(E) = 6$ and e) $\lambda_H(E) = 12$. On the left with the CEV vs. on the right considering THV.

CONCLUSIONS

We have obtained numerical solutions, in the two loop approximation, for the quartic Higgs coupling RGE within the framework of the SM. We have compared the one and two loop running of the λ_H evaluated with two kinds of data: the current experimental values (CEV) and the theoretical hypothetical values (THV) derived from an unification assumption. Using the one loop RGE for the quartic Higgs coupling $\lambda_H(t)$ which is of the Riccati type we studied the zeros and singularities (poles) of $\lambda_H(t)$ as functions of t. This information with the triviality and stability conditions lead to the bounds of the allowed values of the $\lambda_H(t_0)$ and of the running Higgs mass m_H.

The Unification assumption drives to a narrower interval of allowed values for the Higgs running mass and higher bounds. The upper bound increases for about 5.066 GeV and the lower one for about 28.8 GeV at the m_t scale.

To summarize, by the method of the renormalization group we have obtained and compared two different predictions for the Higgs running mass. One considering the current experimental values for the gauge couplings and the other changing the value of the strong gauge coupling for a different one obtained by the assumption of an unification in the SM context.

ACKNOWLEDGMENTS

S.R.J.W. thanks to the "Comisión de Operación y Fomento de Actividades Académicas" (COFAA) and EDD from Instituto Politécnico Nacional. J.H.M.O.Y. thanks to the PIFI-IPN and CONACyT, México.

REFERENCES

1. H. Arason, et. al., Phys. Rev. D **46**, 3945 (1992); Mingxing Luo and Yong Xiao, Phys. Rev. Lett. **90**, 011601 (2003).
2. Particle data group, S. Eidelman et. al., Phys. Lett. B **592**, 1 (2004).
3. S.R. Juárez W., A. Morales S. and P. Kielanowski, Revista Mexicana de Física **50**, 401 (2004); S. Rebeca Juárez W. and Piotr Kielanowski, Proc. IX Reunión Nacional Académica de física y matemáticas, ESFM-IPN (2004).
4. H.G. Solís-Rodríguez, S.R. Juárez W. and P. Kielanowski, X Mexican Workshop on Particles and Fields to be published, AIP CP (2006).
5. L. Wolfenstein, Phys. Rev. Lett. **51**, 1945 (1983).
6. P. Kielanowski, S.R. Juárez W. and H.G. Solís-Rodríguez, Phys. Rev. D **72**, 96003 (2005).
7. T. Hambye, K. Riesselmann, Phys. Rev.D. **55** , 7255 (1997).

III. STATISTICAL PHYSICS

Complex Networks: Statics and Dynamics

Albert Diaz-Guilera

Departament Física Fonamental, Universitat de Barcelona, Marti i Franques 1, 08028 Barcelona, SPAIN

Abstract. We present some of the results obtained during the last 8 years about complex networks. Starting with the collection of data in the form of networks or graphs, we proceed on the characterization at different scales: microscopic, macroscopic, and mesoscopic. We introduce also the basic models incorporating complexity in the pattern of connectivities. Finally we review some results on dynamical features on complex networks.

INTRODUCTION

Complex systems show emergent properties that cannot be understood as a simple superposition of the dynamical behavior of the single units that form it. Emergent properties arise as a collective effect from the interaction between units. For most of the complex systems we can find in nature or in society the interaction patterns of the units are far from being regular or from being completely random. These two extreme patterns of connectivity between units had been the subject of analysis during the last decades, but since the pioneering work of Watts and Strogatz [1] and Barabasi and Albert [2] most of the interest turned to non-trivial patterns of interaction.

Access to large datasets and the power of the new generation of computers have enabled to generate networks from the data and to scrutinize them, finding the relevant statistical properties and the common features shared by many of them. And such common features are not the subject of particular disciplines; this new framework has received a really interdisciplinar support involving disciplines such diverse as economy, social sciences, medicine, engineering, physics, chemistry, biology, just to mention a few (see [3, 4, 5, 6, 7] for reviews, and [8, 9] for contributed essays on various topics by leading researchers).

Although in principle the interest relied on the structural properties, i.e. the topology of the interactions, in the last years a great deal of attention has turned to the dynamical properties. This dynamical properties can involve the growth of the network (in terms of nodes or links or both), the regeneration of links between the nodes, and even the dynamics of the nodes properties.

In this paper we will cover the different aspects of complex networks. Starting with an overview of networks that are found in our environment, then we follow with the characterization of the networks at different scales (microscale, macroscale, and mesoscale) and introducing some simple network models that describe the topological features. Finally we describe some dynamical models on networks.

CP885, *Advanced Summer School in Physics 2006, Frontiers in Contemporary Physics—EAV06,*
edited by O. Miranda, M. Carbajal, L. M. Montaño, O. Rosas-Ortiz, and S. A. Tomás Velázquez

NETWORKS EVERYWHERE

We can currently find structures among data that can be classified as networks in many unrelated fields. The only thing that is needed is some sort of relation between the data. Networks or graphs are formed by nodes that are connected by links or edges. What is a node and a link will a characteristic of the data. For instance in social networks nodes can be individuals and links could be any kind of relation: friendship, coauthoring, trust, ... In technological networks, as for example the Internet, the edges correspond to physical wiring between computers or routers. Or in biology, metabolic networks are formed by metabolites as nodes and the links represent the biochemical reactions. Those are just a few examples of the type of data that can be represented as a network. A very large number of publicly available repositories of huge databases are at our fingertips.

Classification

Here we present a list of examples, mainly from papers published in Physics journals, of data that has been collected in the form of network. They can roughly be classified into four categories (following [6]):

- Social networks
  Actor collaborations [10, 1]
  Boards of directors [11, 12]
  Physics and biology coauthorships [13, 14, 15]
  Email messages [16, 17, 18]
  Sexual contacts [19, 20]
  Jazz bands and musicians [21]
  Pretty Good Privacy trust network [22]
- Information networks
  World Wide Web [23, 2]
  Citation networks [24]
  Word co-occurrence [25, 26]
- Technological and transport networks
  Internet [27]
  Power grid [1]
  Software packages [28]
  Software routine calls [29]
  Electronic circuits [30]
  Airport network [31, 32]
  Railroad network [33]
- Biological networks
  Metabolic networks [34, 35, 36, 37, 38]
  Protein interactions [39, 40, 41, 42, 43]

Food webs [44, 45]
Neural networks [1, 46, 47, 48]
Genetic regulatory networks [49, 50]
Signaling networks [51]

This is by no means a complete list. Nevertheless, it attempts to be a cross section of the various lines of investigation where network analysis has been useful.

How are networks constructed from data

It is then clear that every dataset will give rise to a network in which nodes and links have very different meanings: social agents and social relationships, computers and cables, species and predator-prey relationships, neurons and synapses, web pages and hyperlinks, and so on.

The link can be directed or undirected, depending on the reciprocity of the relation. An example of directed network, in which some links can be directed, is presented in Fig. 1(top-left). Links with an arrow pointing from one node to another one are directed, whereas the links without any arrow are undirected or bidirectional, because the relation holds in the two directions.

Some of the networks that are constructed are called bipartite, those are graphs that contain nodes of two different types, with links only between unlike types. This is what happens for instance in the actors movie database with actors and movies, or with the coauthor-ship databases. In this case networks are constructed by linking those actors that appear in the same movie or those authors that participate in the same paper. An example of such network is presented in Fig. 1(bottom)[15]. This is the network obtained from the coauthorship of presentations in the Spanish Statistical Physics meetings; it has been obtained by accumulating the collaborations along all the editions of the meeting. There are a few nodes that are identified because they correspond to members of the different scientific committees; the role played by these members will be discussed later, in the community identification discussion section.

Another fact that has become very important in the last years is the weight of the edges. If one is just interested in the existence of the relation then we talk about unweighted networks. If, on the contrary, there is some measure for the strength of the relation, as for instance the number of flights or the number of passengers between airports, or the band-width between Internet routers, some weight is associated to the link, and in this case the networks are called weighted [52]. In Fig. 1(top-right) we present a network obtained from the email exchange of the Universitat Rovira i Virgili. We take each node in the original network and measure the number of steps across the e-mail network needed to reach any other node. Then we average over all the nodes in the same center and obtain average distances between centers. This average distance between the centers accounts for the weight of the link. To summarize this information in a new network of centers, we proceed as follows. First, we calculate the distance from one center A to all other centers, d_{AB}, d_{AC}, and so on. Then we compute the average distance from A to the other centers $\langle d_A \rangle$. Finally, node A (that now represents a center, not an individual) is linked to another node B if $d_{AB} < \langle d_A \rangle$. In this case, the network is directed because, in general, $d_{AB} < \langle d_A \rangle$ does not imply $d_{BA} = d_{AB} < \langle d_B \rangle$. [53]

CHARACTERIZING NETWORKS

Microscale

From a microscopical point of view, the interest would lie on the role played by the nodes in the overall context of the whole network. This has been the main issue for decades from the social sciences viewpoint [54]. Several measures of centrality were introduced and the special roles played by the nodes discussed. For instance, the degree of a node corresponds to its number of links or the mean distance is a measure of the average distance, measured as the shortest number of links necessary to reach one node from another, from a node to the rest of the population. Another example is the clustering coefficient of a node, which measures the fraction of links between neighbors of a given node. Finally, another interesting measure is what is called the betweenness, of a node, which corresponds to the number of shortest paths between each pair of nodes in the network that go through the reference node. In many problems related to flow or traffic in the network the betweenness is a good measure for the load of the node [55, 56].

Macroscale

On the other hand, when dealing with very large networks, the roles played by the individual nodes has not meaning at all and the interest is turned to the statistical characterization of the network at the global or macroscopic scale. Now one studies average quantities like the mean degree, the mean distance between nodes, the average clustering coefficient, the diameter of the network (measured as the maximum distance between nodes). Another statistical characterization of the network comes in terms of the distributions of degree, of load, or on the correlations.

It was the initial study of these statistical characterizations of the networks that started the big interest from the Statistical Physics community. In particular, as we will explain shortly, there were two crucial facts that could not be explained by means of known graph models: the small-world effect and the observation that the distribution of degrees followed a power law indicating that there are no characteristic scales in this distribution and hence those networks were called "scale-free" networks. In Fig. 2(right) we plot the in- and out-degree cumulative distributions [1] in the PGP web of trust of Ref. [22], as an example of power-law distribution; in this case, as a directed network, the in- and out-degrees distributions do not need to be identical.

---

[1] The cumulative distribution $P(k)$ is simply related to the probability density function $p(x)$ by $P(k) = \int_{-\infty}^{k} dx\, p(x)$. In particular, if $p(x)$ is a power law $p(x) \sim x^{-\alpha}$, then $P(k) \sim k^{-\alpha-1}$, and if $p(x)$ is an exponential $p(x) \sim \exp(-x/k^*)$, then $P(k) \sim \exp(-k/k^*)$.

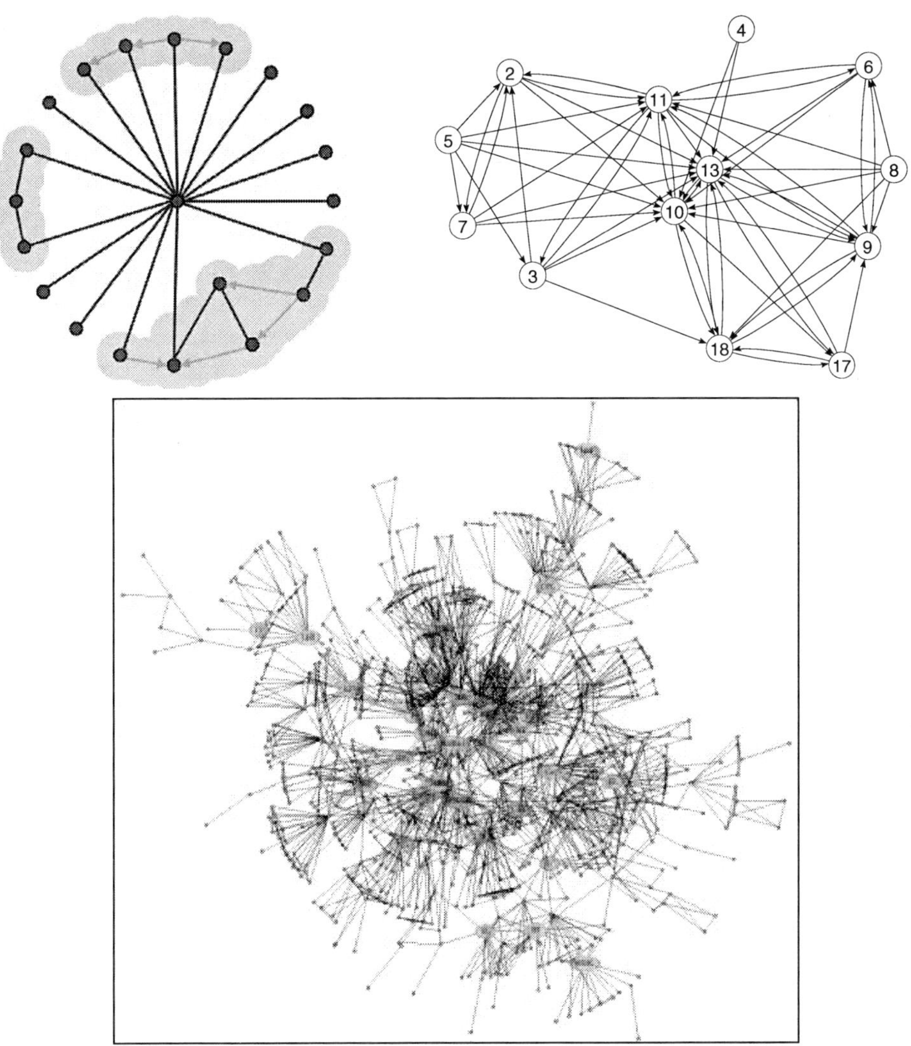

FIGURE 1. (color on-line) Examples of generated networks from data. Top-Left: Directed network. Nodes are shown as grey (blue) circles. Bidirectional links are colored in dark grey (brown) whereas unidirectional links are colored in light grey (orange). The shadowed areas correspond to the robust areas of the network, i.e. they keep connected when the central node is eliminated. Top-Right: Directed and weighted network. Network obtained from the email network at Universitat Rovira i Virgili. The nodes correspond to the centers and the links are related to the distance between centers (see text). Since the distance can take real non-negative values the link, and hence the network, are weighted. Bottom: Collaboration network. Cumulative network of collaborations during Spanish Statistical Physics meetings. Shaded (green) nodes correspond to members of the scientific committee [15].

Later on, different characterizations of the networks have been introduced; for instance in weighted networks, the distribution of weights is also a scale-free [52]. Also, other characterization in the large scale have appeared. For instance, the degree-degree correlation $P(k'|k)$ is the conditional probability that a link of a node with degree k is linked to a node with degree k'; if this probability depends on k we say that the node is correlated, or uncorrelated in the opposite case. In terms of this conditional probability, it is more useful to define the average degree of the nearest neighbors of nodes with degree k

$$k_{\mathrm{nn}}(k) = \sum_{k'} k' P(k'|k). \tag{1}$$

If $k_{\mathrm{nn}}(k)$ is a decreasing function of k then we say that the network is disassortative, as happens in technological or biological networks, whereas if it is a increasing function we call it assortative, as happens in many of the social networks, where clearly the meaning is that most connected nodes tend to be connected between them and less with poorly connected nodes.

MODELS OF NETWORKS

The random graph model of Erdös and Renyi

This is the most simple model of graph[57]. Let us consider a set of N nodes and the probability that every two nodes are connected (form a pair or a link) is p (see Fig. 3(left)). Then the expected value of the connectivity is simply:

$$\bar{k} = p(N-1). \tag{2}$$

When considering very large networks and keeping the average value fixed, the distribution of connectivities approaches a Poisson distribution with mean λ:

$$P(k) = \frac{\lambda^k e^{-k}}{k!} \tag{3}$$

which is sharply peaked at λ, as can be seen in Fig. 2(left).

The small-world model of Watts and Strogatz

In the paper by Watts and Strogatz [1] they realized that many networks in nature had a statistical behavior that could not be fitted to that of the known results up to that time: regular lattices or random graphs. On the one hand regular lattices have very large average distance between nodes, this the so called "small-world" effect, and high average clustering coefficient, due to the high interconnection between neighbors. On the other hand, random graphs have very short average distances, due to the existence of short-cuts and very low clustering due to the random uncorrelated nature of the connections. And the conclusion that the authors got from the analysis of the networks was that they had

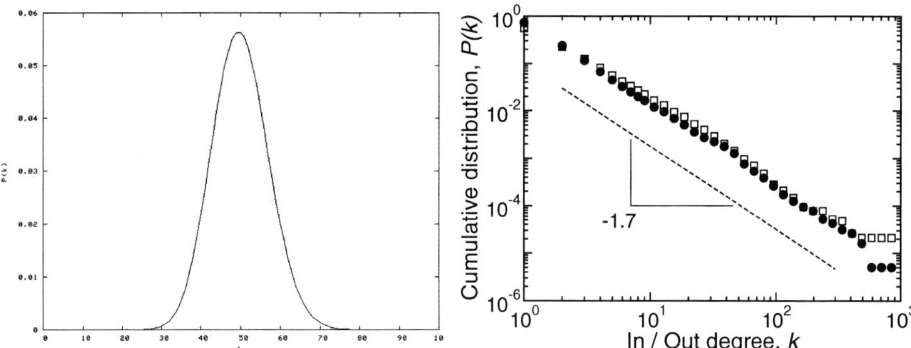

FIGURE 2. Two examples of distribution of connectivities. Left: Poisson distribution with average 50 in a linear-linear scale. Right: Power-law cumulative distribution of in- and out-degrees (incoming and outgoing connections) in the PGP web of trust of Ref. [22].Notice that in this case the scale is log-log.

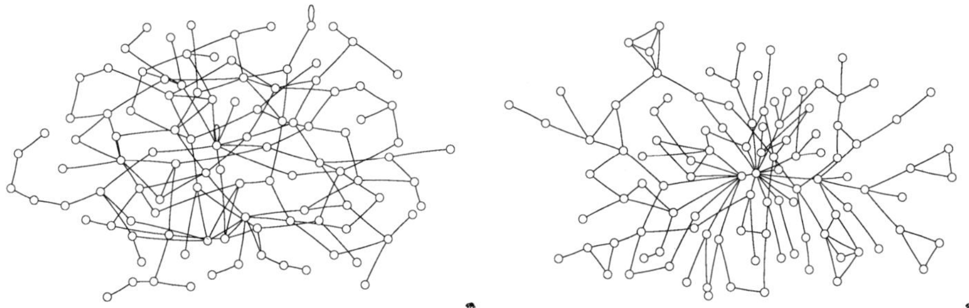

FIGURE 3. Two networks of approximately the same number (100) of nodes and links (130). Left: Erdös-Renyi Random graph. Right: Barabasi-Albert scale free network.

very short distances, like in random networks, and high clustering, like random graphs. This observation opened a completely new field of research since new models that could explain this simultaneous, and in principle opposite, behaviors were needed.

In particular, they already proposed a model, nowadays known as the Watts and Strogatz "small-world" model. To construct the network one starts with a regular one-dimensional lattice (a ring) in which all nodes are linked to their m neighbors in each direction (see Fig. 4(leftmost)); in this case the degree distribution is a delta function $\delta(k-2m)$. Then one proceeds by removing the short range links and substituting them by long-range links between two randomly chosen nodes with probability p. As can be seen in Fig. 4 by increasing p the graph loses the lattice character and resembles every time more to a random graph. Actually, with a relatively small value of p the graph acquires a short average distance between nodes without appreciably changing the clustering. In this way the observation of the simultaneous short distance and high clustering is explained by means of a very simple model.

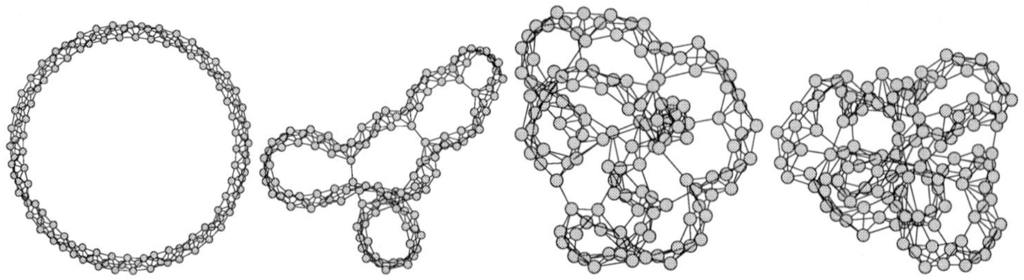

FIGURE 4. Small world model of Watts and Strogatz. Links to neighbors in the original ring are rewired with a probability 0.00, 0.05, 0.10, 0.15 (from left to right).

The scale-free model of Barabasi and Albert

Although the model introduced by Watts and Strogatz could resolve the apparent dichotomy in the observation of some regular and some random characteristics in many networks in nature, it did not change appreciably the distribution of connectivities. Starting from a delta function it rapidly evolves to a Poisson distribution for low values of p, thus the distribution of connectivities resembles that of the Erdos-Renyi model. But, just a few months later than this paper appeared, Barabasi and Albert [2] published their work in which they noticed that again many of the large networks that could be already analyzed at that time (including the Internet or the Web) showed distributions of connectivities that should be fitted to a power-law, instead of a Poisson-like as an Erdos-Renyi random graph.

In order to explain this behavior they also introduced a model, nowadays known as Barabasi-Albert model, in which there were two essential ingredients: growth and preferential attachment (this kind of attachment also gives its name to the model sometimes). On the one hand, networks are not static but are the result of a process of growing, starting from a set of a small number of completely connected nodes. On the other hand, the growth proceeds in such a way that the arriving nodes are linked preferentially to those nodes which already have more connections, as is schematically visualized in Fig. 5 (see also Fig. 3(right) for an example of such network with around 100 nodes). As can be easily interpreted from this simple rule, and also from the kind of distribution showed in Fig. 2(right), one of the main implications of this model is the existence of small fraction of highly connected nodes, named as hubs, whereas the vast majority of nodes have a very low connectivity. These hubs play a crucial role in many aspects of the network; for instance, the network is very sensitive to intentional attacks if the targets are the hubs, but is very robust under random attacks (or failures) in the case that the target is chosen at random [58, 59]. They are also important in the spreading of information or in the dynamics of synchronization, as we will see in next sections.

The finding that networks in natural or technological or social environments were scale free, showing some remarkable similarities of many other phenomena studied in the physical sciences, like critical phenomena or fractals, together with the "small-world" concept introduced by Watts and Strogatz, started the new theory of complex networks with contributions in many different fields, but with a major contribution from

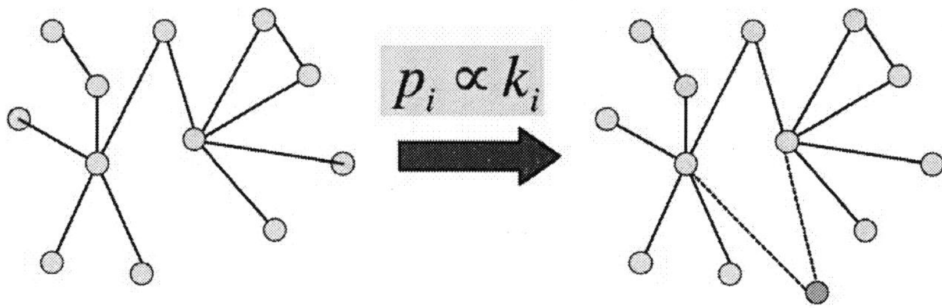

FIGURE 5. Preferential attachment rule of the Barabasi-Albert model. The arriving node is more likely to be connected to those nodes which already have more existing connections, and hence the new links correspond to the dotted lines.

the Statistical Physics community.

DESCRIBING THE MESOSCALE: COMMUNITIES

Clearly, in the previous sections we characterize the networks either from the microscopic or from the macroscopic point of view, but many networks show structures that are important in the intermediate scales, the mesoscale. Those structure can have different meanings depending on the origin of the network: communities in social networks[54], functional groups in biology[60], regional groups in geographically based networks, thematic clusters in the web [61, 62], and so on. Many times theses structures have an important role in their own and they have not been constructed by chance but by an ordered process of growth. For this reason identifying the communities in a network is a process from which we can gain a lot of useful information. Furthermore, dynamics is also affected by this community structure since dynamics is tightly related to the underlying topology of the network. The readers are pointed to Refs. [63, 64] for recent reviews on the subject of communities in complex networks.

Distinct modules or communities within networks can loosely be defined as subsets of nodes which are more densely linked, when compared to the rest of the network. But this is a very simple definition that cannot assure the correct identification of the groups that form the complex network.

The problem of community detection is quite challenging and has been the subject of discussion in various disciplines. A simpler version of this problem, the graph bipartitioning problem has been the topic of study in the realm of computer science for decades. In real complex networks we often have no idea how many communities we wish to discover, but in general it is more than two. This makes the process all the more costly. What is more, communities may also be hierarchical, that is communities may be further divided into sub-communities and so on [16, 21, 15, 65].

Nevertheless, many attempts to tackle these problems have been proposed recently. The proposed methods vary considerably in terms of approach and application, which makes them difficult to compare. Community identification is potentially very useful and researchers from a number of fields may be interested in using one or several of the methods for their own purposes. In [66] we review all these methods comparing their performance and their computational cost.

But community identification is not merely a qualitative problem; actually, the performed comparison between the different algorithms is done in terms of a quantity that measures how good a given partition is. Since communities are sometimes not perfectly defined with clear border-line separation among them, different algorithms to detect communities can give rise to slightly different partitions. Them a measure that quantifies the accuracy of the partition is welcome. A simple approach that has become widely accepted was proposed in [67]. It is based on the intuitive idea that random networks do not exhibit community structure. Let us imagine that we have an arbitrary network and an arbitrary partition of that network into n_c communities. It is then possible to define a $n_c \times n_c$ size matrix $\mathbf{e}$ where the elements e_{ij} represent the fraction of total links starting at a node in partition i and ending at a node in partition j. Then, the sum of any row (or column) of $\mathbf{e}$, $a_i = \sum_j e_{ij}$ corresponds to the fraction of links connected to i.

If the network does not exhibit community structure, or if the partitions are allocated without any regard to the underlying structure, the expected value of the fraction of links within partitions can be estimated. It is simply the probability that a link begins at a node in i, a_i, multiplied by the fraction of links that end at a node in i, a_i. So the expected number of intra-community links is just $a_i a_i$. On the other hand we know that the *real* fraction of links exclusively within a partition is e_{ii}. So, we can compare the two directly and sum over all the partitions in the graph.

$$Q \equiv \sum_i (e_{ii} - a_i^2) \tag{4}$$

This is the measure known as *modularity*, that for a very good partition approaches 1. It is important to say that the network can have a very clear community separation and then a good partition can attain a large value of the modularity.

But sometimes, we are not only interested in the best partition but in the hierarchical organization of the network in nested communities. One of the early methods of community detection, proposed by Girvan and Newman [68], consists in splitting the networks by cutting the links with the highest betweenness. In this case this procedure can be iterated up to the level of individual nodes giving rise then to a hierarchy of nested communities. The application of this procedure is very useful for the understanding on the different levels of organization in a network. We have applied this procedure to the email network of the Universitat Rovira i Virgili [16] finding that the hierachical organization of the community structure maintains many treats of the supposed formal chart of the organization; but, at the same time, we could observe that some nodes are not placed in the supposed community. This is of course very valuable as a tool for the management of a organization [53]. Also as a tool of identifying the working communities and the respective leaders we applied the procedure to the Statistical Physics meetings network shown in Fig. 1. The network in Fig. 6 is the result of such community partition, where we can see that the green nodes, identified as the members of the scientific committees are equally distributed between the different branches and appear mainly at their tips. The former means that members have been chosen in a homogeneous way between the

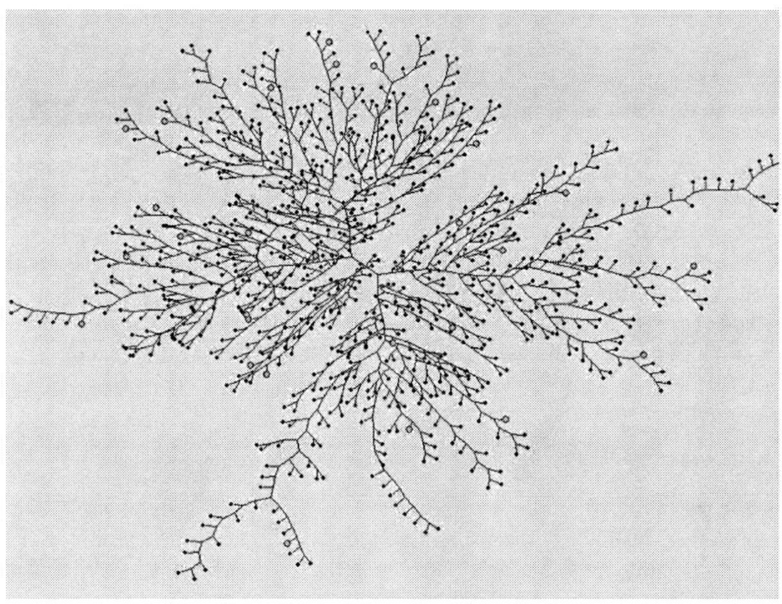

FIGURE 6. (color on-line) Community structure of the collaboration network in the Spanish Statistical Physics meetings. The small branches correspond to the research groups that are grouped into Universities that, at the same time, are closely grouped according to geographical proximity. The grey (green) nodes, that correspond to the members of the scientific committees appear mainly at the tips of the branches, showing their leadership in the respective groups. The homogeneous distribution of grey (green) nodes also shows that they have been chosen uniformly among the different groups.

different groups that form the Statistical Physics community and that these members are the leaders of the respective teams.

Another fact that has been obtained from this hierarchical community structure is that in many networks the distribution of community sizes also shows a power-law, indicating an underlying mechanism of auto-organization in the network and the absence of characteristic community sizes. In this way another scaling of magnitudes within communities can be analyzed and hence, in a language very familiar to physicists, networks can be classified in different universality classes [16, 15].

DYNAMICS ON THE NETWORK

Complex networks have become such widespread analyzed not only because of their universal topological properties, but also because the effect of the topology on the dynamics. Dynamical systems had been largely studied mainly in three different playgrounds: regular lattices, random graphs, and completely connected networks. Thus the evidence of the existence in nature and society of complex patterns of interaction again offered a large number of new possibilities to those studying the dynamical properties of complex systems. And hence, many different types of dynamics have been studied according to different patterns of connectivity. Just to mention a few in different contexts: flow of physical magnitudes or information in communication networks [56], spreading of epidemics [69, 20, 19] or rumors[70], synchronization of dynamical units (mainly oscillators) [71, 7], opinion formation [72], cultural dissemination [73], technological innovations [74], strategic games [75], Boolean dynamics in genetic networks [51], neural networks [1, 76, 77, 48].

Just to present a comprehensive view of these phenomena we will show results on two different types of dynamics: search and congestion as an example of transport in networks, and the dynamics of oscillators towards synchronization since it is a good example on how dynamics can help in elucidating some details of the topology.

Search and congestion

Concerning transport, the flow of information has been one of the mainly discussed issues. Information, in this case, can be understood as packets in a computer network [78], problems in a company that need to be solved [55], passengers in a transportation network [79]. As an example of information flow in [55] we presented a formalism that is able to cope with search and congestion simultaneously in any type of network, allowing the determination of optimal topologies. This formalism avoids the problem of simulating the dynamics of the communication process and provides a general scenario applicable to any communication process.

Let us focus on a single information packet at node i whose destination is node k. The probability for the packet to go from i to a new node j in its next movement is p_{ij}^k. In particular, $p_{kj}^k = 0\,\forall j$ so that the packet is *removed* as soon as it arrives to its destination. This formulation is completely general, and the precise form of p_{ij}^k will depend on the search algorithm and on the connectivity matrix of the network. In particular, when the search is Markovian, p_{ij}^k does not depend on previous positions of the packet. In this case, the probability of going from i to j in n steps is given by

$$P_{ij}^k(n) = \sum_{l_1,l_2,\dots,l_{n-1}} p_{il_1}^k p_{l_1 l_2}^k \cdots p_{l_{n-1}j}^k. \tag{5}$$

This definition allows us to compute the average number of times, b_{ij}^k, that a packet generated at i and with destination at k passes through j.

$$b^k = \sum_{n=1}^{\infty} P^k(n) = \sum_{n=1}^{\infty} \left(p^k\right)^n = (I - p^k)^{-1} p^k. \tag{6}$$

and the effective betweenness of node j, B_j, is then defined as the sum over all possible origins and destinations of the packets,

$$B_j = \sum_{i,k} b_{ij}^k. \tag{7}$$

When the search algorithm is able to find the minimum paths between nodes, the effective betweenness will coincide with the topological betweenness, β_j, as usually defined in the previous sections [80, 81].

Once these quantities have been defined, we focus on the load of the network, $N(t)$, which is the number of floating packets. These floating packets are stored in the nodes that act as queues. In a general scenario where packets are generated at random and independently at each node with a probability ρ, the arrival of packets to a given node j is a Poisson process. In this simple picture, the queues are called M/M/1 in the computer science literature and the average load of the network is [82, 55]

$$\overline{N} = \sum_{j=1}^{S} \frac{\frac{\rho B_j}{S-1}}{1 - \frac{\rho B_j}{S-1}}. \tag{8}$$

There are two interesting limiting cases of equation (8). When ρ is very small, taking into account that the sum of betweennesses is proportional to the average distance, one obtains that the load is proportional to the average effective distance. On the other hand, when ρ approaches ρ_c most of the load of the network comes from the most congested node, and therefore

$$\overline{N} \approx \frac{1}{1 - \frac{\rho B^*}{S-1}} \qquad \rho \to \rho_c, \tag{9}$$

where B^* is the effective betweenness of the most central node. The last results suggest the following interesting problem: to minimize the load of a network it is necessary to minimize the effective distance between nodes if the amount of packets is small, but it is necessary to minimize the largest effective betweenness of the network if the amount of packets is large. The first is accomplished by a *star-like* network, that is, a network with one central node and all the others connected to it. The second, however, is accomplished by a very decentralized network in which all the nodes support a similar load. This behavior is similar to any system of queues provided that the communication depends only on the sender.

It is worth noting that there are only two assumptions in the calculations above. The first one has already been mentioned: the movement of the packets needs to be Markovian to define the jump probability matrices p^k. Although this is not strictly true in real communication networks—where packets are not usually allowed to go through a given node more than once—it can be seen as a first approximation [78, 83, 84]. The second assumption is that the jump probabilities p_{ij}^k do not depend on the congestion state of the network, although communication protocols sometimes try to avoid congested regions, and then $B_j = B_j(\rho)$. However, all the derivations above will still be true in a number of general situations, including situations in which the paths that the packets follow are unique, in which the routing tables are fixed, or situations in which the structure of the network is very homogeneous and thus the congestion of all the nodes is similar. Compared to situations in which packets avoid congested regions, it corresponds to the worst case scenario and thus provide bounds to more realistic scenarios in which the search algorithm interactively avoids congestion.

Equation (8) relates a dynamical variable, the load, with the topological properties of the network and the properties of the algorithm. So we have converted a dynamical

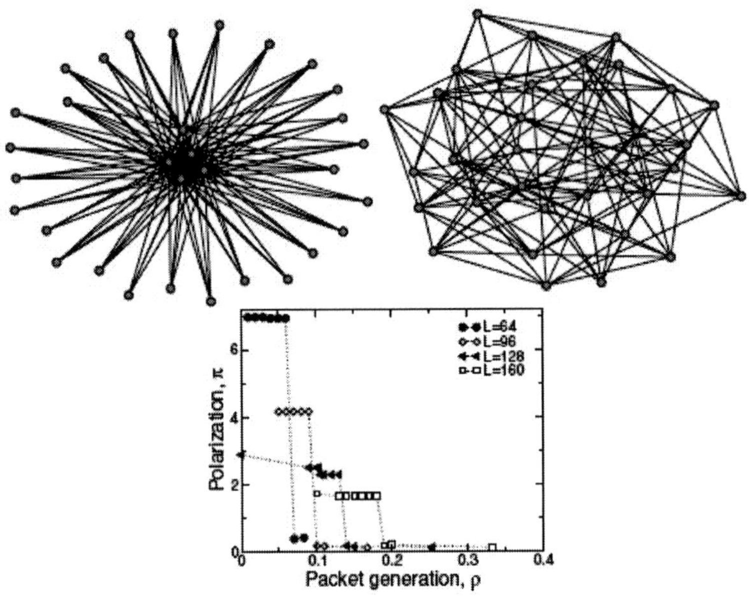

FIGURE 7. Optimal topologies for networks with $S = 32$ nodes, $L = 32$ links.

communication problem into a topological problem. Hence, the dynamical optimization procedure of finding the structure that gives the minimum load is reduced to a topological optimization procedure where the network is characterized completely by its effective betweenness distribution. In [55] we considered the problem of finding optimal structures for a purely local search, using a generalized simulated annealing procedure, as described in [85]. On the one side, we have found (see Fig. 7) that for $\rho \to 0$ the optimal network has a star-like centralized structure as expected, which corresponds to the minimization of the average effective distance between nodes. On the other extreme, for high values of ρ, the optimal structure has to minimize the maximum betweenness of the network; this is accomplished by creating a homogeneous network where all the nodes have essentially the same degree, betweenness, etc. One could expect that the transition centralized-decentralized occurs progressively. Surprisingly, the results of the optimization process reveal a completely different scenario. According to simulations, star-like configurations are optimal for $\rho < \rho^*$; at this point, the homogeneous networks that minimize B^* become optimal. Therefore there are only two type of structures that can be optimal for a local search process: star-like networks for $\rho < \rho^*$ and homogeneous networks for $\rho > \rho^*$.

Beyond the existence of both centralized and decentralized optimal networks, it is significant that the transition from one sort of networks to the other is abrupt, meaning that there are no intermediate optimal structures between total centralization and total decentralization. Our explanation of this fact is the following. Since we are con-

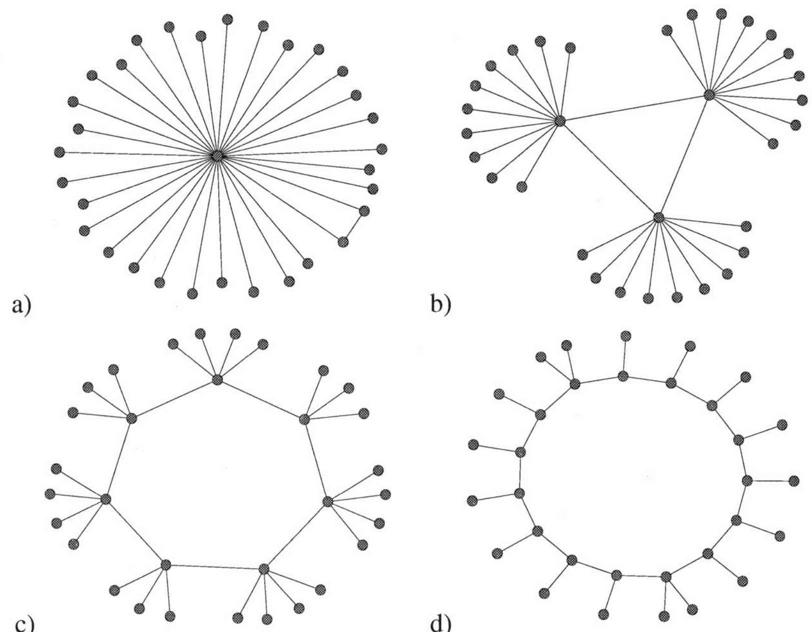

FIGURE 8. Optimal topologies for networks with $S = 32$ nodes, $L = 32$ links and global knowledge. (a) $\rho = 0.010$. (b) $\rho = 0.020$. (c) $\rho = 0.050$. (d) $\rho = 0.080$. In this case of global knowledge, the transition from centralization to decentralization seems smooth.

sidering local knowledge of the network topology, centered star-like configurations are extremely efficient in searching destinations and thus minimizing the effective distance between nodes. This explains that stars are optimal for a wide range of values of ρ, until the central node (or nodes) becomes congested. At this point, structures similar to stars will have the same problem and will be much worse regarding search; at this point, the only alternative is something completely decentralized, where the absence of congestion can compensate the dramatic increase in the effective distance between nodes. If this explanation is correct, one should be able to obtain a smooth transition from centralization to decentralization by considering global knowledge of the network, in such a way that the average effective distance (that in this case coincides with the average path length) is not much larger in an arbitrary network than in the star. Although we do not have extensive simulations in this case, Fig. 8 shows that there is some evidence to think that this is indeed the case.

Dynamics towards synchronization

Physicists have largely studied the dynamics of complex biological systems, and in particular the paradigmatic analysis of large populations of coupled oscillators [86, 87, 88]. The connection between the study of synchronization processes and complex networks is interesting by itself. This synchronization phenomena as many others e.g. asian fireflies flashing at unison, pacemaker cells in the heart oscillating in harmony, etc. have been mainly described under the mean field hypothesis that assumes that all oscillators behave identically and interact with the rest of the population. Recently, the emergence of synchronization phenomena in complex networks has been shown to be closely related to the underlying topology of interactions [89] beyond the macroscopic description.

One of the most successful attempts to understand synchronization phenomena was due to Kuramoto [88], who analyzed a model of phase oscillators coupled through the sine of their phase differences. The model is rich enough to display a large variety of synchronization patterns and sufficiently flexible to be adapted to many different contexts [90]. The Kuramoto model consists of a population of N coupled phase oscillators where the phase of the i-th unit, denoted by $\theta_i(t)$. Here we consider a simplified dynamics in which all units have the same frequency, that can be set to zero without loss of generality. Thus we have

$$\frac{d\theta_i}{dt} = \sum_j K_{ij} \sin(\theta_j - \theta_i) \quad i = 1, ..., N \tag{10}$$

where K_{ij} describes the coupling between units. In absence of noise the only attractor of the dynamics is the complete synchronization, $\theta_i = \theta$, $\forall i$.

Originally, this model had been studied in networks which are complete, but recently these studies have been extended to systems where the patterns of connections is local but non-trivial [7]. In this context the interest concerns not the final synchronized state in itself but the route to the attractor. In particular, it has been shown [7] that high densely interconnected sets of oscillators (motifs) synchronize more easily that those with sparse connections. This scenario suggests that for a complex network with a non-trivial connectivity pattern, starting from random initial conditions, those highly interconnected units forming local clusters will synchronize first and then, in a sequential process, larger and larger spatial structures also will do it up to the final state where the whole population should have the same phase. This process occurs at different time scales if a clear community structure exists. Thus, the dynamical route towards the global attractor reveals different topological structures, presumably those which represent communities. Therefore, it is the complete dynamical process what unveils the whole organization at all scales, from the microscale at a very early stages up to the macroscale at the end of the time evolution. On the contrary, those systems endowed with a regular topological structure displays a trivial dynamics with a single time scale for synchronization.

We have analyzed the dynamics towards synchronization in computer-generated graphs with community structure. For this reason, we define a local order parameter measuring the average of the correlation between pairs of oscillators

$$\rho_{ij}(t) = < cos(\theta_i(t) - \theta_j(t)) > \tag{11}$$

where the brackets stand for the average over initial random phases. The main advantage of this approach is that it allows to trace the time evolution of pairs of oscillators and therefore to identify compact clusters reminiscent of the existence of communities.

The paradigmatic model of network with a well defined community structure that has been used as a benchmark for different community detection algorithms [66], was proposed by Girvan and Newman [68]. In that model the authors construct a network of 128 nodes as a set of 4 communities, each one formed by 32 nodes. Fixing the mean number of links per node at a value of 16, the parameter describing the sharpness of the community distribution is z_{in}, the average number of links within the community. In Fig. 9 we show the time evolution of one of these networks, $z_{in} = 15$ and hence a very clearly defined community structure, averaging over random initial phases.

Dealing only with topological information we can, from the connectivity matrix, construct the Laplacian matrix and compute their eigenvalue spectrum. This spectrum gives information on the time scales involved in the dynamical process. We plot the eigenvalues spectrum of this matrix in the following way: in the horizontal axis we represent the inverse of the eigenvalue, which in a dynamical process accounts for the time, and in the vertical axis we represent the index of the eigenvalue which accounts for the number of groups along the dynamics. This picture is useful because it can be compared with the way groups (clusters or communities) are formed along the synchronization process, obtaining a very striking similarity, meaning that these eigenvalues control the formation of the synchronized communities. We also plot, for completion, the dendogram of the synchronization process (Fig. 9c): In this picture we show how the groups merge according to the synchronization dynamics along time (vertical axis). Finally we also plot (d) the relative time to achieve synchronization for each pair of oscillators. This synchronization is understood as a correlation being larger than some threshold value. The characterization is completely independent of the threshold, as is shown in [91], since it only changes the absolute time scale not the relative one. Nodes are ordered in the same way than in the picture of the dendogram just to get together those nodes that synchronize earlier.

In this way we have been able to relate topology, in terms of the eigenvalue spectrum of the Laplacian matrix, with dynamics, in terms of the appearance of synchronized groups of oscillators. Topologically these groups correspond to the communities, but there can be some cases where communities are not so well defined and this informations keep being useful. There can be some occasions where synchronized groups of oscillators do not fit exactly with topological communities. Synchronization is a global dynamical process that can identify the relevant structures (perhaps hierarchical) along its evolution. Also the effect of hubs in the dynamical evolution is interesting, since hubs are sometimes above the community structure.

For more information on this issue the reader is pointed to [91, 92] and to the website `http://www.ffn.ub.es/albert/synchro.html`.

CONCLUSIONS AND OPEN PROBLEMS

Complex patterns of interactions are so often found in any natural, technological or social environment that it has been widely accepted that new tools are needed. From Statistical Physics, many valuable existing tools have been applied to this new emergent field. Researchers in many different subjects are generating new repositories of data, very large networks are generated and these tools need new implementations. A network

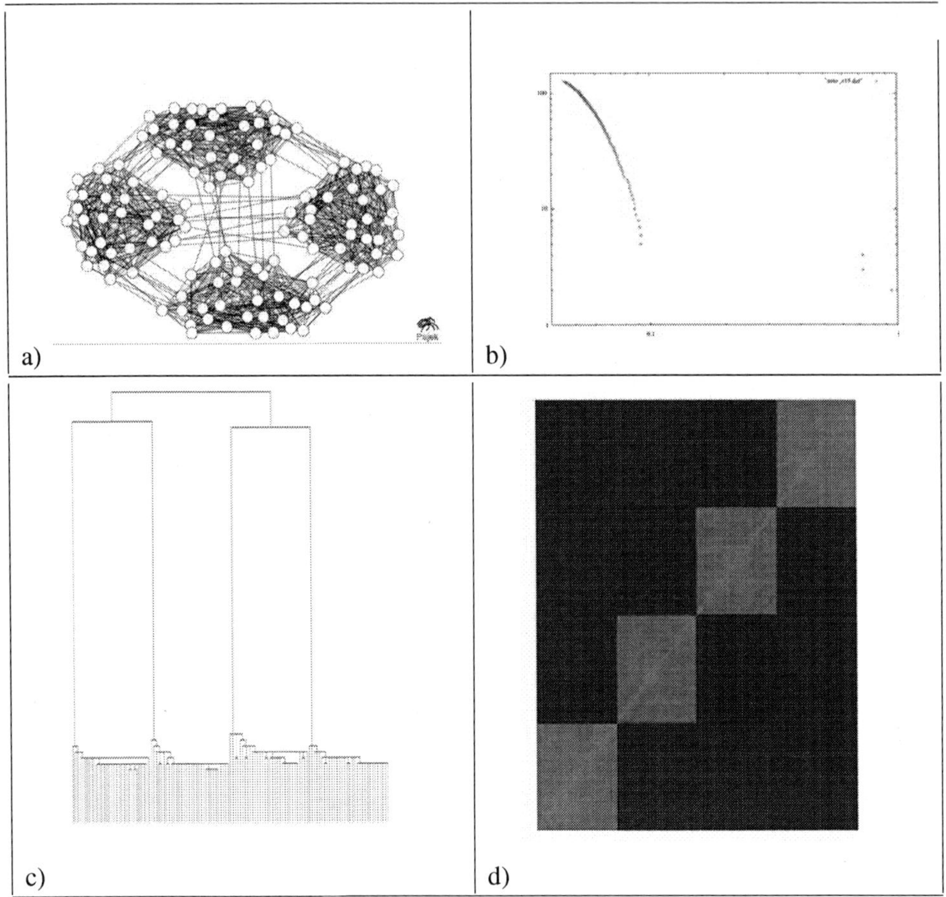

FIGURE 9. (color on-line) Synchronization process in a network with a homogeneous distribution of communities. a) the network structure; b) eigenvalue spectrum; c) dendogram of the community merging; d) time needed for each pair of oscillators to synchronize. Light grey (red) for shorter times, dark grey (blue) for larger times.

is not just a collection of nodes and binary relationships between those nodes. Nodes and links can be anything, depending on the considered data, but nodes can have weights, links can have weights as well, and hence new theories have appeared to deal with this additional degree of complexity.

Usually, networks are characterized either from the microscopic level or from the macroscopic level. From a microscopic point of view we are mainly interested in node properties: degree, different measures of centrality, clustering, and so. However, from a macroscopic point of view we deal with statistical properties of the set of nodes and/or links; which are the distributions of connectivities, of load, of distances, and how the different measures are, on average, correlated. These characterizations enable to classify the networks into different universality classes, which is quite common in physics grounds. We also know that in many problems in physics we have descriptions that are scale invariant and hence we can move from the microscopic to the macroscopic scale. Here we have reviewed some concepts and methods in the intermediate scale, the mesoscale, where the definition and identification of communities or functional groups play a crucial role. Up to now, there has been a large amount of work on methods of community identification. Which are the most efficient in terms of accuracy or which are the more economic in terms of computer resources needed. These properties have also turned out to show some degree of universality.

Nevertheless, this identification based solely on topological properties needs to be related with the exact relations between the nodes of the different groups. Nodes can belong topologically to a given group but their functionality can be quite different. Understanding these relations, why topological communities are or are not related with functional groups, social communities, or some sort of thematic clusters, is still one of the open problems related with the mesoscale properties. Another interesting point that needs more clarification is the community structure at different scales, why are they ordered in some kind of hierarchical or nested way and their relation again with some ordering in this scales that can be related with some dynamical properties of the processes taking place on the network. This hierarchical structure goes far beyond many of the current methods to identify community partitions in networks; all this methods try to find the optimal value of a kind of cost function, called modularity, which is a property of the network and of the partition, then the best partition is that with the highest modularity, but there can be partitions that, even with a high value of modularity, are very unlikely from a physical point of view. Hence a proper understanding of the precise location and the neighboring areas in the partition space of special configurations can be of great help in understanding the functionality of networks.

But, at the same time, networks are not formed by static objects; nodes (social agents, computers, companies, ...) evolve in time and they can change their status and this evolution is strongly correlated with the evolution of the links (social relationships, hard rewirings, new business strategies, ...), All these new evolving, rewiring, updating, growing, removing, ... open many new problems that will be faced in the next future. Also, as stated in the previous paragraph, we need a proper understanding of the topologies and its relation with the dynamics of the node properties. We have presented here just two examples on how the topological structure affects dynamics. First, a problem of transport in which the nodes are agents that process and deliver information that has to arrive to the right destination. Here we have found the characteristics of the optimal network depending on the external load. Second, the time evolution of synchronized populations of oscillators shows a striking degree of community ordering that reflects the topological structure; furthermore, we have highlighted the relations between topological properties of the connectivity matrix with dynamical properties of the synchronization. This is just to get a glance on the wide applicability of these ideas in physical, economical, social, biological, or even engineering problems.

In any case, we are dealing with a subject, Complex Networks, that is very young, but that in such a short period of time has given so many relevant contributions (in the form of reviews, technical books, popularization books, ...) that we have to think that the future has just started and many new players are welcome to the ground.

ACKNOWLEDGMENTS

The author gratefully acknowledges fruitful and enlightening discussions with L.A.N. Amaral, A. Arenas, M. Boguña, L. Danon, J. Duch, X. Guardiola, R. Guimera, M. Llas, Y. Moreno, and C.J. Pérez-Vicente. I would like also to thank L. Danon and J. Duch for providing some of the data and figures and A. Arenas for a critical reading of the manuscript.

REFERENCES

1. D. J. Watts, and S. Strogaz, *Nature* **393**, 440–442 (1998).
2. A. L. Barabási, and R. Albert, *Science* **286**, 509–512 (1999).
3. S. H. Strogatz, *Nature* **410**, 268–276 (2001).
4. A. L. Barabási, and R. Albert, *Review of Modern Physics* **74**, 47–97 (2002).
5. S. Dorogovtsev, and J. F. F. Mendes, *Advances in Physics* **51**, 1079–1187 (2002).
6. M. E. J. Newman, *SIAM Review* **45**, 167–256 (2003).
7. S. Boccaletti, V. Latora, Y. Moreno, M. Chavez, and D.-U. Hwang, *Physics Reports* **424**, 175–308 (2006).
8. S. Bornholdt, and H. G. Schuster, editors, *Handbook of Graphs and Networks - From the Genome to the Internet*, Wiley-VCH, Berlin, 2002.
9. R. Pastor-Satorras, M. Rubí, and A. Díaz-Guilera, editors, *Statistical Mechanics of Complex Networks*, Springer, 2003.
10. L. Amaral, A. Scala, M. Barthelemy, and H. Stanley, *Proceedings of the National Academy of Sciences, USA* **97**, 11149–11152 (2000).
11. G. F. Davis, M. Yoo, and W. E. Baker, *preprint, University of Michigan Business School* (2001).
12. M. E. J. Newman, S. Strogatz, and D. J. Watts, *Physical Review E* **64**, 026118 (2001).
13. M. E. J. Newman, *Physical Review E* **64**, 016132 (2001).
14. M. E. J. Newman, *Proceedings of the National Academy of Sciences, USA* **98**, 404–409 (2001).
15. A. Arenas, L. Danon, A. Diaz-Guilera, P. M. Gleiser, and R. Guimerà, *European Physical Journal B* **38**, 373–380 (2004).
16. R. Guimerà, L. Danon, A. Díaz-Guilera, F. Giralt, and A. Arenas, *Physical Review E* **68**, 065103 (2003).
17. H. Ebel, L. I. Mielsch, and S. Bornholdt, *Physical Review E* **66**, 035103 (2002).
18. M. E. J. Newman, S. Forrest, and J. Balthrop, *Physical Review E* **66**, 035101 (2002).
19. F. Liljeros, C. R. Edling, and L. A. N. Amaral, *Microbes and Infections* **5**, 189–196 (2003).
20. F. Liljeros, C. Edling, L. A. N. Amaral, H. E. Stanley, and Y. Aberg, *Nature* **411**, 907–908 (2001).
21. P. Gleiser, and L. Danon, *Advances in Complex Systems* **6**, 565–573 (2003).
22. X. Guardiola, R. Guimerà, A. Arenas, A. Diaz-Guilera, D. Streib, and L. Amaral, *preprint* pp. cond–mat/0206240 (2002).
23. R. Albert, H. Jeong, and A.-L. B. ., *Nature* **401**, 130 (1999).
24. S. Redner, *European Physical Journal B* **4**, 131–134 (1998).
25. S. N. Dorogovtsev, and J. F. F. Mendes, *Proceedings of the Royal Society, London B* **268**, 2603–2608 (2001).
26. R. F. i Cancho, and R. Solé, *Proceedings of the Royal Society London B* **268**, 2261–2265 (2001).
27. M. Faloutsos, P. Faloutsos, and C. Faloutsos, *Comp. Comm. Rev.* **29**, 251–262 (1999).
28. M. E. J. Newman, *Physical Review E* **67**, 026126 (2003).
29. S. Valverde, R. F. i Cancho, and R. Solé, *Europhysics Letters* **60**, 512–517 (2002).
30. R. F. i Cancho, C. Janssen, and R. Solé, *Physical Review E* **64**, 046119 (2001).
31. R. Guimerà, S. Mossa, A. Turtschi, and L. A. N. Amaral, *Proceedings of the National Academy of Sciences, USA* **102**, 7794–7799 (2005).
32. V. Colizza, A. Barrat, M. Barthelemy, and A. Vespignani, *Proceedings of the National Academy of Sciences, USA* **103**, 2015 (2006).

33. P. Sen, S. Dasgupta, A. Chatterjee, P. A. Sreeram, G. Mukherjee, and S. S. Manna, *Physical Review E* **67**, 036106 (2003).
34. H. Jeong, B. Tombor, R. Albert, Z. N. Oltvai, and A. L. Barabási, *Nature* **407**, 651–654 (2000).
35. S. M. Gomez, S. H. Lo, and A. Rzhetsky, *Genetics* **159**, 1291–1298 (2001).
36. O. Ebenhoh, and R. Heinrich, *Bulletin of Mathematical Biology* **65**, 323–57 (2003).
37. S. Schuster, T. Pfeiffer, F. M. I. Koch, and T. Dandekar., *Bioinformatics* **18**, 351ñ61 (2002).
38. A. Wagner, and D. A. Fell, *Proceedings of the Royal Society London B* **268**, 1803–10 (2001).
39. H. Jeong, S. Mason, A. L. Barabási, and Z. N. Oltvai, *Nature* **411**, 41–42 (2001).
40. D. S. Goldberg, and F. P. Roth, *Proceedings of the National Academy of Sciences, USA* **100**, 4372–76 (2003).
41. M. Vendruscolo, N. V. Dokholyan, E. Paci, and M. Karplus, *Physical Review E* **65**, 061910 (2002).
42. A. Wagner, *Molecular Biology and Evolution* **18**, 1283–92 (2001).
43. S. Wuchty, *Proteomics* **2**, 1715–23 (2002).
44. J. A. Dunne, R. J. Williams, and N. D. Martinez, *Proceedings of the National Academy of Sciences, USA* **99**, 12917–22 (2002).
45. J. M. Montoya, and R. V. Solé, *Journal of Theoretical Biology* **214**, 405–12 (2002).
46. S. Morita, K. Oshio, Y. Osana, Y. Funabashi, K. Oka, and K. K. ., *Physica A* **298**, 553–61 (2001).
47. O. Shefi, I. Golding, R. Segev, E. B.-J. E, and A. Ayali, *Physical Review E* **66**, 021905 (2002).
48. V. M. Eguíluz, D. Chialvo, G. Cecchi, M. Baliki, and A. Apkarian, *Physical Review Letters* **92**, 028102 (2005).
49. A. Bhan, D. J. Galas, and T. G. Dewey, *Bioinformatics* **18**, 1486–1493 (2002).
50. N. Guelzim, S. Bottani, P. Bourgine, and F. Kepes, *Nature Genetics* **31**, 60–63 (2002).
51. L. A. N. Amaral, A. Díaz-Guilera, A. A. Moreira, A. L. Goldberger, and L. A. Lipsitz, *Proceedings of the National Academy of Science* **101**, 15551–15555 (2004).
52. A. Barrat, M. Barthelemy, R. Pastor-Satorras, and A. Vespignani, *Proceedings of the National Academy of Science* **101**, 3747 (2004).
53. R. Guimerà, L. Danon, A. Arenas, A. Díaz-Guilera, and F. Giralt, *Journal of Economic Behavior and Organization* (2007).
54. S. Wasserman, and K. Faust, *Social Network Analysis, Methods and Applications*, Cambridge University Press, 1994.
55. R. Guimerà, A. Díaz-Guilera, F. Vega-Redondo, A. Cabrales, and A. Arenas, *Physical Review Letters* **89**, 248701 (2002).
56. B. Tadic, G. Rodgers, and S. Thurner, *preprint* (2006).
57. P. Erdos, and A. Renyi, *Publ. Math. Debrecen* **6**, 290–297 (1959).
58. R. Albert, H. Jeong, and A.-L. Barabasi, *Nature* **406**, 378 (2000).
59. R. Cohen, K. Erez, D. ben Avraham, and S. Havlin, *Physical Review Letters* **85**, 4626 (2000).
60. H. Zhou, and R. Lipowsky, *preprint* (2005).
61. G. W. Flake, S. Lawrence, C. L. Giles, and F. M. Coetzee, *IEEE Computer* **35**, 66 – 71 (2002).
62. J.-P. Eckmann, and E. Moses, *Proceedings of the National Academy of Sciences, USA* **99**, 5825–5829 (2002).
63. M. E. J. Newman, *European Physical Journal B* **38**, 321–330 (2004).
64. L. Danon, J. Duch, A. Arenas, and A. Díaz-Guilera, *COSIN project*, World Scientific, 2005, chap. Community structure identification.
65. M. E. J. Newman, *Physical Review E* **69**, 066133 (2004).
66. L. Danon, A. Díaz-Guilera, J. Duch, and A. Arenas, *J. Stat. Mech* p. P09008 (2005).
67. M. E. J. Newman, and M. Girvan, *Physical Review E* **69**, 026113 (2004).
68. M. Girvan, and M. E. J. Newman, *Proceedings of the National Academy of Sciences USA* **99**, 7821–7826 (2002).
69. R. Pastor-Satorras, and A. Vespignani, *Physical Review Letters* **86**, 3200–3203 (2001).
70. D. H. Zanette, *Physical Review E* **64**, 050901 (2001).
71. L. Donetti, P. I. Hurtado, and M. A. Muñoz, *Physical Review Letters* **95**, 188701 (2005).
72. F. A. Rodrigues, and L. da F. Costa, *International Journal of Modern Physics C* **16**, 1785–1792 (2005).
73. K. Klemm, V. M. Eguíluz, R. Toral, and M. San Miguel, *Physical Review E* **67**, 026120 (2003).
74. M. Llas, P. M. Gleiser, A. Díaz-Guilera, and C. J. Pérez, *Physica A* **326**, 567–577 (2003).
75. H. Ebel, and S. Bornholdt, *Physical Review E* **66**, 056118 (2002).

76. L. F. Lago-Fernández, R. Huerta, F. Corbacho, and J. A. Sigüenza, *Physical Review Letters* **84**, 2758–2761 (2000).
77. M. Aldana, and H. Larralde, *Physical Review E* **70**, 066130 (2004).
78. T. Ohira, and R. Sawatari, *Physical Review E* **58**, 193 (1998).
79. M. Barthelemy, and A. Flammini, Optimal traffic networks (2006).
80. L. C. Freeman, *Sociometry* **40**, 35–41 (1977).
81. M. E. J. Newman, *Physical Review E* **64**, 016133 (2001).
82. O. Allen, *Probability, Statistics and Queueing Theory with Computer Science Application*, Academic Press, New York, 2nd edition,, 1990.
83. A. Arenas, A. Diaz-Guilera, and R. Guimera, *Physical Review Letters* **86**, 3196–3199 (2001).
84. R. Sole, and S. Valverde, *Physica A* **289**, 595–605 (2001).
85. C. Tsallis, and D. A. Stariolo, *Annual Rev. Comp. Phys. II*, World Sci. Singapore, 1994.
86. A. Winfree, *The geometry of biological time*, Springer, 2001.
87. S. H. Strogatz, *Sync: The Emerging Science of Spontaneous Order*, Hyperion, 2003.
88. Y. Kuramoto, *Chemical oscillations, waves, and turbulence*, Dover, 2003.
89. F. M. Atay, T. Biyikoglu, and J. Jost., *IEEE Trans. Circuits and Systems* **53** (2006).
90. J. A. Acebrón, L. L. Bonilla, C. J. Pérez Vicente, F. Ritort, and R. Spigler, *Reviews of Modern Physics* **77**, 137–185 (2005).
91. A. Arenas, A. Diaz-Guilera, and C. J. Perez-Vicente, *Physical Review Letters* **96**, 114102 (2006).
92. A. Arenas, A. Diaz-Guilera, and C. J. Perez-Vicente, *Physica D* **96**, (submitted) (2007).

Microrheology of Polyacrylamide Solutions from Depolarized Dynamic Light Scattering

José Luis Arauz-Lara

Instituto de Física, Universidad Autónoma de San Luis Potosí, Alvaro Obregón 64, 78000 San Luis Potosí, SLP, Mexico
Departamento de Física, Cinvestav, Av. IPN 2508, Colonia Zacatenco, 07360 México D. F., Mexico

Abstract. The microrheology of viscoelastic fluids such as high molecular weight polyacrylamide aqueous solutions is determined from a Fourier analysis of both the translational and the rotational diffusion of optically anisotropic spherical colloidal probes dispersed in the system. Depolarized dynamic light scattering is used to measure the rotational and translational mean squared displacements from which the storage and loss moduli are obtained. We found both microrheology measurements to be in excellent agreement between them and with those determined by mechanical measurements. This extents the capabilities of the microrheological methods based on the diffusional motion of colloidal probes.

INTRODUCTION

Soft materials abound in nature and in industrial products, from biological materials (for instances, protein or DNA solutions) to gels, polymeric solutions, paints, cosmetics, foods, etc. Thus, the development of methods to characterize their mechanical properties (among others) is of wide interest. The diffusive motion of colloidal particles embedded in a complex fluid such as those mentioned above, probes the local mechanical response of the host medium to oscillatory perturbations at frequencies and space scales not attainable by conventional mechanical instruments [1, 2, 3, 4, 5]. This method to determine the mechanical properties is referred to as microrheology [6, 7]. The main advantages of microrheology is the possibility to measure the mechanical properties of very small systems (e. g., in living cells with a volume of only few picoliter) and the accessibility to much higher shear frequencies than those attained mechanically. The basic idea behind this method, is the use of a generalization of the Stokes-Einstein relation between the translational motion of the probe particles, described by the mean squared displacement $W(t) = \langle \Delta \mathbf{r}^2(t) \rangle /6$, with $\Delta \mathbf{r}(t)$ being the particle's displacement at time t, and the mechanical properties of the medium described by the stress relaxation modulus $G(t)$ [8]. For non-interacting spherical particles in a simple viscous fluid, the mean squared displacement is a linear function of time, i. e., $W(t) = Dt$. Here D is the free-particle self-diffusion coefficient, related to the shear viscosity η of the fluid by the Stokes-Einstein relation, i. e., $\eta = k_BT/6\pi aD$, with k_BT being the thermal energy and a the particle's radius. In the absence of interactions with neighbor particles and external fields, deviations of $W(t)$ from linearity would reflect a more complex (viscoelastic) local mechanical response of the host medium. This is expressed in the generalized Stokes-Einstein (GSE) relation, which in the Fourier space reads [6],

CP885, *Advanced Summer School in Physics 2006, Frontiers in Contemporary Physics—EAV06*, edited by O. Miranda, M. Carbajal, L. M. Montaño, O. Rosas-Ortiz, and S. A. Tomás Velázquez

$$G^*(\omega) = \frac{k_B T}{i\omega\pi a \langle \Delta \mathbf{r}^2(\omega) \rangle}, \tag{1}$$

where $\langle \Delta \mathbf{r}^2(w) \rangle$ is the Fourier transform of the mean squared displacement, ω is the frequency of shearing and $G^*(\omega)$ is the complex shear modulus whose real part $G'(\omega)$ is the elastic or storage modulus and the imaginary part $G''(\omega)$ is the viscous or loss modulus. These quantities can be obtained from eq. (1), using the method devised by Dasgupta *et al* [9]. In that method, one assumes a local power law for the mean squared displacement, leading to the evaluation of its first and second logarithmic derivatives.

Here we show that the rotational diffusion of probe colloidal particles can also be used to determine the microrheological properties of viscoleastic media. For this, we use spherical particles with internal optical anisotropy and depolarized dynamic light scattering to measure simultaneously their rotational and translational mean squared displacements. The rotational diffusion, due to fluctuating torques exerted on the particle by the medium, also probes the material's response in a frequency domain similar to that of the translational diffusion and can be used to characterize that response. The rotational motion is analogous to the translational motion, and can be described by the mean squared angular displacement $\Omega(t) \equiv \langle [\Delta\theta(t)]^2 \rangle / 2$, where $\Delta\theta(t)$ is the one dimensional angular displacement of the particle. For freely rotating spherical particles $\Omega(t) = \Theta t$, where $\Theta = k_B T / 8\pi\eta a^3$ is the rotational diffusion coefficient [10]. Here too, in the absence of interparticle interactions, deviations of $\Omega(t)$ from linearity would be due to the complex response of the medium. The GSE relation for rotational diffusion of spherical particles, expressed in the Fourier space, is given by [11],

$$G^*(\omega) = \frac{k_B T}{i\omega 4\pi a^3 \langle \Delta\theta^2(\omega) \rangle}. \tag{2}$$

In this case too, the microrheological moduli are obtained by applying the method used in dealing with eq. (1).

SAMPLE PREPARATION

Polymeric solutions were prepared by dissolving polyacrylamide (Sigma), molecular weight $M_W = 5 - 6 \times 10^6$ g/mol, in deionized water of resistivity 17.0 MΩ·cm. The probes are optically anisotropic spherical particles of diameter 340 nm and a size polydispersity of 7.25%. These particles were dispersed in the solution at a very low volume fraction $\phi_{lc} = 10^{-5}$ to avoid interaction between them. The probes were made by emulsification of liquid crystal (RM257, Merck) in water at the temperature of the nematic phase as follows: 0.2 mg of reactive monomer RM257 (Merck) and 0.1 mg of photoinitiator Darocur 1173 (Ciba) were dissolved in 20 g of ethanol at room temperature. The mixture is then injected into stirring water, producing a polydisperse dispersion of liquid crystalline (with photo-initiator) droplets in water. The system is allowed to equilibrate at 80 °C, above the temperature of the crystal-nematic phase transition of the monomer. The photo-initiator is activated by irradiation with UV light, which polymerize the monomer freezing the internal nematic order in the droplet [12, 13]. Suspensions of the anisotropic particles with low polydispersity are obtained by fractionation following the depletion crystallization method. Mechanical measurements of the viscoelastic moduli were carried out using the concentric cylinders geometry in a Paar-physica Mcr300 rheometer. All experiments were carried at a constant temperature of 23°C. The mesh size ξ of the polymeric network is estimated as $\xi = R_G(c^*/c)^{3/4}$ [14], where R_G is the radius of gyration, c and c^* are the polymer concentration and its critical value, respectively. The latter was determined as the intercept of the linear and the power law behavior of the low shear viscosity vs. c. We found ξ to be in the range of 1-9 nm for matrices with polymer concentration in the range of .8% to .1% w/w studied here. Thus, the mesh size is much smaller than the particle's size.

DEPOLARIZED DYNAMIC LIGHT SCATTERING

The sample is placed in a goniometer (Brookhaven) in the center of a optical vat filled with index matching fluid (decalin). A polarized laser beam of wavelength $\lambda = 488$ nm is focused onto the sample cell. The light scattered at an angle θ_s with respect to the incident beam is collected by a monomode optical fiber located in the scattering plane. A second polarizer (the analyzer) is located before the optical fiber to make sure that only one mode reaches the detector. The scattered light is split and directed to two photon detectors (ALV/SO-SIPD), and the signal is processed by a time correlator (ALV 6010/160) operated in the pseudo-cross correlation mode. The optically anisotropic particles have a nematic-like internal structure, whose director $\hat{\mathbf{n}}(t)$ changes direction randomly due to fluctuating torques exerted by the solvent molecules. Thus, the time correlation function $g^{(2)}(k,t) \equiv \langle I(k,0)I(k,t)\rangle / \langle I^2(k,0)\rangle$ of the light scattered $I(k,t)$ by the particles can be measured in two different geometries of the polarizers: both vertical (VV) and one vertical and the other (the analyzer) horizontal (VH). Here $k = (4\pi n_s/\lambda)\sin(\theta_s/2)$ is the magnitude of the scattering wavevector and n_s the refraction index of the medium. In each polarizers geometry, the corresponding correlation function $g^{(1)}(k,t)$ of the scattered electric field $\vec{E}(k,t)$, is obtained via the Siegert relation, i. e., $g^{(2)}(k,t) = 1 + b \mid g^{(1)}(k,t) \mid^2$, where b is an experimental constant of order 1 [15]. Thus, the measured quantities of interest are: $g^{(1)}_{VV}(k,t) = \langle E^*_{VV}(k,0)E_{VV}(k,t)\rangle / \langle |E_{VV}(k,0)|^2\rangle$ and $g^{(1)}_{VH}(k,t) = \langle E^*_{VH}(k,0)E_{VH}(k,t)\rangle / \langle |E_{VH}(k,0)|^2\rangle$. These correlation functions describe the particle's dynamics and can be written as [15],

$$g^{(1)}_{VV}(k,t) = \left[A + B f_R(t)\right] f(k,t) \tag{3}$$

and

$$g^{(1)}_{VH}(k,t) = f_R(t) f(k,t), \tag{4}$$

where $f_R(t)$ and $f(k,t)$ are the dynamic correlation functions describing single particle rotational and translational motion in the host matrix, respectively. A and B are constants depending only on the components of the particle's polarizability tensor, which are

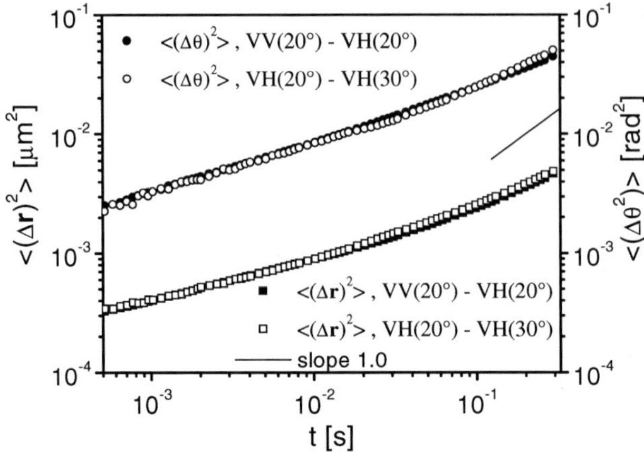

FIGURE 1. Translational and rotational mean squared displacements of optically anisotropic particles, measured in the *VV-VH* geometry (closed symbols) and in the *VH-VH* geometry at two different angles (open symbols). As shown here, the slope of both mean squared displacements is lower than 1 (the slope of the solid line is 1). Thus, the motion of the particles in the polymeric matrix is sub-diffusive.

intrinsic particle properties, and can be measured in a known host fluid. In this work, those constants are determined by light scattering from particles dispersed in water [13]. Although the rotational and translational dynamics are mixed up in the field correlation functions, one can see from equations (3) and (4) that the dynamic correlation functions can be obtained either, by measuring both $g^{(1)}_{VV}(k,t)$ and $g^{(1)}_{VH}(k,t)$ at the same scattering angle, or by measuring $g^{(1)}_{VH}(k,t)$ at two different scattering angles. As we show below, both possibilities can be realized and lead to the same results at low polymer concentrations when the light scattered by the polymer matrix is not more than a few percent of the light scattered by the probe particles. For more concentrated polymeric matrices, with higher scattering power, the second option provides a method to block the light scattered from the polymeric matrix from reaching the detector. The mean squared displacements, rotational and translational, can be obtained from the dynamic correlation functions by using the Gaussian approximation, i. e., by assuming $f(k,t) = \exp(-k^2W(t))$ and $f_R(t) = \exp(-6\Omega(t))$.

RESULTS

Figure 1 shows the translational and rotational mean squared displacements measured in a system with a polyacrylamide concentration of 0.5% w/w, using both procedures mentioned above. Open symbols represent the mean squared displacements extracted from measurements in the *VH* optical configuration at two different scattering angles, whereas the closed symbols correspond to measurements in the *VV* and in the *VH* con-

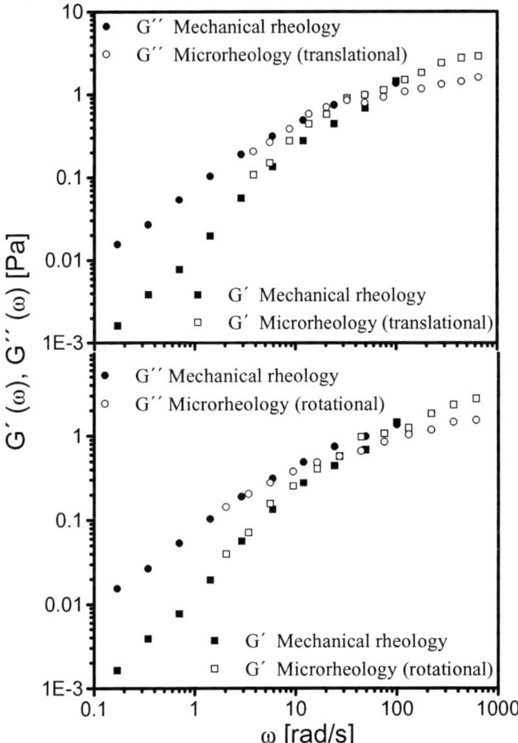

FIGURE 2. Comparison of the storage G' and loss G'' moduli of the sample in fig. (1), measured by translational (a) and rotational (b) probe diffusion microrheology and mechanical rheology.

figurations at the same scattering angle. As one can see here, both procedures lead to identical results for most of the time window accessible to the experiment. Furthermore, curves for the rotational motion are pretty similar to those corresponding to the translational motion. In fact, if we normalize $\langle \Delta \mathbf{r}^2(t) \rangle$ with $4a^2$, the curve superimpose with the curve of $\langle \Delta\theta^2(t) \rangle$. This indicates already that the microrheology obtained from rotational diffusion will coincide with that from translational motion, see eqs. (1) and (2).

Finally, in figure 2 we show the storage (open squares) and loss (open circles) moduli for the system in fig. 1. Closed symbols correspond to mechanical measurements. Fig. (a) shows microrheology results from translational diffusion, whereas fig. (b) shows those from rotational diffusion. As one can see here, there is an excellent agreement between microrheology, from both diffusional motions, and mechanical measurements. This nice agreement is observed in the range of polymer concentration studied here, 0.1

to 0.8 % w/w. For higher concentrations, the particle's dynamics becomes too slow and the intensity correlation functions do not reach their asymptotic values of 1 within the time scale of the experiment. Investigations in that concentration regime may require the introduction of a non-ergodic treatment to the light scattering data [13].

ACKNOWLEDGMENTS

Work supported by the Consejo Nacional de Ciencia y Tecnología, México, Grant SEP-2004-C01-46121.

REFERENCES

1. T. G. Mason, K. Ganesan, J. H. van Zanten, D. Wirtz and S. C. Kuo, Phys. Rev. Lett. **79**, 3282 (1997)
2. Y. Tseng, T. P. Kole and D. Wirtz, Biophysical J. **83**, 3162 (2002).
3. M. L. Gardel, M. T. Valentine, J. C. Crocker, A. R. Bausch and D. A. Weitz, Phys. Rev. Lett. **91**, 158302 (2003).
4. A. W. C. Lau, B. D. Hoffman, A. Davies, J. C. Crocker and T. C. Lubensky, Phys. Rev. Lett. **91**, 108301 (2003).
5. T. P. Kole, Y. Tseng, L. Huang, J. L. Katz and D. Wirtz, Mol. Biol. Cell **15**, 3475 (2004).
6. T. G. Mason and D. A. Weitz, Phys. Rev. Lett. **74**, 1250 (1995).
7. F. C. Mackintosh and C. F. Schmidt, Curr. Opin. Colloid Interface Sci. **4**, 300 (1999).
8. J. D. Ferry, *Viscoelastic Properties of Polymers* John Wiley and Sons, New York, 1980.
9. B. R. Dasgupta, S. Y. Tee, J. C. Crocker, B. J. Frisken, and D. A. Weitz, Phys. Rev. E **65**, 51505 (2002).
10. J. K. G. Dhont, *An Introduction to Dynamics of Colloids* (Elsevier Science, Amsterdam, 1996).
11. Z. Cheng and T. G. Mason, Phys. Rev. Lett. **90**, 018304 (2003).
12. A. Mertelj, J. L. Arauz-Lara, G. Maret, T. Gisler and H. Stark, Europhys. Lett. **59**, 337 (2002).
13. P. Díaz-Leyva, E. Pérez and J. L. Arauz-Lara, J. Chem. Phys. **121**, 9103 (2004).
14. E. C. Cooper, P. Johnson and A. M. Donald, Polymer **32**, 2815 (1991).
15. B. J. Berne and R. Pecora, *Dynamic Light Scattering: With Applications to Chemistry, Biology and Physics* (John Wiley, New York, 1976).

Multiple Light Scattering Probes of Soft Materials

Frank Scheffold

Soft Condensed Matter Group and Fribourg Centre for Nanomaterials, Department of Physics, University of Fribourg, CH-1700 Fribourg, Switzerland; Frank.Scheffold@unifr.ch

Abstract. I will discuss both static and dynamic properties of diffuse waves. In practical applications the optical properties of colloidal systems play an important role, for example in commercial products such as sunscreen lotions, food (drinks), coatings but also in medicine for example in cataract formation (eye lens turbidity). It is thus of importance to know the key parameters governing optical turbidity from the single to the multiple scattering regime. Temporal fluctuations of multiply scattered light are studied with photon correlation spectroscopy (Diffusing Wave Spectroscopy). This DWS method and its various implementations will be treated.

INTRODUCTION

Static transmission or reflection measurements on turbid media provide information about the scattering strength and also contain some information about the microstructural order. From the analysis of the fast fluctuations of the laser speckle pattern, it is possible to access local dynamic processes. Based on a combination of single and multi-speckle detection schemes, it is nowadays possible to cover an extended range of relaxation times from a few nanoseconds to minutes or hours. These advanced light scattering methods can be used to study colloid dynamics, foams, granular media and many other complex systems. In a completely different domain, biomedical imaging, the dynamic analysis of multiply scattered light is called laser speckle imaging (LSI). Finally, I will address the fundamental analogy between electron transport in metals and alloys and photon transport in dielectrics. Transport of (coherent) light in strongly scattering colloidal systems is a very active field of research with topics such as conductance fluctuations, weak and strong (Anderson) localization of light, photonic crystals, random laser or (quantum) statistics of multiply scattered light.

Opaque systems and diffuse light analysis

In many dense systems, scattering is so strong that the propagation of light can be described as a diffusion process with light being scattered in random directions thus smearing out the q-dependence of scattering. In the absence of absorption, the scattering properties are now characterized by the transport cross section σ^* which, for colloidal particles, can be derived from Mie-theory. The diffuse transmission coefficient for a slab of thickness L is given by $T \approx l^*/L$ where $l^* \propto 1/\sigma^*$ denotes the transport mean free path. Information about structural order or particle sizes in this regime is

CP885, *Advanced Summer School in Physics 2006, Frontiers in Contemporary Physics—EAV06,* edited by O. Miranda, M. Carbajal, L. M. Montaño, O. Rosas-Ortiz, and S. A. Tomás Velázquez

therefore only accessible in an indirect way by comparison of $l*$ with Mie-theory and structure modelling. The $l*$ parameter can be extracted both from transmission [1] and backscattering experiments [3]. A more refined method to determine the scattering properties in opaque media is frequency domain photon migration (FDPM). It allows to determine the absorption and isotropic scattering properties separately with high precision and accuracy by modulating the laser beam intensity (for details see [2]).

In principle, model independent q-resolved information can be recovered if the wavelength dependent transmission is analyzed. This approach has been named diffuse transmission spectroscopy (DTS) [1]. It exploits the fact that the maximum momentum transfer $q_{max} = 2k_0$ is wavelength dependent. In turn, the wavelength dependence of l^* can be related to the q-dependence of the scattering cross section or $I(q)$. In practice, the method suffers, however, a number of limitations. The transmission coefficient T can only be recorded over a limited range of wavelengths, the wavelength dependent refractive index of the sample has to be known, the sample must be in the diffusive regime for all wavelengths, the accuracy of the measurements is limited and taking the derivative further reduces the quality of the data.

PROBING DYNAMIC PROPERTIES OF SOFT MATERIALS

Dynamic light scattering techniques are widely applied in different areas of science and technology where the microscopic movement of small particles has to be studied. Light scattering is certainly one of the best methods to study these materials. The technique offers convenient access to microscopic properties such as diffusion coefficients. The accuracy of remote optical sensing makes light scattering one of the most valuable tools for optical characterization and monitoring. The surging interest in slowly relaxing and arrested colloidal systems such as gels or glasses [4, 5] has created a need to monitor dynamic properties on time scales of seconds and minutes. Traditionally, a single speckle mode of scattered light is detected and fluctuations are recorded over a time much longer that the relaxation time. However, this time averaging scheme is not applicable to rigid, nonergodic systems. For these systems, the ensemble average can be obtained by summing a collection of consecutive experiments conducted on different sample realizations. Usually, the sample is translated or rotated and a scan over a large number of independent speckles is performed. A major drawback of this approach is the extensive duration of measurements. It is not unusual today to investigate relaxation process on time scales of seconds and minutes with a corresponding measurement time of hours and days. As a matter of fact, several authors have reported data collection times of more than a day for a single intensity correlation function [6].

Besides being tedious and time consuming this approach is restricted to the systems in (quasi-)equilibrium. Only the advent of multi-speckle detection schemes made it possible to conveniently monitor very slow relaxation processes. Dynamic light scattering using a digital (CCD/CMOS) camera as a detector offers the possibility to perform simultaneously a large number of independent experiments thus achieving ensemble averages in real time [5]. But unfortunately with a time resolution of typically 1-10 ms digital camera based detection is restricted to rather long correlation times. Thus, traditional photon correlation spectroscopy has to be done as well in a separate experiment if

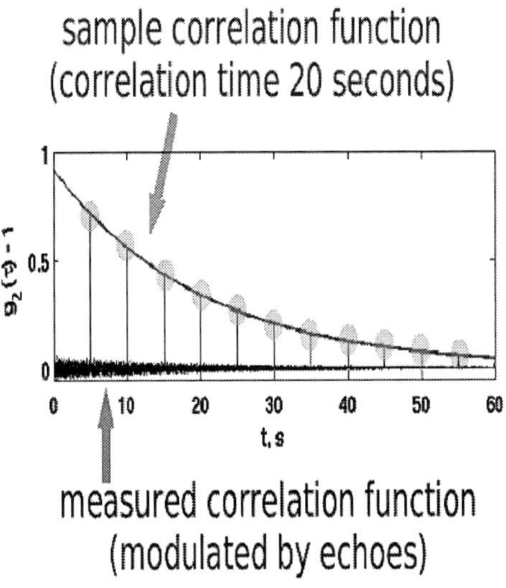

FIGURE 1. DWS echoes in backscattering for a sample of TiO_2 in glycerol.

access to the full range of correlation times is required [7, 8]. Due to the strong multiple scattering, DWS offers the flexibility to modify the experimental design, which has been exploited in a recent approach. A new two-cell detection scheme for diffusing wave spectroscopy (DWS) provides an effective multi-speckle averaging using single mode detection. To obtain an ensemble averaged signal the sample is illuminated with laser light scattered from a rotating diffuser. Correlation peaks (echoes) in the recorded correlation function appear at any revolution while the correlation function of the sample remains finite. The echo signal is generated by a large number of independent speckles thus efficient ensemble averaging is performed almost in real time. [9]

DWS BASED MICRORHEOLOGY

A particular interesting application is called "optical microrheology" . The underlying idea of optical microrheology is to study the thermal response of small (colloidal) particles embedded in the system under study [10, 11, 12, 13]. The particles are either artificially introduced, which is then called "tracer-microrheology", or can be part of the system itself, e.g. like in the case of yoghurt. By analyzing the thermal motion of the particle it is possible to obtain quantitative information about the storage and loss moduli over an extended range of frequencies. Thereby, it is possible to obtain quantitative information about the loss and storage moduli, $G'(\omega)$ and $G''(\omega)$ over an extended range of frequencies [14, 15]. This technique has been introduced some years ago when Mason and Weitz suggested a quantitative relation between the tracer mean squared displacement and the complex shear modulus [12, 16]

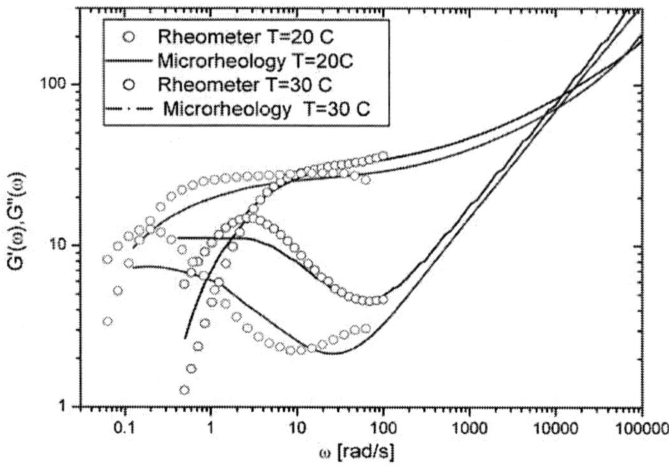

FIGURE 2. Classical rheometry and optical microrheology of a dense surfactant solution at two different temperatures T = 20° C and 30° C. Measurement time for the optical microrheology was approximately 5 min for each temperature.[21]

$$\tilde{G}(s) = \frac{s}{6\pi a}\left[\frac{6k_BT}{s^2\langle\Delta\tilde{r}^2(s)\rangle}\right] \quad (1)$$

One of the most popular techniques to study the thermal motion of the tracer particles is diffusing wave spectroscopy (DWS) since it allows relatively easy access to small particle displacements over an extended range of distances from less than $1nm$ to about $50nm$. Thereby, elastic moduli from less than $1Pa$ to $1MPa$ can be accessed in a two-cell or multispeckle DWS experiment [17, 18].

Figure 2 shows some recent measurements where tracer particles have been introduced in an otherwise transparent matrix consisting of a concentrated surfactant solution: the surfactant Cetylpyridinium Chloride (CPy 100mM) with Sodium Salicylate (60mM). Under these conditions the surfactant molecules self-assemble and form large elongated aggregates leading to a strongly viscoelastic fluid [20]. Polystyrene Particles (diameter 720nm) were added during preparation (final volume fraction ca. 1 %). The sample is filled in a 5mm standard rectangular glass cell and kept at constant temperature. From the intensity autocorrelation function of the diffusely transmitted laser light, the motion profile of the small tracer particles is calculated with nanometer resolution . Using this information the full frequency spectrum is determined (Figure 2). The measurement nicely reveals the typical Maxwell-type relaxation of these fluids: liquid like for low frequencies and solid-like for frequencies above 10 rad/sec. A comparison of mechanical and optical measurements in the high frequency regime is currently underway [21].

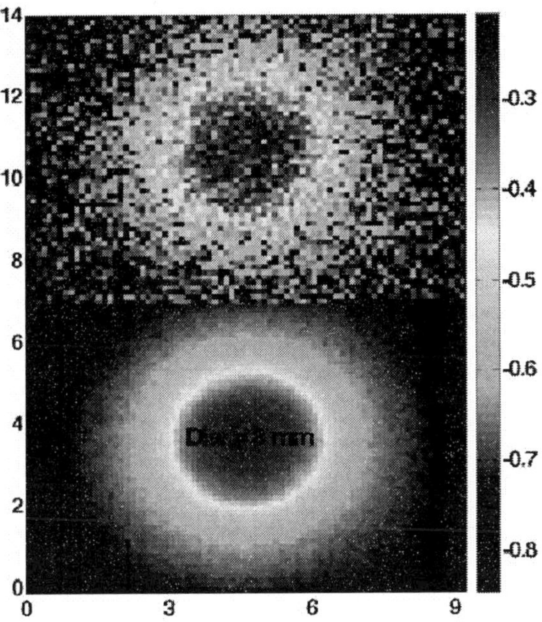

FIGURE 3. Laser speckle image taken from a medical phantom (a mm-size liquid volume in a white Teflon block) For the picture on the left side the incident optical laser pattern is created by a slowly rotating diffuser which substantially reduces the speckle noise [26].

LASER SPECKLE BIOMEDICAL IMAGING

In a completely different domain, biomedical imaging, the dynamic analysis of multiply scattered light is called laser speckle imaging (LSI). This technique, a variant of DWS, is an efficient and simple method for full-field monitoring of dynamics in heterogeneous media[22, 23, 24, 25, 26]. It is widely used in biomedical imaging of blood flow [23, 24, 25] since it provides access to physiological processes *in vivo* with excellent temporal and spatial resolution.

In a more general context LSI can be considered a simplified version of the DWS approach, which analyzes the temporal intensity fluctuations of scattered laser light in order to derive the microscopic properties of the scatterers position and motion. In LSI, an image of dynamic heterogeneities is obtained by analyzing the spatial statistics of speckles produced by scattered light and thus allows a full-field monitoring of dynamics. The original method consists of analyzing the local speckle contrast K in the image plane. K is defined by the variance of the intensity fluctuations for a given integration time T : $K^2(T) = \langle I^2 \rangle / \langle I \rangle^2 - 1$. In the absence of scatterers motion, the contrast takes a maximum value while motion decreases the contrast[27].

WAVES IN RANDOM MEDIA

Since the Anderson's discovery that the propagation of a quantum particle can be blocked by disorder [28] and subsequent realization that this 'Anderson localization' can also take place for electromagnetic waves (photons) [29], the quest for observing it has become a very active field of research [30, 31, 32, 33, 34, 35, 36, 37, 38, 39, 40, 41, 42]. Although the observation of microwave localization in quasi-one dimensional disordered samples [30] now seems to be accepted by the scientific community, the localization of visible light in strongly scattering, three-dimensional (3D) semiconductor powders [31] has been questioned [32]. For both light and microwaves, an efficient way of revealing localization effects is to study fluctuations and correlations (upon varying frequency for static scatterers [33, 34, 35] or time for mobile scatterers [35, 36, 42]) of the transmission coefficient T of the random sample, or even the full probability distribution of T [37, 38, 30, 39]. Statistics of T, as well as many other mesoscopic optical phenomena (coherent backscattering and weak localization [40], long-range spatial intensity correlations [33], universal conductance fluctuations [41], etc.) can be understood and discussed.

REFERENCES

1. P. D. Kaplan, A. D. Dinsmore, A. G. Yodh, and D. J. Pine, Phys. Rev. E **50**, 4827 (1994).
2. S. M. Richter, E. M. Sevick-Muraca, Colloids Surf. A **172**, 163 (2000).
3. Bavarian et al. Phys. Rev. E **71**, 066603 (2005).
4. F. Scheffold and P. Schurtenberger, Soft Materials, 2002.
5. L. Cipelletti, S. Manley, R.C. Ball, and D.A. Weitz, Phys. Rev. Lett. **84**, 2275 (2000).
6. W. van Megen, S. M. Underwood, and P. N. Pusey, Phys. Rev. Lett. **67**, 1586 (1991).
7. S. Romer, F. Scheffold and P. Schurtenberger, Phys. Rev. Lett., **85**, 4980 (2000).
8. F. Scheffold, S. E. Skipetrov, S. Romer and P. Schurtenberger, Phys. Rev. E, **63**, 061404 (2001).
9. P. Zakharov, F. Cardinaux and F. Scheffold, Phys. Rev E, **75**, 011413 (2006).
10. T. Gisler and D.A. Weitz, Curr. Opin. Coll. Int. Sci. **3,** 586 (1998); M.L. Gardel, M.T. Valentine, and D.A. Weitz, in: *Microscale Diagnostic Techniques*, K. Breuer (Ed.) (Springer Verlag, Berlin, 2002), in press.
11. N.J. Wagner and R.K. Prud'homme (Eds.), Curr. Opin. Coll. Int. Sci. **6** (2001).
12. T.G. Mason and D.A. Weitz, Phys. Rev. Lett. **74,** 1250 (1995); T.G. Mason *et al.,* J. Opt. Soc. Am. A **14,** 139 (1997); T.G. Mason *et al.,* Phys. Rev. Lett. **79,** 3282 (1997).
13. F. Gittes, B. Schnurr, P.D. Olmsted, F.C. MacKintosh, and C.F. Schmidt, Phys. Rev. Lett. **79**, 3286 (1997).
14. C.W. Macosko, *Rheology, Priciples, Measurements, and Applications* (Wiley, New York, 1994)
15. J.D. Ferry, *Viscoelastic Properties of Polymers* (Wiley, New York, 1980)
16. Recently a rigorous theoretical derivation (albeit certain constraints) for this generalized Stokes-Einstein relation has been reported: A.J. Levine and T.C. Lubensky, Phys. Rev. Lett. **85,** 1774 (2000); A.J. Levine and T.C. Lubensky, Phys. Rev. E **63,** 041510 (2000).
17. B.R. Dasgupta, S.Y. Tee, J.C. Crocker, B.J. Frisken, and D.A. Weitz, Phys. Rev. E. **65,** 051505 (2002).
18. C. Heinemann, F. Cardinaux, F. Scheffold, P. Schurtenberger, F. Escher, and B. Conde-Petit, Carbohydrate Polym. **55**, 155 (2004).
19. F. Cardinaux, L. Cipelletti, F. Scheffold, and P. Schurtenberger, Europhys. Lett. **57**, 738 (2002).
20. P. Fischer and H. Rehage, Langmuir **13**, 7012 (1997).
21. N. Willenbacher, F. Scheffold, F. Cardinaux, P. Fischer, C. Oelschlaeger, M. Schopferer, F. Nettesheim, N. Wagner, in preparation.
22. J. D. Briers, Physiological Measurement **22** : R35-R66 (2001).

23. B. Weber, C. Burger, M. T. Wyss, G. K. von Schulthess, F. Scheffold, and A. Buck, Eur. J. Neurosci. **20**, 2664 (2004).
24. T. Durduran, M. G. Burnett, C. Zhou G. Yu, D. Furuya, A. G. Yodh, J. A. Detre, and J. H. Greenberg, J. of Cerebral Blood Flow & Metabolism, **24** 518 (2004).
25. A. Dunn, A. Devor, M. Andermann, H. Bolay, M. Moskowitz, A. Dale, and D. Boas, Optics Lett. **28** 28 (2003).
26. A.C. Völker, P. Zakharov, B. Weber, F. Buck, and F. Scheffold Opt. Exp., **13** 9782 (2005).
27. P. Zakharov, A.C. Völker, A. Buck, B. Weber, and F. Scheffold, Quantitative modeling of laser speckle imaging, Optics Lett. to be published.
28. P.W. Anderson, Phys. Rev. **109**, 1492 (1958).
29. S. John, Phys. Rev. Lett. **53,** 2169 (1984); P.W. Anderson, Phil. Mag. B **52,** 505 (1985); S. John, Phys. Today **44**(5), 32 (1991).
30. A.A. Chabanov, M. Stoytchev, A.Z. Genack, Nature **404,** 850 (2000); A.A Chabanov and A.Z. Genack, Phys. Rev. Lett. **87**, 153901 (2001).
31. D.S. Wiersma *et al.,* Nature **390,** 671 (1997).
32. Comment on Ref. [31]: F. Scheffold *et al.,* Nature **398**, 206 (1999); Reply to comment: D.S. Wiersma *et al., ibid.* **398**, 207 (1999).
33. A.Z. Genack, N. Garcia, and W. Polkosnik, Phys. Rev. Lett. **65,** 2129 (1990).
34. M.P. Van Albada, J.F. de Boer, and A. Lagendijk, Phys. Rev. Lett. **64,** 2787 (1990); J.F. de Boer, M.P. van Albada, and A. Lagendijk, Phys. Rev. B **45,** 658 (1992).
35. E. Akkermans and G. Montambaux, *Physique mésoscopique des électrons et des photons* (EDP Sciences/CNRS Editions, 2004).
36. F. Scheffold *et al.,* Phys. Rev. B **56**, 10942 (1997).
37. J.F. de Boer *et al.,* Phys. Rev. Lett. **73**, 2567 (1994).
38. M. Stoytchev and A.Z. Genack, Phys. Rev. Lett. **79**, 309 (1997); Opt. Lett. **24**, 262 (1999).
39. A.Z. Genack and A.A. Chabanov, J. Phys. A: Math. Gen. **38**, 10433 (2005).
40. M.P. van Albada and A. Lagendijk, Phys. Rev. Lett. **55,** 2692 (1985); P.-E. Wolf and G. Maret, *ibid.* **55,** 2696 (1985); E. Akkermans, P. E. Wolf, and R. Maynard, *ibid.* **56**, 1471 (1986).
41. F. Scheffold and G. Maret, Phys. Rev. Lett. **81**, 5800 (1998).
42. S. Balog, S. E. Skipetrov, and F. Scheffold, Phys. Rev. Lett. **97**, 103901 (2006).

Depolarization of Multiple Scattered Reflected Light

L. F. Rojas-Ochoa[*,†], D. Lacoste[**], R. Lenke[‡], P. Schurtenberger[*] and F. Scheffold[*]

[*]*Department of Physics, University of Fribourg, CH-1700 Fribourg, Switzerland*
[†]*Departamento de Física, Cinvestav-IPN, Av. IPN 2508, 07360 México D.F., Mexico*
[**]*Physico-Chimie Théorique, Ecole Supérieure de Physique et de Chimie Industrielles (ESPCI), 10 Rue Vauquelin, 75231 Paris Cedex 05, France*
[‡]*Carl Zeiss Laser Optics GmbH, D-73446 Oberkochen, Germany,*

Abstract. We formulate a quantitative description of backscattered linearly polarized light using an extended photon diffusion formalism that explicitly takes into account the scattering anisotropy parameter, g, of the medium. We investigate the path length distribution on the scattering anisotropy parameter g spanning an extended range from 0 (isotropic scattering) to 1 (forward scattering). Good agreement is found with Monte Carlo simulations of multiple scattered light.

INTRODUCTION

Polarized light scattered many times in a random medium leaves the sample partially depolarized. Unfortunately, despite its importance in areas like biomedical optical imaging, coherent backscattering or dynamic spectroscopy [1, 2, 3, 4], the depolarization of light in a random medium is still not completely understood due to the complexity of vector wave multiple scattering (as compared to the much simpler problem of scalar wave propagation). Previous attempts have mainly focussed on isotropic (Rayleigh) scattering [5], or on the depolarization of circularly polarized light [4, 6]. Only few studies have discussed specifically the mechanism of depolarization of linearly polarized light in the case where the anisotropy parameter $g =< \cos\theta >$ is different from 0 and furthermore a detailed comparison with simulations has been lacking [7, 8, 9, 10, 11, 12, 13]. An accurate description for arbitrary scattering anisotropy is however crucial to analyze the information contained in backscattered light if progress is to be made in applications like remote sensing, photon correlation spectroscopy or optical imaging of biological tissues [3, 14, 15].

In this work, we formulate a quantitative description of backscattered linearly polarized light using an extended photon diffusion formalism taking explicitly into account the scattering anisotropy parameter, g. The details of our model are adjusted by comparison with Monte Carlo simulations of multiply scattered light. We can distinguish the following limiting situations for the transport of light and its polarization: isotropic scattering $g \simeq 0$ and forward-peaked scattering $g \simeq 1$. The situation of forward-peaked scattering $g \simeq 1$ is typical of Mie scattering [16] with large particles and of biological tissues. The situation of $g < 0$ has only been made possible experimentally recently by

CP885, *Advanced Summer School in Physics 2006, Frontiers in Contemporary Physics—EAV06,* edited by O. Miranda, M. Carbajal, L. M. Montaño, O. Rosas-Ortiz, and S. A. Tomás Velázquez

tuning the interaction of the light using mesostructured colloidal liquids [17]. Here we show that with our additional correction the simple photon diffusion picture successfully describes the distribution of path lengths in the backscattering geometry.

PATH LENGTHS DISTRIBUTION FOR BACKSCATTERING

On length scales much larger than the transport mean free path l^*, the transport of light in a turbid medium can be described by the diffusion approximation. This approximation is connected to the idea of treating the transport of photons as a random walk, characterized by a distribution of path lengths [5, 19, 20, 21]. An exact solution of the diffusion equation applied to light transport can be obtained using the method of images. This method takes into account the boundary conditions through two lengths which are both of the order of a transport mean free path: the extrapolation length z_e; where the flux of photons vanishes outside the sample, and z_p; which is the location of the photon source (for a more detailed interpretation of z_e, z_p based on a random walk model see reference [4]). The method leads to

$$P(s) = \frac{\sqrt{3}}{4\sqrt{\pi \ell^*}\, s^{3/2}} \left[z_p e^{-\frac{3}{4}\frac{z_p^2}{\ell^* s}} + (z_p + 2z_e)\, e^{-\frac{3}{4}\frac{(z_p+2z_e)^2}{\ell^* s}} \right], \qquad (1)$$

which obeys the normalization condition $\int_0^\infty P(s)ds = 1$. Note that the path length is simply related to the number of scattering events n by $s/\ell = n-1$, so that the path length is 0 for single scattering. Here l is the scattering mean free path. Both quantities, l and l^*, are related by $l^*/l = 1/(1-g)$. The scattering anisotropy parameter g is defined as the average of the cosine of the scattering angle $g = \langle \cos\Theta \rangle$.

To check the validity of Eq. 1, we have performed Monte-Carlo simulations of linearly polarized light reflected from a semi-infinite turbid medium (details about the simulation method can be found in refs. [4, 22, 23]). These simulations use the Mie scattering cross section in the range $0 \leq g \leq 1$ and are able to evaluate numerically an exact path length distribution as a function of the number of scattering events n and polarization. The simulations were done for uncorrelated spherical scatterers (structure function $S(q) \equiv 1$) and for a non-reflecting interface. Values of $\lambda_0/n_s = 532\,nm$ for the incident wavelength, $n_p = 1.59$ for the refractive index of the particle and $n_s = 1.332$ for solvent refractive index were used. In Fig. 1 we compare the results of the simulations with the prediction of the method of images according to Eq. 1. We see clearly in this figure that the method of images provides an excellent description of the path length distribution for the case of isotropic scattering ($g \equiv 0$), but that the method fails to give an equally good description for the case of anisotropic scattering corresponding to $g \simeq 0.806$.

The disagreement in the latter case is not surprising as the diffusion approximation is known to overestimate the contribution from the short paths of the distribution, the error becoming more and more severe as the anisotropy of scattering increases. One way to improve the distribution of paths length of Eq. 1, is by introducing a cutoff in the distribution as suggested by Mackintosh and John [7]. Here we extend their approach by taking into account explicitly the scattering anisotropy factor g (with $\int_0^\infty P_{corr}(s)ds = 1$):

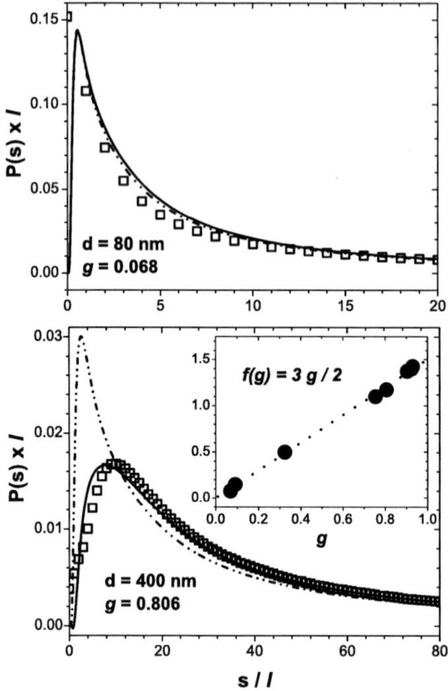

FIGURE 1. Normalized path length distribution $P(s) \times l$ for backscattered light from a semi-infinite medium. Symbols: Monte Carlo simulations. Lines: calculations based on Eq. 2 with $f(g) = 3g/2$ (solid) and $f(g) \equiv 0$ (dashed). Inset: $f(g)$ obtained from Eq. 1 adjusted to fit the simulation results. Wavelength $\lambda_0 = 532nm$, refractive index of the particle $n_p = 1.59$ and of the solvent $n_s = 1.332$. Non-reflecting boundary conditions were used.

$$P_{corr}(s) \propto P(s) \cdot \left[1 - f(g)\, e^{-s/\ell^*}\right] . \tag{2}$$

According to Mackintosh and John [7] $f(1)$ is of order unity and for isotropic scattering ($g \rightarrow 0$) there is no correction $f(0) = 0$. Using the corrected distribution of path lengths to fit the simulation results, we have found that the function f is well approached by a linear dependence $f(g) = 3g/2$ which we assume to be valid also for $g < 0$. As can be seen in figure 1, the use of the correction factor significantly improves the prediction of Eq. 1. We note that, alternatively, for the case of forward-peaked scattering $g \simeq 1$, other schemes of approximations have been suggested. For example the recent reference [13] reports that the Fokker-Plank equation provides a better description than does the (uncorrected) diffusion approximation for forward-peaked scattering.

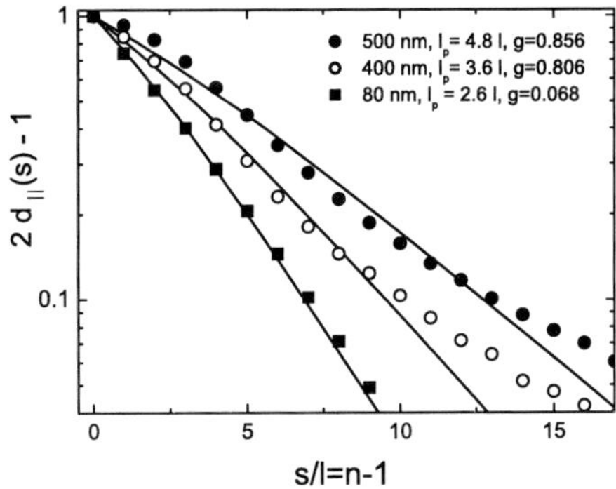

FIGURE 2. Depolarization of multiply scattered light. $2d_{\|}(s) - 1$ from theory (Eq.(4) solid lines) and simulation (symbols). Excellent agreement if found for Rayleigh scatterers while for larger particles the agreement becomes somewhat less good.

DEPOLARIZATION LENGTH FOR LINEAR POLARIZATION

Incident polarized light looses its polarization in random multiple scattering [5, 7, 8, 9, 12]. For linearly polarized light, only two configurations ($\|$ and $\perp$) need to be considered (for isotropic samples). Physically, in the $\|$ geometry more photons are detected for short paths as compared to the unpolarized case. In a seminal paper Akkermanns et al. [5] found that the path length distribution for the two configurations can be written as

$$P_{\|,\perp}(s) = d_{\|,\perp}(s) \cdot P(s), \tag{3}$$

with the depolarization ratio given by

$$d_{\|}(s) = \frac{1 + 2e^{-s/\ell_p}}{2 + e^{-s/\ell_p}}, \tag{4}$$

$$d_{\perp}(s) = \frac{1 - e^{-s/\ell_p}}{2 + e^{-s/\ell_p}}, \tag{5}$$

in terms of the characteristic length of depolarization for linearly polarized light ℓ_p. For point-like scatterers ($g = 0$) Akkermans et al. obtained $\ell_p = \ell / \ln(10/7) \cong 2.804\ell$ [5]. We find good agreement between Eq.3 and our numerical simulations with ℓ_p as an adjustable parameter (Fig. 2). For large particles (and therefore large g) the agreement is somewhat less good. However polarization effects in DWS usually are found weak for $g \approx 1$ and therefore we did not attempt to improve the accuracy of Eqs. (4) and (5)(It is worthwhile to note that close to the sample surface very interesting polarization effects, such as a butterfly pattern, persist [24, 25]).

In the limit $s/\ell \gg 1$ Eq. 4 and 5 reduce to :

$$P_{\parallel,\perp}(s) \cong \left[\frac{1}{2} \pm \frac{3}{4} e^{-s/\ell_p}\right] \cdot P(s) \,. \tag{6}$$

We consider this expression the most simple generalization since it captures well intermediate path lengths $s/\ell > 3$, where polarization effects are important, but at the same time the number of scattering events is already sufficiently large to apply the diffusion approximation.

CONCLUSION

In this paper, we have shown how to describe the effect of the scattering anisotropy on the depolarization of linearly polarized light. By means of numerical simulations, we checked the limit of validity of the diffusion approximation when the scattering anisotropy g is increased, and we have shown how to correct the predictions by means of an anisotropy dependent cutoff for the path length distribution $P(s)$. We discuss the path length distribution dependence on the scattering anisotropy parameter g over an extended range of values of g. Since our description only uses a single adjustable parameter, ℓ_p, it is now possible to fully characterize backscattered light with polarization resolved measurements. We think that this approach can strongly benefit applications in the field of soft and bio material analysis, as well as diffuse light imaging techniques [3, 13]. An extension of the model to describe DWS experiments in the backscattering geometry is presented in [18].

ACKNOWLEDGMENTS

Financial support from the Swiss National Science foundation is gratefully acknowledged.

REFERENCES

1. G. Maret, and P. E. Wolf, *Z. Phys. B* **65**, 409–431 (1987).
2. D. J. Pine, D. A. Weitz, P. M. Chaikin, and E. Herbolzheimer, *Phys. Rev. Lett.* **60**, 1134–1137 (1988).
3. D. A. Weitz, and D. J. Pine, "Diffusing-wave spectroscopy," in *Dynamic Light Scattering*, edited by W. Brown, Oxford U. Press, New York, 1993, pp. 652–720.
4. R. Lenke, and G. Maret, "Multiple Scattering of Light: Coherent Backscattering and Transmission," in *Scattering in Polymeric and Colloidal Systems*, edited by W. Brown and K. Mortensen, Gordon and Breach Science Publishers, London, 2000, pp. 1–72.
5. E. Akkermans, P. E. Wolf, R. Maynard, and G. Maret, *J. Phys. France* **49**, 77–98 (1988).
6. E. E. Gorodnichev, A. I. Kuzovlev, and D. B. Rogozkin, *JETP Letters* **68**, 22–28 (1998).
7. F. C. MacKintosh, and S. John, *Phys. Rev. B* **40**, 2383–2406 (1989).
8. F. C. MacKintosh, J. X. Zhu, D. J. Pine, and D. A. Weitz, *Phys. Rev. B* **40**, 9342–9345 (1989).
9. D. Bicout, C. Brosseau, A. S. Martinez, and J. M. Schmitt, *Phys. Rev. E* **49**, 1767–1770 (1994).
10. V. L. Kuzmin, and V. P. Romanov, *Phys. Rev. E* **56**, 6008–6019 (1997).
11. D. Lacoste, V. Rossetto, F. Jaillon, and H. Saint-Jalmes, *Opt. Lett.* **29**, 2040–2042 (2004).

12. D. A. Zimnyakov, Y. P. Sinichkin, P. V. Zakharov, and D. N. Agafonov, *Waves Random Media* **11**, 395–412 (2001).
13. A. D. Kim, and J. B. Keller, *J. Opt. Soc. Am. A* **20**, 92–98 (2003).
14. P. Sebbah, *Waves and imaging through complex media*, Kluwer Academic Publishers, Dordrecht; Boston, 2001.
15. M. Moscoso, J. B. Keller, and G. Papanicolaou, *J. Opt. Soc. Am. A* **18**, 948–960 (2001).
16. H. C. van de Hulst, *Light Scattering by Small Particles*, Dover, New York, 1981; C. F. Bohren, and D. R. Huffman, *Absorption and Scattering of Light by Small Particles*, Wiley, New York, 1983.
17. L. F. Rojas-Ochoa, J. M. Mendez-Alcaraz, J. J. Saenz, P. Schurtenberger, and F. Scheffold, *Phys. Rev. Lett.* **93**, 073903 (2004).
18. L. F. Rojas-Ochoa, D. Lacoste, R. Lenke, P. Schurtenberger, and F. Scheffold, *J. Opt. Soc. Am. A* **21**, 1799–1804 (2004).
19. A. Ishimaru, *Wave propagation and scattering in random media*, Academic Press, New York, 1978.
20. A. Lagendijk, R. Vreeker, and P. DeVries, *Phys. Lett. A* **136**, 81–88 (1989).
21. J. X. Zhu, D. J. Pine, and D. A. Weitz, *Phys. Rev. A* **44**, 3948–3959 (1991).
22. R. Lenke, and G. Maret, *Eur. Phys. J. B* **17**, 171–185 (2000).
23. R. Lenke, R. Tweer, and G.Maret, *J. Opt. A: Pure Appl. Opt.* **4**, 293–298 (2002).
24. A. C. Maggs, and V. Rossetto, *Phys. Rev. Lett.* **87**, 253901 (2001).
25. A. H. Hielscher et. al., *Optics Exp.* **1**, 441–453 (1997).

Entropic Interactions in Complex Fluids

J. M. Méndez-Alcaraz[1,2]

[1]*Departamento de Física, Cinvestav, Av. IPN 2508, Col. San Pedro Zacatenco, 07360 México, D.F., Mexico*
[2]*On sabbatical stay in Centro de Investigación en Energía, UNAM, 62580 Temixco, Morelos, Mexico.*

Abstract. When in a complex fluid the volume exclusion between the component particles is the leading interaction, then depletion forces of entropic nature may become able to drive the physics of the system. In this paper we quantify the corresponding depletion interaction potential and evaluate some of its structural and thermodynamic effects in model colloidal mixtures.

It is quite common to find entropically driven complex fluids. Mixtures of hard particles are perhaps the best example, and therefore all the systems which can be accurately modeled as such a mixture, as it is the case of charge-stabilized colloids with high salt concentrations, or steric-stabilized colloids, etc. Actually, every liquid, provided that the thermal energy scale k_BT dominates over all interactions, behaves like a mixture of hard particles. In such systems, the internal energy U is only of kinetic origin, i.e., $U = \sum_{i=1}^{p} a_i k_B T N_i$, and therefore does not depend on the volume V (the value of a_i depends on the form of the N_i particles of species i in the mixture of p components). The volume effects in the free energy F must then arise from the entropy S, i.e., $F(T,N_1,\cdots,N_p,V) = U(T,N_1,\cdots,N_p) - TS(T,N_1,\cdots,N_p,V)$. This is why this kind of systems are called entropic liquids.

At a microscopic level the volume exclusion interaction between the component particles is responsible for the properties of the system. This kind of interactions lead to depletion effects which become so relevant for size-asymmetric mixtures that they may be able to drive the physics of the system. These effects can be understood as follows: Let us assume a colloidal suspension and pay attention to any two particles moving in the bulk and approaching each other by chance. Simultaneously other particles are expelled from the gap between the two approaching particles, giving rise to an imbalance between the osmotic pressure inside the gap and outside it. This imbalance results in the depletion interaction between approaching particles, which can be most easily evaluated in the case of extremely dilute binary mixtures of spherical colloidal particles.

Let us assume a very dilute binary mixture of spherical particles with diameters σ_1 and σ_2, with $\sigma_1 > \sigma_2$, and let us observe the bigger particles. When two of them approach each other, as described in the previous paragraph, the volume forbidden to the smaller particles decreases by ΔV, as shown in Fig. 1, where the case of a binary mixture in front of a hard wall is also displayed. In the case that we have only two particles of species 1, the corresponding change in the free energy of the system is roughly given by $\Delta F \approx -p_2 \Delta V \approx -(k_B T n_2)\Delta V$, where we have approximated the partial pressure of species 2 by the ideal gas pressure ($n_i = N_i/V$ is the number density of species i),

CP885, *Advanced Summer School in Physics 2006, Frontiers in Contemporary Physics—EAV06,* edited by O. Miranda, M. Carbajal, L. M. Montaño, O. Rosas-Ortiz, and S. A. Tomás Velázquez

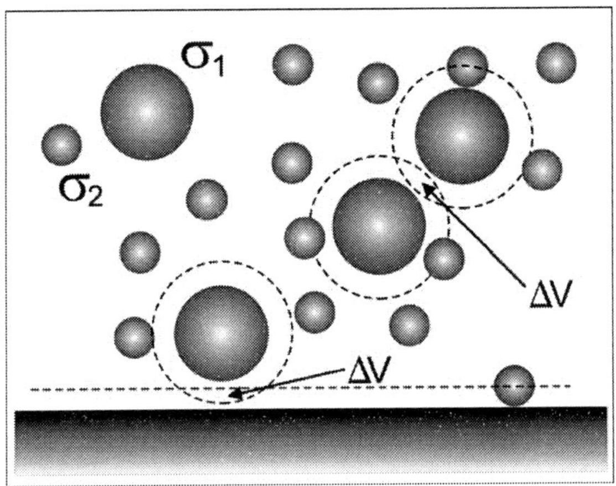

FIGURE 1. The figure schematically shows a binary mixture of spherical hard particles of diameters σ_1 and σ_2 in front of a hard wall. The dashed lines represent the volume forbidden to the particles of species 2 around the particles of species 1, and in front of the wall. When two particles of species 1 approach each other, or when a particle of species 1 approaches the wall, that volume decreases by ΔV.

which is compatible with our assumption of a very dilute system. We do not consider the contributions to ΔF arising from the depletion of particles of any species from the gap between any two approaching particles of species 2, or of species 1 and 2, since we only want to calculate the depletion interaction potential between both particles of species 1. We will come back to the analysis of the validity of this approximation, but for the moment let us assume that it is correct.

The quantity ΔV can be straightforwardly evaluated by means of a geometrical sketch, as first done by Asakura and Oosawa [1], and by Vrij [2]. The result leads to the effective interaction potential $u_{11}^{eff}(r)$ between both particles of species 1:

$$\beta u_{11}^{eff}(r) \approx -\varphi_2 \left[(\eta+1)^3 - \frac{3}{2}(\eta+1)^2 \frac{r}{\sigma_2} + \frac{1}{2}\frac{r^3}{\sigma_2^3} \right], \tag{1}$$

where we have identified ΔF with $u_{11}^{eff}(r)$, $\varphi_i = \pi n_i \sigma_i^3/6$ is the volume fraction of species i, $\eta = \sigma_1/\sigma_2$ is the size ratio, r is the distance between the centers of the approaching particles, and $\beta = 1/k_BT$ is the Boltzmann factor. From this result we can see that the depletion interaction ranges from $r = \sigma_1$ up to $r = \sigma_1 + \sigma_2$, and that it is attractive. The latter can be straightforwardly understood because the volume accessible to the particles of species 2, which are much more numerous than the particles of species 1, increases when both particles of species 1 approach each other (ΔV increases). This means that the maximum entropy (minimum free energy) is reached when both particles of species 1 stay together, leaving free as much volume as possible for the particles of species 2. In other words, nature favors the order of the larger particles (under order we understand that they are put together) and the disorder of the smaller and more numerous particles, because the entropy of the system becomes maximum by this way.

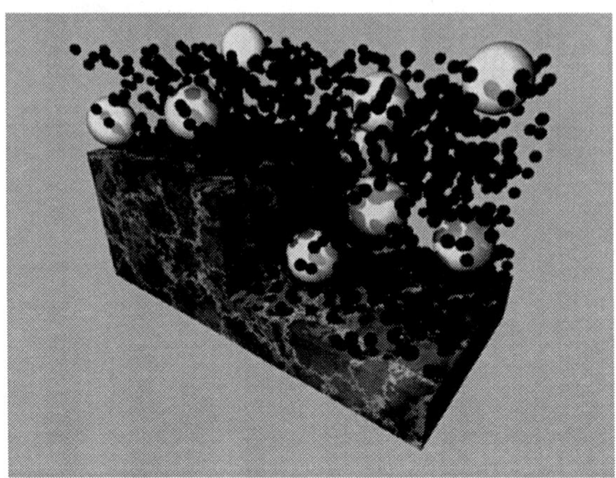

FIGURE 2. The figure schematically shows a binary mixture of spherical hard particles of diameters σ_1 and σ_2 in front of a planar hard wall with a linear step edge of height h along the y-axis.

In the case of a binary mixture in front of a planar hard wall, also shown in Fig. 1, the quantity ΔV can also be straightforwardly evaluated along the same lines followed to obtain Eq. (1) by means of a geometrical sketch [3]. This time, however, we assume only one particle of species 1 approaching the wall. The result for the wall-particle effective interaction potential $u_{w1}^{eff}(x)$ is given by

$$\beta u_{w1}^{eff}(x) \approx -\varphi_2 (1+3\eta) \left(\frac{3}{2} - \frac{x}{\sigma_2}\right), \tag{2}$$

where x is the distance between the surface of the wall and the center of the approaching particle. The interaction ranges from $x = \sigma_1/2$ up to $x = \sigma_1/2 + \sigma_2$, and it is still more attractive than in the case of two spheres (compare with Eq. (1)). Equation (2) can actually be obtained from Eq. (1) by just taking the limit $\sigma_1 \to \infty$ for only one of the two approaching particles of species 1 [4]. Once again we can claim that nature favors the order of the larger particles (under order we understand that they are put on the surface of the wall) and the disorder of the smaller and more numerous particles. Moreover, we can also expect that the induction of such order in the particles of species 1 depends not only on the parameters of the suspension, like φ_i and η, but also on the geometry of the particles and of the wall. Let us take a closer look at the latter.

In the derivation of Eq. (1) the quantity ΔV can be interpreted as the volume of the region in the gap between the particles of species 1, separated by the distance r, from which the particles of species 2 are excluded due to their simultaneous overlap with both particles of species 1. Therefore, Eq. (1) can be rewritten as [5]

$$\beta u_{11}^{eff}(r) \approx -n_2 \int_V c_{12}^{(0,0)}(r') c_{21}^{(0,0)}(|\mathbf{r}-\mathbf{r}'|) d\mathbf{r}', \tag{3}$$

with $c_{ij}^{(0,0)}(r) = -1$ for $r < (\sigma_i + \sigma_j)/2$, and 0 elsewhere. We will discuss later about the convenience of the notation. In a similar way, in the derivation of Eq. (2) the quantity

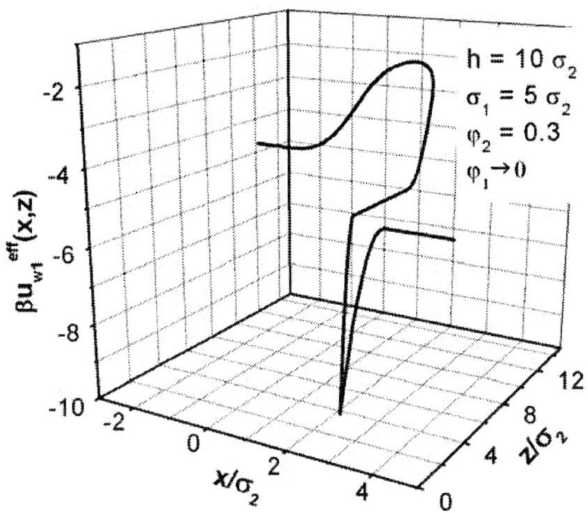

FIGURE 3. The figure shows the contact wall-particle depletion potential for a binary mixture of hard spheres in front of a hard wall with a linear step edge of height $h = 10\sigma_2$ along the y-axis. The upper level of the step is in the region with $x < 0$ and $z = h$, on a plane parallel to the xy-plane, and the lower level in the region with $x > 0$ and $z = 0$, on the xy-plane. The parameters of the mixture are $\varphi_1 \to 0$, $\varphi_2 = 0.3$, and $\eta = 5$.

ΔV can be interpreted as the volume of the region in the gap between the particle of species 1 and the wall, separated by the distance x, from which the particles of species 2 are excluded due to their simultaneous overlap with both the wall and the particle of species 1. Therefore, Eq. (2) can be rewritten as [5]

$$\beta u_{w1}^{eff}(r) \approx -n_2 \int_V c_{w2}^{(0,0)}(\mathbf{r}') c_{21}^{(0,0)}(|\mathbf{r} - \mathbf{r}'|) d\mathbf{r}', \tag{4}$$

with $c_{wi}^{(0,0)}(\mathbf{r}) = -1$ when the wall and a particle of species i overlap, and 0 elsewhere. The advantage of these equations is that they can be numerically evaluated in more general cases by constructing an spatial grid around the overlapping region. We check the simultaneous overlapping condition in every point of the grid. If the condition is fulfilled we add a volume element to the corresponding discretization of the integral in Eqs. (3) and (4), and pass to the next point of the grid. In the opposite case, we only pass to the next point of the grid, and so forth. Let us now try this idea in the case of a binary mixture of hard spheres in front of a planar hard wall with a linear step edge of height h along the y-axis, as shown in Fig. 2.

Let us consider a hard wall with a step edge of height $h = 10\sigma_2$, and, in front of it, a binary mixture of hard spheres with parameters $\varphi_1 \to 0$, $\varphi_2 = 0.3$, and $\eta = 5$. The upper level of the step is in the region with $x < 0$ and $z = h$, on a plane parallel to the xy-plane, and the lower level in the region with $x > 0$ and $z = 0$, on the xy-plane. Note that the meaning of the variable x differs from the one previously used in this paper. The wall-particle depletion potential is given by Eq. (4) with the appropriate overlapping conditions for $c_{w2}^{(0,0)}(\mathbf{r})$ and $c_{21}^{(0,0)}(r)$. In order to evaluate this equation at contact we construct a cubic grid of step $\Delta = \sigma_2/10$ over the wall. Then, we check in every point of the grid the condition of simultaneous overlapping of a particle of species 2 with the wall and with a particle of species 1 always in contact with the wall. The latter is then moved along the x-axis in order to scan the region around the step edge. The results we found for $\beta u_{w1}^{eff}(x,z)$ are shown in Fig. 3. The line represents a contact scanning of the wall, beginning on the upper level and ending on the lower level of the step. Far away from the edge the scanning particle only sees a flat wall and, therefore, the value $\beta u_{w1}^{eff}(x \ll 0, h) = \beta u_{w1}^{eff}(x \gg 0, 0) = -4.8$ can be also obtained from the contact limit of Eq. (2).

Closing the edge from the left, the scanning particle first feels a force opposite to its motion, parallel to the wall, and a some weaker attraction perpendicular to the wall. Those features arise from the collisions with the smaller spheres in front of the lower level of the step and in the neighborhood of its convex edge. After crossing this region, the scanning particle falls into a very attractive well located on the concave edge of the step. In order to leave this well the particle has to move against a force, parallel to the wall, pushing it back to the concave edge. Some of these predictions have been already observed in the experiment [6].

In order to analyze the validity of the previous approximations, we invite the reader to take a look of references [4] and [5], where a theoretical scheme for depletion forces is worked out in the framework of the integral equations theory of simple liquids. The authors assume that depletion forces are a special case of the more general effective interactions resulting from a contraction of the description of liquid mixtures. Therefore, if certain components of a mixture are not explicitly considered, their influence on the structure of the remaining particles has to be included in the effective interaction potential between the latter ones. This is obtained by demanding the spatial distribution of the explicitly described particles to be the same as in the original mixture. Technically, this is done by rewriting the Ornstein-Zernike equation for the original mixture as an effective Ornstein-Zernike equation for the remaining particles, and connecting it with the effective interaction potential by means of an appropriate closure relation. Along those lines it is found that the effective interaction potential between particles of species 1 in our binary mixtures is given by

$$\beta u_{11}^{eff}(r) = -n_2 \mathscr{F}^{-1}\left\{\frac{\widetilde{c}_{12}^{2}(q)}{1 - n_2 \widetilde{c}_{22}(q)}\right\} + \left[b_{11}^{eff}(r) - b_{11}(r)\right], \tag{5}$$

where $\mathscr{F}^{-1}\{X\}$ denotes the inverse Fourier transform of X, $\widetilde{c}_{ij}(q)$ is the Fourier transform of the direct correlation function $c_{ij}(r)$ between particles of species i and j, $b_{11}(r)$ is the bridge function between particles of species 1 in the original mixture, and $b_{11}^{eff}(r)$ is the bridge function between the same particles but in the contracted system where the particles of species 2 are not longer explicitly considered. Equation (5) is exact.

Considering that the leading terms of the bridge functions are of quadratic order in the density [7], and performing the virial-like expansion $c_{ij}(r) = \sum_{\alpha,\beta=0}^{\infty} n_1^{\alpha} n_2^{\beta} c_{ij}^{(\alpha,\beta)}(r)$ of the direct correlation functions in Eq. (5), we reproduce Eq. (3) for $n_1 \to 0$, up to linear terms in n_2. This demonstrates that the equations we were using above are exact in the

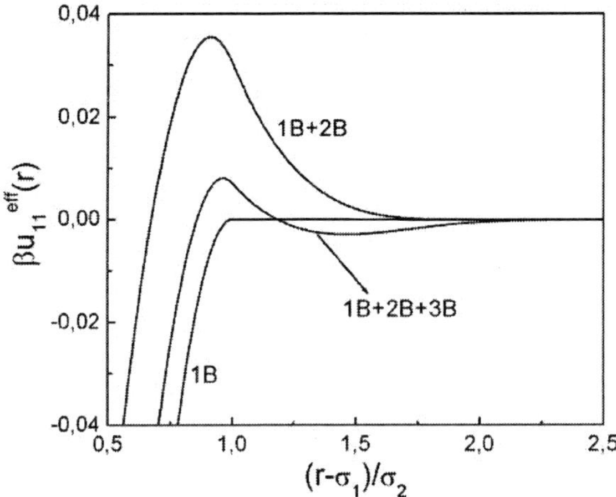

FIGURE 4. The figure shows the single bridges, double bridges, and triple bridges contributions to the depletion potential in a binary mixture of hard spheres with $\sigma_1 = 10\sigma_2$, $\varphi_2 = 0.05$, and $\varphi_1 \to 0$.

dilute limit, as we have actually assumed, as well as the convenience of our previous notation. Equation (4) can also be straightforwardly obtained along the same lines, as it can be seen in references [5] and [8]. Moreover, if we neglect the difference between the bridge functions in Eq. (5), the following approximate expression for the depletion potential can be obtained still in the infinite dilution limit of species 1:

$$\beta u_{11}^{eff}(r) \approx -n_2 \mathscr{F}^{-1}\left\{\tilde{c}_{12}^{(0,0)}(q)\tilde{c}_{21}^{(0,0)}(q)\right\} - n_2^2 \mathscr{F}^{-1}\left\{\tilde{c}_{12}^{(0,0)}(q)\tilde{c}_{22}^{(0,0)}(q)\tilde{c}_{21}^{(0,0)}(q)\right\}$$
$$- n_2^3 \mathscr{F}^{-1}\left\{\tilde{c}_{12}^{(0,0)}(q)\tilde{c}_{22}^{(0,0)}(q)\tilde{c}_{22}^{(0,0)}(q)\tilde{c}_{21}^{(0,0)}(q)\right\} + \cdots, \tag{6}$$

where we neglected the functions $c_{12}^{(0,1)}$ and $c_{12}^{(0,2)}$ in the terms of quadratic and cubic order in n_2. The first term, which corresponds to Eq. (3), can be interpreted as representing single bridges between two particles of species 1 formed by one particle of species 2. The second term represents the bridges between two particles of species 1 formed by two particles of species 2. The third term represents triple bridges formed by two particles of species 1 and three particles of species 2 in between, and so forth.

We evaluate Eq. (6) in order to get an idea about how the depletion potential changes when the concentration of depleting particles increases. In Fig. 4 we show, for $\sigma_1 = 10\sigma_2$, $\varphi_2 = 0.05$, and $\varphi_1 \to 0$, that the single bridges lead to the depletion attraction already predicted by Eq. (1) and displayed by the line with the legend 1B. The double bridges add a repulsive barrier in front of the attractive well at contact, as displayed by the line with the legend 1B+2B corresponding to the first two terms in Eq. (6). The triple bridges produce a secondary attraction in front of that repulsive barrier, as shown by the line with the legend 1B+2B+3B corresponding to the first three terms in Eq. (6). These results show that the depletion interactions become richer when the system concentrates, for this paper however we leave our calculations at this point.

The previous results indicate that the suspended particles in front of a wall are pushed to the wall due to the depletion interaction, and that the amplitude of such attraction can be manipulated by changing the parameters of the suspension, such as particle size and concentration, and of the wall, such as its curvature. This occurs although we are working with hard particles and hard walls, whose interaction is only repulsive and short-ranged. The origin of the long-ranged depletion forces is that the system increases its entropy when the larger particles are placed close to the wall, letting free the volume of the bulk for the much more numerous smaller particles. Moreover, the entropy of the smaller particles may become still larger when the larger particles are put together. Indeed, this phenomenon has been already used in order to generate self-assembled structures, as in the crystallization of protein suspensions [9], or in the selective adsorption of colloidal particles on walls with a relief pattern alternating convex and concave edges [10].

Entropy plays a very important role in determining the properties of complex fluids in a large variety of situations. The ideas presented above could help by understanding the way in which the entropy drives the physics of the system, but also to design methods of manipulating the entropy in order to reach the desired result. When the entropy increases two opposite effects occur; a part of the system becomes more ordered and, simultaneously, the rest of the system becomes more disordered. The trick is to find the way to get the order in the form we want to have it, and the disorder where it is irrelevant.

ACKNOWLEDGMENTS

Useful discussions with P. González-Mozuelos, R. Castañeda-Priego, and A. Rodríguez-López are gratefully acknowledged. This work was possible thanks to financial support provided by Conacyt (Grants 33815-E, 46373/A-1, and 47200/A-1).

REFERENCES

1. S. Asakura and F. Oosawa, J. Chem. Phys. **22**, 1255 (1954).
2. A. Vrij, Pure Appl. Chem. **48**, 471 (1976).
3. R. Castañeda-Priego, *Structure and Effective Interactions in Colloidal Suspensions* (PhD Thesis, Cinvestav, Mexico, 2003).
4. J. M. Méndez-Alcaraz and R. Klein, Phys. Rev E **62**, 5360 (2000).
5. R. Castañeda-Priego, A. Rodríguez-López, and J. M. Méndez-Alcaraz, Phys. Rev. E **73**, 051404 (2006).
6. A. D. Dinsmore, A. G. Yodh, and D. J. Pine, Nature (London) **383**, 239 (1996).
7. J. P. Hansen and I. R. McDonald, *Theory of Simple Liquids* (Academic, London, 1986)
8. P. González-Mozuelos and J. M. Méndez-Alcaraz, Phys. Rev. E **63**, 021201 (2001).
9. S. Fraden and R. D. Kamien, Biophys. J. **78**, 2189 (2000).
10. A. D. Dinsmore and A. G. Yodh, Langmuir **15**, 314 (1998).

Counterions Released from Oppositely Charged Surfaces: a Liquid Theory Study

A. Flores-Amado and M. Hernández-Contreras

Departamento de Física, Centro de Investigación y de Estudios Avanzados del IPN, A.P. 14-740, México D.F., Mexico

Abstract. The strength of the attractive interaction of two oppositely charged surfaces in an electrolyte solution, and the counterion release from the slit between them was determined within the hypernetted chain approximation and compared with the Poisson-Boltzmann equation. Liquid theory predicts small deviations, for the attraction between the plates and release of ions, than mean field theory at moderate and low added salt. At higher salt concentration both theories agree on their predictions for these thermodynamic properties.

The release of counterions upon association of macromolecules of opposed electrical sign is a mechanism of self-assembly formation [1]. Here we present a liquid theory study of this phenomenon and its effects on the interaction between two charged flat plates of contrary electrical sign and same strength embedded in an aqueous electrolyte solution. The counterion release has been studied on this system by Fleck *et al* [2] and Safran [3] by means of the Poisson-Boltzmann (PB) mean field theory. On the other hand, our investigation relies on the use of the anisotropic hypernetted chain approximation (AHNC) [4] which has the capability of taking into account, the charge-charge density and ion's size correlations which are neglected in PB theory. We considered the cases of smeared out and discrete distribution of surface charges on the plates. It is found that in all cases PB theory always predicts a more enhanced, pressure between the plates, and number of counterions released, than liquid theory does. We found that systems with the least concentration of added salt display a stronger attractive interaction than those cases where the electrolyte concentration was higher. Similar effects are also noted on the excess number of monovalent counterions. Thus, the strength of released counterions is bigger at low salt concentrations than at larger values of bulk electrolyte concentration. These findings will be described below in the next paragraph.

The model system consists of two identical surfaces with density of electric surface charge $\sigma_s = 0.267$ C/m^2, each one of opposite sign separated the distance L by an 1:1 NaCl electrolyte solution where the two species of ions are formed by spheres of the same size of hydrated diameter $d = 4.25$Å dispersed in an aqueous solvent of dielectric constant $\varepsilon = 78.5$. The ions direct interaction is represented by the long range Coulomb electrostatic potential plus a hard core to avoid overlapping between the spheres. We shall not consider image charge effects. The theoretical approach we use is based on the Ornstein-Zernike (OZ) integral equation of inhomogeneous electrolytes for the: total, and direct correlation functions and the profile distribution of both species of ions $n^{\mp}(z)$ in the space between the plates [5].

Such integral equation is solved numerically for the total correlation function and

CP885, *Advanced Summer School in Physics 2006, Frontiers in Contemporary Physics—EAV06,* edited by O. Miranda, M. Carbajal, L. M. Montaño, O. Rosas-Ortiz, and S. A. Tomás Velázquez

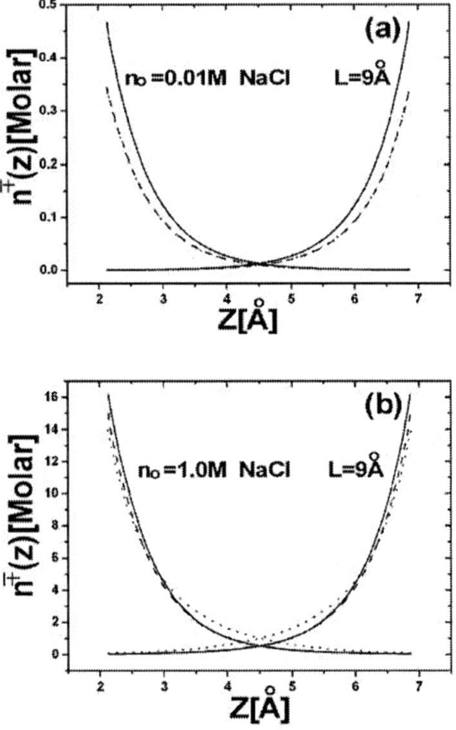

FIGURE 1. Profile distribution of positive and negative ions $n^{\mp}(z)$ between two asymmetrically charged surfaces separated the distance $L = 9$Å. Figure (a) for three cases of $\sigma_s = 0.267$ C/m^2: discrete (black continuous line, results of AHNC theory), smeared out (dashed line, and AHNC theory), and smeared out (dotted line, PB theory results using pointlike ions), respectively, with salt concentration of $n_0 = 0.01$M NaCl. Figure (b) same as Figure (a) but salt concentration $n_0 = 1.0$M NaCl.

$n^{\mp}(z)$ by imposing the AHNC closure relation between the total and direct correlation functions, and the direct pair potential of two particles. The converged solution of the OZ equation corresponds to the thermodynamic equilibrium of the whole system and therefore leads to the profile distribution of positive and negative ions in the slit between the plates $n^{\mp}(z)$ as a function of the perpendicular distance z from the origin of coordinate located on the left surface. These structural properties are plotted in Fig. 1 for two bulk electrolyte concentrations $n_0 = (0.1, 0.01)$M of NaCl, wall to wall separation $L = 9$Å, and absolute temperature $T = 300$K. In those figures, black continuous line corresponds to discrete σ_s on the surfaces and use was made of the OZ+AHNC equations, dashed line are results of OZ theory for smeared out σ_s, whereas dotted line is the PB result of pointlike ions with σ_s being a smooth distribution of surface charge.

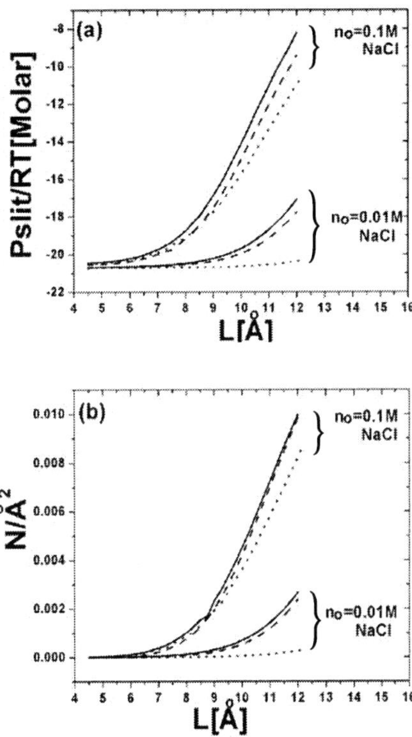

FIGURE 2. Figure (a) provides the comparison of P_{slit} calculated with AHNC and PB theories as a function of surfaces separation L at two bulk electrolyte concentrations $n_0 = (0.1, 0.01)$M. Figure (b) depicts the comparison of the excess number of monovalent counterions $N/\text{Å}^2$ as a function of surfaces separation at the same electrolyte concentrations of Figure (a). In both cases same symbols for lines as in Fig. 1 and $\sigma_s = 0.267$ C/m^2.

According to Fig. 1, the OZ equation always predicts for the case of discrete density of surface charge σ_s a higher ion's concentration close and in contact with the surfaces, while for a continuous density of charge σ_s both OZ and PB theories lead to profiles $n^{\mp}(z)$ roughly of the same order of magnitude when the salt concentration is low, thus, for instance at $n_0 = 0.01$M (see Fig. 1(a)). However, OZ y PB approaches lead to different numerical values for $n^{\mp}(z)$ at contact when the concentration of salt is raised up to $n_0 = 1.0$M (see Fig. 1(b)). In previous work [6] we reported also of an observed increase in the ions contact densities $n^{\mp}(z_{\text{cont}})$ for the case of equally charged symmetric surfaces due to the positional correlations among the bulk ions densities with those charges located at the surfaces when there are present discrete surface charge distributions. This high contact values of $n^{\mp}(z_{\text{cont}})$ has a direct consequence in another structural property, namely, the pressure between the two walls, which is given by the contact theorem as $P_{\rm slit} = k_{\rm B}T[n^{+}(z_{\rm cont}) + n^{-}(z_{\rm cont})] - \frac{\sigma_s^2}{2\varepsilon}$, where $z_{\rm cont}$ is the distance of closest approach of the center of an ion to one of the charged surfaces.

Figure 2(a) depicts the pressure $P_{\rm slit}$ between the walls given in Molar units as a function of wall's separation L for two concentrations of NaCl, $n_0 = (0.1, 0.01)$M. In such a figure, black continuous and dashed lines correspond to results of OZ theory for discrete and smeared out charge σ_s on the surfaces, respectively, while dotted line is the PB theory result. We note that the value of the slit pressure between the walls is found to have higher values at the PB level of approximation than those of OZ theory with σ_s being a smooth distribution of charge, and also than the OZ equation prediction for σ_s discrete, consecutively. And the strength of the slit pressure is in any case larger for $n_0 = 0.01$M of NaCl than the case at hight salt concentration of $n_0 = 0.1$M. Yet the maximun number of counterions released $N/\text{Å}^2$ takes place at the lowest concentration of salt and it occurs in the same order of their relative magnitudes as for the pressure curves shown in Fig. 2(a) obtained according to the different theories quoted above. Both Fleck *et al* [2] and Safran [3] found that the $P_{\rm slit}$ pressure value is maximun at the largest counterion release with low salt concentrations in solution. However, we found quantitative differences among the predictions made by the different theories, namely, PB and OZ equation, on the structural properties such as $P_{\rm slit}$ and $N/\text{Å}^2$. Particularly, those differences are larger for wall to wall separations on the order of two or more ion's diameters (see Fig. 2).

ACKNOWLEDGMENTS

This work was supported by CONACyT Grant No. 48794-F, México.

REFERENCES

1. K. Wagner, D. Harries, S. May, V. Kahl, J.O. Rädler, A. Ben-Shaul, *Langmuir* **26**, 303-306 (2000).
2. A.A. Meier-Koll, C.C. Fleck, and H.H. von Grünberg, *J. Phys.:Condens. Matter* **16**, 6041-6052 (2004).
3. S.A. Safran, *Europhys. Lett.* **69**, 826-831 (2005).
4. R. Kjellander and S. Marcelja, *J. Chem. Phys.* **82**, 2122-2135 (1985).
5. R. Kjellander, T. Åkesson, B. Jönson, and S. Marcelja *J. Chem. Phys.* **97**, 1424-1431 (1992).
6. O. González-Amezcua, M. Hernández-Contreras, and P. Pincus *Phys. Rev. E* **64**, 041603(1-3) (2001).
7. A. Flores-Amado and M. Hernández-Contreras, submitted to *Phys. Rev. E*, August (2006).

IV. MEDICAL PHYSICS

Complex Dynamic Behavior in Simple Gene Regulatory Networks

Moisés Santillán Zerón[1]

Centro de Investigación y Estudios Avanzados del IPN
Unidad Monterrey
Av. Cerro de las Mitras No. 2565
Col. Obispado, 64060 Monterrey NL, MÉXICO

Abstract. Knowing the complete genome of a given species is just a piece of the puzzle. To fully unveil the systems behavior of an organism, an organ, or even a single cell, we need to understand the underlying gene regulatory dynamics. Given the complexity of the whole system, the ultimate goal is unattainable for the moment. But perhaps, by analyzing the most simple genetic systems, we may be able to develop the mathematical techniques and procedures required to tackle more complex genetic networks in the near future.

In the present work, the techniques for developing mathematical models of simple bacterial gene networks, like the tryptophan and lactose operons are introduced. Despite all of the underlying assumptions, such models can provide valuable information regarding gene regulation dynamics. Here, we pay special attention to robustness as an emergent property.

These notes are organized as follows. In the first section, the long historical relation between mathematics, physics, and biology is briefly reviewed. Recently, the multidisciplinary work in biology has received great attention in the form of systems biology. The main concepts of this novel science are discussed in the second section. A very slim introduction to the essential concepts of molecular biology is given in the third section. In the fourth section, a brief introduction to chemical kinetics is presented. Finally, in the fifth section, a mathematical model for the lactose operon is developed and analyzed..

A BRIEF HISTORICAL REVIEW

Since the classification of sciences during the Enlightenment, we have the generalized impression that physics and mathematics, on the one hand, and biology, on the other hand, evolved following distinct and separated pathways. The truth, however, is different. Since the origins of modern science, there have been people making important contributions at the very interface between these sciences. William Harvey discovered blood circulation with the aid of a mathematical model. Electrodynamics started with the works of Galvani and Volta (both physicians) on animal electricity. Later, Boltzmann (also a physician, but better known for his contributions to physics) invented the myograph and the ophtalmoscope, recorded for the first time the velocity of a nervous impulse, discovered the first law of thermodynamics (based on metabolic considerations), and helped to settle the foundations of all modern theories of resonance with his

[1] Also at: **Instituto Politécnico Nacional,** Esc. Sup. de Física y Matemáticas, Edif. 9, UP Zacatenco, 07738, MÉXICO DF, México, and **McGill University,** Centre for Nonlinear Dynamics, 3655 Promenade Sir William Osler, H3G 1Y6 Montreal, QC, CANADA

CP885, *Advanced Summer School in Physics 2006, Frontiers in Contemporary Physics—EAV06,*
edited by O. Miranda, M. Carbajal, L. M. Montaño, O. Rosas-Ortiz, and S. A. Tomás Velázquez

studies on hearing physiology.

During the 20th century, electrophysiology (the science that studies the interactions between biological tissues and electromagnetic fields) advanced greatly. Archibald V. Hill, Bernard Katz, Max Planck, Walter Nernst, Kenneth S. Cole, Alan L. Hodgkin, Andrew, F. Huxley, Erwin Neher, and Haldan K. Hartline, among others, made important contributions to its progress. Some landmarks in the history of electrophysiology were: the explanation for the action potential, discovered by Hodgkin y Huxley with the aid of mathematical models; the Huxley model for muscle contraction; and the invention of the patch clamp technique by Neher.

Charles Darwin published his theory of evolution through natural selection in 1859. Since the beginning, it was clear that such theory lacked adequate proper statistical foundations, and this was its main weakness. Indeed, an apparent contradiction arose between Darwinism and Mendel's laws of inheritance, just after they were rediscovered in 1900. This gap was closed through the work of many mathematicians who, between 1860 and 1940, developed the necessary statistical tools to fuse genetics and Darwinism. The result is what we now know as population genetics or neo-Darwinism. Some of the most important contributors to this success were: Fleeming Jenkin, Francis Galton, Karl Pearson, Raphael Weldon, Godfrey H. Hardy, Ronald A. Fisher, Sewall Wright, and Theodosius Dobzhansky.

Molecular biology consolidated between 1940 and 1960, and until 1970 two different schools were recognized: the structural and the informatics schools.

The physicists W. H. Bragg y W. L. Bragg (father and son) founded the structural school in Cambridge; they invented X-ray crystallography, and the structural analysis of biological molecules soon started in their lab. Some of the most notorious structuralists were W. T. Astbury , John D. Bernal, Max Perutz, and John C. Kendrew (all of them from Cambridge), as well as Linus Pauling from Caltech. The structuralists were convinced that no new physical laws were required to explain vital phenomena. They were aimed at explaining the function of biomolecules (and so of tissues and organs) from their structure. The secondary proteinic structure known as alpha helix, and the structures of hemoglobin y la myoglobin are some of the most important discoveries of this school.

Inspired on the uncertainty principle of quantum mechanics, Niels Bohr proposed that new principles from physics may be necessary to understand life. With this assertion, Bohr founded the informatics school of molecular biology. Max Delbrück and Erwin Schrödinger were two of the most important spokesmen for this school. Delbrück moved to the USA were he started a very successful collaboration with Salvador Luria on bacteriophage research. On its own, Schrödinger had to move to Dublin due to the Nazis, and there we published a little book entitled *What is life?*, which was tremendously influential.

Some of Schrödinger's recruits were James Watson and Francis Crick (who discovered the structure of DNA in 1953), Maurice Wilkins (who provided essential physical data to Wanton and Crick) Seymour Benzer (who sequenced the first gene ever), and François Jacob (who discovered mRNA and, together with Jacques Monod, the regulatory mechanisms of the *lac* operon.

SYSTEMS BIOLOGY

A new area of interdisciplinary research in biology emerged a few year ago, which continues with the long tradition described in the previous section; its name is systems biology. Conceptually, the closest ancestors to systems biology are systems theory and cybernetics. Since the first one inherits part of the philosophy and the goals of the former ones, it is interesting to briefly review them.

Systems theory is an interdisciplinary field which studies relationships of systems as a whole. It was founded in the 1950s and focuses on organization and interdependence of relationships. Part of systems theory, system dynamics is a method for understanding the dynamic behavior of complex systems. The basis of the method is the recognition that the structure of any system —the many circular, interlocking, sometimes time-delayed relationships among its components— is often just as important in determining its behavior as the individual components themselves. Examples are chaos theory and social dynamics. In some cases it is not possible to explain the behavior of the whole in terms of the behavior of the parts.

Cybernetics is the study of communication and control, typically involving regulatory feedback, in living organisms, in machines, and in combinations of the two. It is an earlier but still-used generic term for many of the subject matters that are increasingly subject to specialization under the headings of adaptive systems, artificial intelligence, complex systems, complexity theory, control systems, decision support systems, dynamical systems, information theory, learning organizations, mathematical systems theory, operations research, simulation, and systems engineering.

Contemporary cybernetics began as an interdisciplinary study connecting the fields of control systems, electrical network theory, logic modeling, and neuroscience in the 1940s. The name cybernetics was coined by Norbert Wiener to denote the study of "teleological mechanisms" (i.e. machines with corrective feedback) and was popularized through his book Cybernetics, or Control and Communication in the Animal and Machine (1948). The emphasis is on the functional relations that hold between the different parts of a system, rather than the parts themselves. These relations include the transfer of information, and circular relations (feedback) that result in emergent phenomena such as self-organization.

What is systems biology?

Systems biology is an academic field that seeks to integrate different levels of information to understand how biological systems function. By studying the relationships and interactions between various parts of a biological system (e.g., gene and protein networks involved in cell signalling, metabolic pathways, organelles, cells, physiological systems, organisms etc.) it is hoped that eventually an understandable model of the whole system can be developed. In contrast to much of molecular biology, systems biology does not seek to break down a system into all of its parts and study one part of the process at a time, with the hope of being able to reassemble all the parts into a whole.

Systems biology begins with the study of genes and proteins in an organism using high-throughput techniques to quantify changes in the genome and proteome in response to a given perturbation. These techniques include microarrays to measure the changes in mRNAs, and mass spectrometry, which is used to identify proteins, detect protein modifications, and quantify protein levels.

Using knowledge from molecular biology, the systems biologist can propose hypotheses that explain a system's behavior. Importantly, these hypotheses can be used to mathematically model the system. Models are used to predict how different changes in the system's environment affect the system and can be iteratively tested for their validity. New approaches are being developed by quantitative scientists, such as computational biologists, statisticians, mathematicians, computer scientists, engineers, and physicists, to improve our ability to make these high-throughput measurements and create, refine, and retest the models until the predicted behavior accurately reflects the phenotype seen.

Recent analysis has revealed that cell signals do not necessarily propagate linearly. Instead, cellular signalling networks can be used to regulate multiple functions in a context dependent fashion. Because of the magnitude and complexity of the interactions in the cell, it is often not possible to understand intuitively the *systems behavior* of these networks. Rather, it has become necessary to develop mathematical models and analyze the behavior of these models both to develop a systems-level understanding and to obtain experimentally testable predictions.

Recent computer simulations of partial or whole genetic networks have demonstrated network behaviors (commonly called systems, or emergent, properties) that were not apparent from examination of only a few isolated interactions alone. One possible pathway to advance in this direction is to start analyzing the simplest genetic networks, which are found in bacteria. Among them, the most widely studied and best known are the tryptophan (*trp*) operon —paradigm for repressible operons, the lactose (*lac*) operon —paradigm for inducible operons, and the phage lambda switch —paradigm for molecular switches.

A FEW CONCEPTS FROM MOLECULAR BIOLOGY

The molecule of life

Deoxyribonucleic acid (DNA) is a nucleic acid —usually in the form of a double helix— that contains the genetic instructions specifying the biological development of all cellular forms of life, and most viruses.

In complex eukaryotic cells such as those from plants, animals, fungi and protists, most of the DNA is located in the cell nucleus. By contrast, in simpler cells called prokaryotes, including the eubacteria and archaea, DNA is not separated from the cytoplasm by a nuclear envelope. The cellular organelles known as chloroplasts and mitochondria also carry DNA.

DNA is often referred to as the molecule of heredity as it is responsible for the genetic propagation of most inherited traits. In humans, these traits can range from hair color to disease susceptibility. During cell division, DNA is replicated and can be transmitted to offspring during reproduction.

DNA is not a single molecule, but rather a pair of molecules joined by hydrogen bonds: it is organized as two complementary strands, head-to-toe, with hydrogen bonds between them. Each strand of DNA is a chain of chemical *building blocks*, called nucleotides, of which there are four types: adenine (abbreviated A), cytosine (C), guanine (G) and thymine (T). These allowable base components of nucleic acids can be polymerized in any order giving the molecules a high degree of uniqueness.

Between the two strands, each base can only *pair up* with one single predetermined other base: A+T, T+A, C+G and G+C are the only possible combinations; therefore, naming the bases on the conventionally chosen side of the strand is enough to describe the entire double-strand sequence. Two nucleotides paired together are called a base pair. On rare occasions, wrong pairing can happen. The double-stranded structure of DNA provides a simple mechanism for DNA replication: the DNA double strand is first *unzipped* down the middle, and the *other half* of each new single strand is recreated by exposing each half to a mixture of the four bases. An enzyme makes a new strand by finding the correct base in the mixture and pairing it with the original strand. In this way, the base on the old strand dictates which base will be on the new strand, and the cell ends up with an extra copy of its DNA.

DNA contains the genetic information that is inherited by the offspring of an organism; this information is determined by the sequence of base pairs along its length. A strand of DNA contains genes, areas that regulate genes, and areas that either have no function, or a function yet unknown. Genes can be loosely viewed as the organism's *cookbook* or *blueprint*.

Ribonucleic acid

RNA is primarily made up of four different bases: adenine, guanine, cytosine, and uracil. The first three are the same as those found in DNA, but uracil replaces thymine as the base complementary to adenine. Unlike DNA, RNA is almost always a single-stranded molecule and has a much shorter chain of nucleotides. RNA contains ribose, rather than the deoxyribose found in DNA.

The central dogma of molecular biology

The central dogma of molecular biology was first enunciated by Francis Crick in 1958 and re-stated in a *Nature* paper published in 1970:

> "The central dogma of molecular biology deals with the detailed residue-by-residue transfer of sequential information. It states that such information cannot be transferred from protein to either protein or nucleic acid."

In other words, *once information gets into protein, it can't flow back to nucleic acid.*

The central dogma is often misunderstood. It is frequently confused with the standard pathway of information flow from DNA to RNA to protein. There are notable exceptions to the normal pathway of information flow and these are often mistakenly referred to as exceptions to the central dogma.

The standard information flow pathway consists of three different steps: transcription, translation, and replication. By new knowledge of the RNA processing in eukaryotic cells, a fourth step must be included: splicing. Below, the different processes accounted for by the central dogma are briefly reviewed.

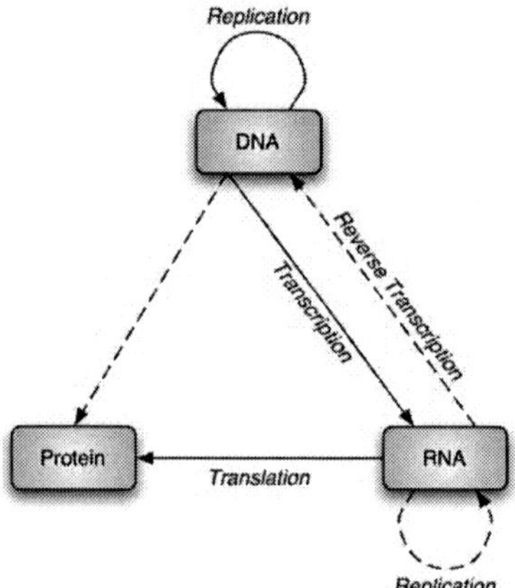

FIGURE 1. The 1970 version of the Central Dogma. The arrows represent the flow of information. Solid arrows represent *probable* information flow, while dotted arrows represent *possible* information flow. Note that information flow from proteins to RNA or DNA is regarded as impossible

Transcription

Transcription is the process through which a DNA sequence is enzymatically copied by an RNA polymerase to produce a complementary RNA. Or, in other words, the transfer of genetic information from DNA into RNA. In the case of protein-encoding DNA, transcription is the beginning of the process that ultimately leads to the translation of the genetic code (via the mRNA intermediate) into a functional peptide or protein.

In prokaryotic cells transcription occurs in the cytoplasm alongside translation. In this cells, transcription initiation takes place through the following steps:

- RNA polymerase (RNAP) recognizes and specifically binds to the promoter region on DNA. At this stage, the DNA is double-stranded (*closed*). This RNAP/wound-DNA structure is referred to as the closed complex.
- The DNA is unwound and becomes single-stranded (*open*) in the vicinity of the initiation site (defined as +1). This RNAP/unwound-DNA structure is called the open complex.
- The RNA polymerase transcribes the DNA, but produces about 10 abortive (short, non-productive) transcripts which are unable to leave the RNA polymerase because the exit channel is blocked by the σ-factor.
- The σ-factor eventually dissociates from the holoenzyme, and elongation proceeds.

Eukaryotes have evolved much more complex transcriptional regulatory mechanisms than prokaryotes. For instance, in eukaryotes the genetic material (DNA), and therefore transcription, is primarily localized to the nucleus, where it is separated from the cytoplasm (where translation occurs) by the nuclear membrane. DNA is also present in mitochondria. Adding to this complexity, eukaryotes have three nuclear RNA polymerases, each with distinct roles and properties:

- RNA Polymerase I transcribes ribosomal RNA (rRNA).
- RNA Polymerase II transcribes messenger RNA (mRNA) and most small nuclear RNAs (snRNAs).
- RNA Polymerase III transcribes transfer RNA (tRNA) and other small RNAs (including the small 5S rRNA).

Further complexity is added by the multitude of transcription factors and signaling pathways that may interact in combination to mediate cell-type and developmental transcriptional regulation. The basal eukaryotic transcription complex includes the RNA polymerase and additional proteins that are necessary for correct initiation and elongation.

Some genes also have enhancer elements that can be thousands of bases upstream or downstream of the transcription initiation site. Combinations of these upstream control elements and enhancers regulate and amplify the formation of the basal transcription complex. Once a polymerase finishes assembling the mRNA molecule, the transcription process is terminated.

Splicing

In eukaryote cells, the primary transcript (pre-mRNA) is processed prior to translation by removing one or more intermediate sequences (introns). This process is known as splicing. The mechanism of alternative splicing makes it possible to produce different mature mRNA molecules, depending on what sequences are treated as introns and what remain as exons. However, not all living cells have mRNA that undergoes splicing; splicing is absent in prokaryotes.

Translation

Eventually, mature mRNA finds its way to a ribosome, where it is translated. In prokaryotic cells, which have no nuclear compartment, the process of transcription and translation may be linked together. In eukaryotic cells, the site of transcription (the cell nucleus) is usually separated from the site of translation (the cytoplasm), so the mRNA must be transported out of the nucleus into the cytoplasm, where it can be bound by ribosomes. mRNA is read by the ribosome as triplet codons, usually beginning with an AUG. Complexes of initiation factors and elongation factors bring aminoacylated transfer RNAs (tRNAs) into the ribosome-mRNA complex, matching the codon in the mRNA to the anti-codon in the tRNA, thereby adding the correct amino acid in the sequence encoding the gene. As the amino acids are linked into the growing peptide chain, they begin folding into the correct conformation. This folding continues until the nascent polypeptide chains are released from the ribosome as a mature protein. In some cases the new polypeptide chain requires additional processing to make a mature protein.

Replication

Finally, as the final step in the Central Dogma, to transmit the genetic information between parents and progeny, DNA must be replicated faithfully. Replication is carried out by a complex group of proteins that unwind the superhelix, unwind the double-stranded DNA helix, and, using DNA polymerase and its associated proteins, copy or replicate the master template itself.

Alterations to the central dogma

In his 1970 paper on the subject, Crick pointed out that the central dogma, while useful as a theory to guide experiment, was not to be taken as dogma:

> "Although the details of the classification proposed here are plausible, our knowledge of molecular biology, even in one cell – let alone for all the organisms in nature – is still far too incomplete to allow us to assert dogmatically that it is correct." - Francis Crick

Since then, a number of facts have emerged suggesting the need for it to be restated:

- **Flow from RNA to DNA:** After the central dogma was expounded, retroviruses were discovered. These transcribe RNA into DNA through the use of a special enzyme called reverse transcriptase. This was thought to occur only in viruses, but, more recently, examples of RNA to DNA flow have been shown in higher animals, including humans.
- **Viruses with RNA-only genomes:** Some virus species have their entire genome encoded in the form of RNA. Thus, their information flow consists only of RNA $\rightarrow$ Protein
- **Non-coding RNAs:** Many RNAs in an organism achieve a functional state capable of affecting the phenotype of the organism without ever being translated into a protein.
- **Prions:** Prions are proteins that propagate themselves by making conformational changes in other molecules of the same type of protein. This change affects the behavior of the protein. In fungi this change can be passed from one generation to the next. This is not considered an exception to the central dogma since the sequence of the protein is unchanged.

GENE REGULATION

Regulation of gene expression (gene regulation) is the cellular control of the amount and timing of appearance (induction) of the functional product of a gene. Any step of gene expression may be modulated, from the DNA-RNA transcription step to post-translational modification of a protein. Gene regulation gives the cell control over structure and function, and is the basis for cellular differentiation, morphogenesis and the versatility and adaptability of any organism.

Any step of gene expression may be modulated, from the DNA-RNA transcription step to post-translational modification of a protein. Following is a list of stages where gene expression is regulated:

- Modification of DNA
- Transcription
- Translation
- Post-transcriptional modification
- RNA transport
- mRNA degradation
- Post-translational modifications

Regulation of transcription

Transcription of a gene by RNA polymerase can be regulated by at least three types of proteins:

- Specificity factors alter the specificity of RNA polymerase for a given promoter or set of promoters, making it more or less likely to bind to them.
- Repressors bind to non-coding sequences on the DNA strand, impeding RNA polymerase's progress along the strand, thus impeding the expression of the gene.
- Activators enhance the interaction between RNA polymerase and a particular promoter, encouraging the expression of the gene.

In prokaryotes, repressors and activators bind to regions called operators that are generally located near the promoter.

In eukaryotes, transcriptional regulation tends to involve combinatorial interactions between several transcription factors, which allow for a sophisticated response to multiple conditions in the environment. This permits spatial and temporal differences in gene expression. Eukaryotes also make use of enhancers, distant regions of DNA that can loop back to the promoter.

Examples:

- When a cell contains a surplus amount of the amino acid tryptophan, the acid binds to a specialized repressor protein (tryptophan repressor). The binding changes the structural conformity of the repressor such that it binds to the genes that help synthesize tryptophan, preventing their expression and thus suspending production. This is a form of negative feedback.

- In bacteria, the lac repressor protein blocks the synthesis of enzymes that digest lactose when there is no lactose to feed on. When lactose is present, it binds to the repressor, causing it to detach from the DNA strand.

The operon

An operon is a group of key nucleotide sequences including an operator, a common promoter, and one or more structural genes that are controlled as a unit to produce messenger RNA (mRNA). Operons occur primarily in prokaryotes and nematodes. They were first described by François Jacob and Jacques Monod in 1961.

An operon contains one or more structural genes which are transcribed into RNA. Upstream of the structural genes lies a promoter sequence which provides a site for RNA polymerase to bind and initiate transcription. Close to the promoter lies an operator sequence. The operon may also contain regulator genes such as a repressor gene which codes for a protein that binds to the operator and inhibits transcription.

An operator is a segment of DNA that regulates the activity of the structural genes of an operon it is linked to, by interacting with a specific repressor or activator. It is a regulatory sequence for shutting a gene down or turning it *on*.

Control of operon genes is a type of gene regulation that enables organisms to regulate the expression of various genes depending on environmental conditions. Operon regulation can be either negative or positive. Negative regulation involves the binding of a repressor to the operator to prevent transcription.

In negative inducible operons, a regulatory repressor protein is normally bound to the operator and it prevents the transcription of the genes on the operon. If an inducer molecule is present, it binds to repressor and changes its conformation so that it is unable to bind to the operator. This allows for the transcription of the genes on the operator.

In negative repressible operons, transcription of the genes on the operon normally takes place. Repressor proteins are produced by a regulator gene but they are unable to bind to the operator in their normal conformation. However certain molecules called corepressors can bind to the repressor protein and change its conformation so that it can bind to the operator. The activated repressor protein binds to the operator and prevents transcription.

Operons can also be positively controlled. With positive control, an activator protein stimulates transcription by binding to DNA (usually at a site other than the operator).

In positive inducible operons, activator proteins are normally unable to bind to the pertinent DNA. However, certain substrate molecules can bind to the activator proteins and change their conformations so that they can bind to the DNA and enable transcription to take place.

In positive repressible operons, the activator proteins are normally bound to the pertinent DNA segment. However, certain molecules can bind to the activator and prevent it from binding to DNA. This prevents transcription.

CHEMICAL KINETICS AND PROKARYOTIC GENE REGULATION

The reaction rate

The reaction rate for a reactant or product in a particular reaction is defined as the amount of the chemical that is formed or removed (in moles or mass units) per unit time per unit volume.

There are several factors that affect the rate of reaction:

- Temperature: Conducting a reaction at a higher temperature puts more energy into the system and increases the reaction rate.
- Concentration: As reactant concentration increases, the frequency of collision increases and so therefore does the frequency of collisions having sufficient energy to cause reaction.
- Pressure: The rate of gaseous reactions usually increases with an increase in pressure. Increase in pressure in fact is equivalent to an increase in concentration of the gas.
- Light: Light is a form of energy. It may affect the rate or even course of a reaction.
- Order: The order of the reaction has a major effect on its rate. The order of a reaction is found experimentally, and, for most basic reactions, is an integer value.
- A catalyst: The presence of a catalyst increases the reaction rate in both the forward and reverse reactions by providing an alternative pathway with a lower activation energy.
- The nature of the reactants: If a reaction involves the breaking and reforming of bonds (complex) compared to just the forming of bonds (simple) then it generally takes longer. The reactants position in the reactivity series also affects reaction rate.
- Surface Area: In reactions on surfaces, which take place during heterogeneous catalysis, the rate of reaction increases as the surface area does.

For a chemical reaction $nA + mB \underset{k_-}{\overset{k_+}{\rightleftharpoons}} C + D$, the forward and backward reaction rates are given by:

$$v_+ = k_+[A]^n[B]^m, \quad \text{and} \quad v_- = k_-[C][D].$$

For a reaction between liquids, gases, or solutes, $[X]$ stands the concentration of X. For a reaction taking place at a boundary it would denote something like moles of X per area. When gases are involved the rate law can also be expressed in pressure units.

The forward and backward rates determine the rate of change of the concentration of the different chemical species involved in the reaction. For instance:

$$\frac{d[A]}{dt} = -nv_+ + nv_-, \quad \frac{[dC]}{dt} = v_+ - v_-.$$

Constants k_+ and k_- are know as the reaction rate coefficients or rate constants, although they are not really constant; they include everything that affects the reaction rate outside concentration: mainly temperature but also ionic strength or light irradiation.

Each reaction rate coefficient k has a temperature dependency, which is usually given by the Arrhenius equation:

$$k = Ae^{-\frac{E_a}{RT}}$$

E_a is the activation energy and R is the Gas constant. Since at temperature T the molecules have energies given by a Boltzmann distribution, one can expect the number of collisions with energy greater than E_a to be proportional to $\exp(-E_a/RT)$. A is the pre-exponential factor or frequency factor. The values for A and E_a are dependent on the reaction.

If the chemical reaction is in equilibrium, the forward and backward reaction rates are equal and therefore:

$$K_D = \frac{k_-}{k_+} = \frac{[A]^n[B]^m}{[C][D]}.$$

This last equation is known as the mass action law, and K_D is the reaction dissociation constant.

Michaelis-Menten kinetics

Michaelis-Menten kinetics describe the rate of enzyme mediated reactions for many enzymes. It is valid only when the concentration of enzyme is much less than the concentration of substrate (i.e., when enzyme concentration is the limiting factor).

The enzymatic reaction is supposed to be irreversible, and the product does not rebind the enzyme:

$$E + S \underset{k_-}{\overset{k_+}{\rightleftharpoons}} E_S \overset{k_2}{\rightharpoonup} E + P.$$

Assume that the E_S-complex formation and dissociation reactions are much faster than the rest, and make the following quasi-steady state approximation accordingly: $d[ES]/dt = k_+[E][S] - k_-[ES] - k_2[E_S] = 0$,

$$[E_S] = \frac{k_1[E][S]}{k_{-1} + k_2}.$$

Define

$$K_m = \frac{k_- + k_2}{k_+}$$

From this it follows that:

$$[E_S] = \frac{[E][S]}{K_m}.$$

The rate (velocity) of the enzymatic reaction is

$$\frac{d[P]}{dt} = k_2[E_S].$$

On the other hand, the total concentration of enzyme is:

$$[E_+] = [E] + [E_S].$$

Hence:

$$[E] = [E_T] - [E_S],$$

and

$$[E_S] = \frac{([E_T] - [E_S])[S]}{K_m}.$$

Rearranging gives:

$$[E_S] = [E_T]\frac{1}{1+\frac{K_m}{[S]}}.$$

Finally:

$$\frac{d[P]}{dt} = k_2[E_T]\frac{[S]}{K_m + [S]} = V_{max}\frac{[S]}{K_m + [S]}.$$

This last equation is known as the Michaelis-Menten equation, from which the next remarks follow.

- E_T is the total or starting amount of enzyme. It is not practical to measure the amount of the enzyme substrate complex during the reaction, so the reaction must be written in terms of the total (starting) amount of enzyme, a known quantity.
- $d[P]/dt$ is the product rate of formation.
- $k_2[E_T] = V_{max}$ is the maximum velocity or maximum rate.
- Notice that if $[S]$ is large compared to K_m, $[S]/(K_m + [S])$ approaches 1. Therefore, the rate of product formation is equal to $k_2[E_T]$ in this case.
- When $[S]$ equals K_m, $[S]/(K_m + [S])$ equals 0.5. In this case, the rate of product formation is half of the maximum rate $(1/2V_{max})$.

Transcription initiation

Transcription initiation takes place in two sequential steps. First, the closed DNA-mRNAP complex forms:

$$D + P \rightleftharpoons D_c.$$

Later, the closed complex transforms into the open complex an the mRNAP starts transcribing the genes ahead:

$$D_c \rightharpoonup D_o.$$

Formation of the open complex is the rate limiting step of transcription initiation. Therefore, we can make a quasi-steady state assumption for the closed complex formation reaction and so, we obtain the following reaction equations:

$$[D]\,[P] = K_D\,[D_c], \quad [\dot{D_o}] = k_M\,[D_c],$$

where K_D is the dissociation constant for the closed complex formation reaction, while k_M is the reaction rate for the closed-to-open complex transformation reaction.

The total DNA concentration ($[D_{tot}]$) remains constant during the whole transcription process:

$$[D] + [D_c] = [D_{tot}].$$

Finally, after a little algebra, the transcription initiation rate comes out to be:

$$k_M [D_c] = k_M [D_{tot}] \frac{[P]/K_D}{1 + [P]/K_D}.$$

Transcription repression

Polymerases and repressors compete for the promoter, and this can be accounted for by the following reactions:

$$D + P \rightleftharpoons D_P, \quad \text{and} \quad D + R_{2T} \rightleftharpoons D_R,$$

where R_{2T} represents an active repressor molecule, and D_R is the operon-repressor complex. The equilibrium equations for the above reactions are:

$$[D][P] = K_P [D_P], \quad \text{and} \quad [D][R_{2T}] = K_R [D_R].$$

Consider again DNA conservation:

$$[D] + [D_P] + [D_R] = [D_{tot}].$$

Then, the transcription initiation rate is:

$$k_M [D_P] = k_M [D_{tot}] \frac{[P]/K_D}{1 + [P]/K_D + [R_{2T}]/K_R}.$$

Repressor activation

Assume that repressor molecules, R, become active when they are bound by a couple of T molecules in two sequential reactions:

$$R + T \rightleftharpoons R_T \quad \text{and} \quad R_T + T \rightleftharpoons R_{2T}.$$

The equilibrium equations for these reactions are:

$$[R][T] = \frac{K_T}{2} [R_T], \quad \text{and} \quad [R_T][T] = 2K_T [R_{2T}].$$

The total repressor concentration R_{tot} is conserved during the above reactions:

$$[R] + [R_T] + [R_{2T}] = [R_{tot}].$$

From the above equations, the concentration of active repressors is:

$$[R_{2T}] = [R_{tot}]\left(\frac{[T]}{[T]+K_T}\right)^2.$$

Finally, with this last result, the rate of transcription initiation (which equals the rate of mRNA production) becomes:

$$\left|\frac{d[M]}{dt}\right|_{\text{prod}} = k_M\,[D_{tot}]\,\frac{\dfrac{[P]}{K_D}}{1+\dfrac{[P]}{K_D}+\dfrac{[R_{tot}]}{K_R}\left(\dfrac{[T]}{[T]+K_T}\right)^2}.$$

THE LACTOSE OPERON

The *lac* operon is an operon required for the transport and metabolism of lactose in *Escherichia coli* and some other enteric bacteria. It consists of three adjacent structural genes, a promoter, a terminator, and an operator. The lac operon is regulated by several factors including the availability of glucose and of lactose. Gene regulation of the lac operon was the first genetic regulatory mechanism to be elucidated and is often used as the canonical example of prokaryotic gene regulation.

The three structural genes in the *lac* operon are: *lacZ*, *lacY*, and *lacA*. *lacZ* encodes β-galactosidase (LacZ), an intracellular enzyme that cleaves the disaccharide lactose into glucose and galactose. Gene *lacY* encodes β-galactoside permease (LacY), a membrane bound transport protein that pumps lactose into the cell. *lacA* encodes β-galactoside transacetylase (LacA), an enzyme that transfers an acetyl group from acetyl-CoA to β-galactosides. Only *lacZ* and *lacY* appear to be necessary for lactose catabolism.

Transcription of *lac* structural genes starts with the binding of the enzyme RNA polymerase (RNAP) to the *lac* promoter. From this position RNAP proceeds to transcribe all three genes (*lacZYA*) into mRNA.

The regulatory response to lactose requires an intracellular regulatory protein called the lactose repressor. The *lacI* gene encoding repressor lies nearby the lac operon and it is always expressed. If lactose is missing from the growth medium, the repressor binds very tightly to a short DNA sequence just downstream of the promoter near the beginning of *lacZ* called the *lac* operator. Repressor bound to the operator interferes with binding of RNAP to the promoter, and therefore mRNA encoding LacZ and LacY is only made at very low levels. When cells are grown in the presence of lactose, a lactose metabolite called allolactose binds to the repressor, causing a change in its shape. Thus altered, the repressor is unable to bind to the operator, allowing RNAP to transcribe the lac genes and thereby leading to high levels of the encoded proteins.

Gene *cya* encodes adenylate cyclase, which produces cyclic AMP (cAMP). Production of cAMP is inhibited by external glucose. A second gene, *crp*, encodes a protein called catabolite repressor protein (CRP), which can be bound by cAMP. The cAMP-CRP complex receives the name of catabolite activator protein (CAP). Protein CAP bind DNA close to the *lac* promoter and acts as an activator of transcription. In conclusion, the presence of external glucose inhibits expression of the lac operon genes.

This dual regulation causes the lactose metabolism enzymes to be made in small quantities in the presence of both glucose and lactose (sometimes called leaky expression) due to lactose inhibiting LacI binding to the operator, but at high cAMP concentrations and in the presence of lactose there are high levels of expression. Leaky expression is necessary in order to allow for metabolism of some lactose after the glucose source is expended, but before lac expression is fully activated.

In summary:

- When lactose is absent then there is no Lac enzyme production (the operator has LacI bound to it).
- When lactose is present but a preferred carbon source (like glucose) is also present then a small amount of enzyme is produced (LacI is not bound to the operator).
- When lactose is the favored carbon source (for example in the absence of glucose) CAP binds to the promoter and Lac enzyme production is maximized.

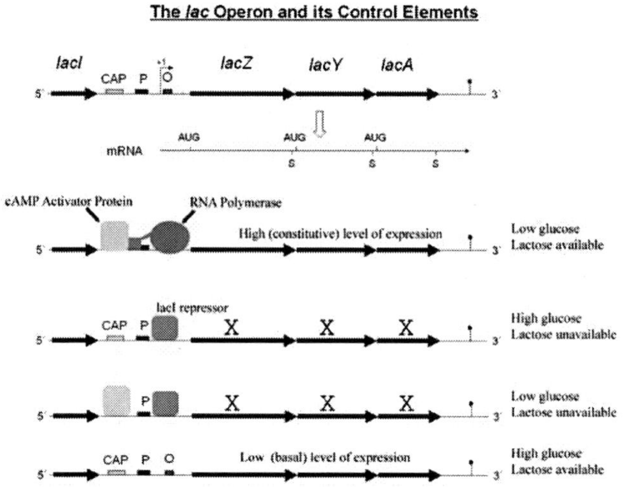

FIGURE 2. Regulation of the *lac* operon.

Lac repressor is a tetramer of identical subunits. Each subunit contains a helix-turn-helix (HTH) motif capable of binding to DNA. The operator site where repressor binds is a DNA sequence with inverted repeat symmetry. The two DNA half-sites of the operator together bind to two of the subunits of the tetrameric repressor. Although the other two subunits of repressor are not doing anything in this model, this property was not understood for many years.

Eventually it was discovered that two additional (minor) operators are involved in *lac* regulation. One (O3) lies in the end of the lacI gene and the other (O2) is about 400 bp downstream in the early part of lacZ. These two sites were not found in the early work because they have redundant functions and individual mutations do not affect repression very much. Single mutations to either O2 or O3 have only 2 to 3-fold effects. However, their importance is demonstrated by the fact that a double mutant defective in both O2 and O3 is dramatically de-repressed (by about 70-fold).

In the current model, repressor is bound simultaneously to both the main operator O1 and to either O2 or O3. The intervening DNA loops out from the complex. The redundant nature of the two minor operators suggests that it is not a specific looped complex that is important. One idea is that the system works through tethering. If bound repressor releases from O1 momentarily, binding to a minor operator keeps it in the vicinity, so that it may rebind quickly. This would increase the affinity of repressor for O1.

The repressor is an allosteric protein, i.e. it can assume either one of two slightly different shapes, which are in equilibrium with each other. In one form the repressor is capable of binding to the operator DNA, and in the other form it cannot bind to the operator. According to the classical model of induction, binding of the inducer, either allolactose or IPTG, to the repressor affects the distribution of repressor between the two shapes. Thus, repressor with inducer bound is stabilized in the non-DNA-binding conformation.

Mathematical model

Bellow, we present a mathematical model for the *lac* operon developed elsewhere. All of the model equations are tabulated in Table 1 and they are briefly explained below. The reader may find useful to refer to Figure 3, where the *lac* operon regulatory pathway is schematically represented.

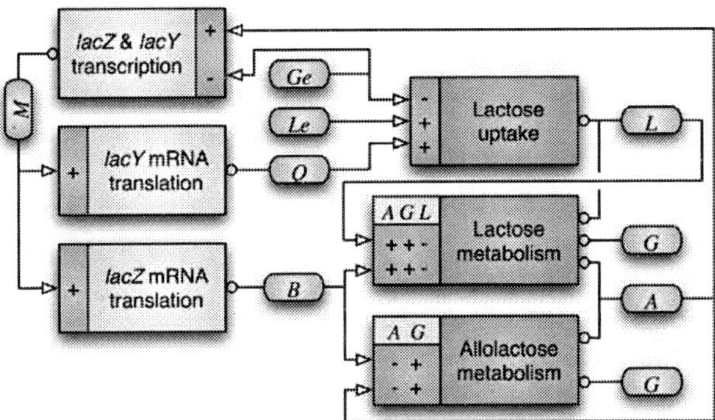

FIGURE 3. Schematic representation of the *lac* operon regulatory mechanisms. The meaning of the variables appearing in this figure is as follows: *Ge* and *Le* stand for external glucose and lactose concentrations; *M*, *Q*, and *B* denote mRNA, permease, and β-galactosidase, molecule concentrations, respectively; finally, *L*, *A*, and *G* correspond to the intracellular lactose, allolactose, and galactose molecule concentrations. All of the processes underlying the *lac* operon regulatory pathway are represented by rectangles. Inputs (outputs) are denoted with empty arrow heads (circles). Finally, plus and minus signs stand for the effect each input variable has on every output variable.

The model consists of three differential equations (1)-(3), that respectively account for the time evolution of mRNA (M), *lacZ* polypeptide (E), and internal lactose (L) concentrations. mRNA is produced via transcription of the *lac* operon genes, and its concentration decreases due to active degradation and dilution due to cell growth. γ_M

TABLE 1. Full set of equations for the lactose operon model. Differential equations (1)-(3) govern the time evolution of the intracellular concentration of mRNA (M), polypeptide (E), and lactose (L) molecules. Ge and Le respectively stand for the extracellular glucose and lactose concentrations. A, Q, and B represent the intracellular allolactose, permease and β-galactosidase molecule concentrations. The functions $\mathscr{P}_D$, $\mathscr{P}_R$, β_L, and β_G account for the negative effect of external glucose on the initiation rate of transcription (via catabolite repression), the probability that the lactose promoter is not repressed, the positive effect of external lactose on its uptake rate, and the negative effect of external glucose on lactose uptake (inducer exclusion), respectively. $2\phi_M\mathscr{M}$ is the rate of lactose molecules metabolized per β-galactosidase. Finally, p_c and ρ represent internal auxiliary variables.

$$\dot{M} = Dk_M \mathscr{P}_D(Ge)\mathscr{P}_R(A) - \gamma_M M \tag{1}$$

$$\dot{E} = k_E M - \gamma_E E \tag{2}$$

$$\dot{L} = k_L \beta_L(Le)\beta_G(Ge)Q - 2\phi_M \mathscr{M}(L)B - \gamma_L L \tag{3}$$

$$A = L \tag{4}$$

$$Q = E \tag{5}$$

$$B = E/4 \tag{6}$$

$$\mathscr{P}_D(Ge) = \frac{p_p\left(1 + p_c(Ge)\left(k_{pc} - 1\right)\right)}{1 + p_p p_c(Ge)\left(k_{pc} - 1\right)} \tag{7}$$

$$p_c(Ge) = \frac{K_G^{n_h}}{K_G^{n_h} + Ge^{n_h}} \tag{8}$$

$$\mathscr{P}_R(A) = \frac{1}{1 + \rho(A) + \dfrac{\xi_{123}\rho(A)}{(1 + \xi_2\rho(A))(1 + \xi_3\rho(A))}} \tag{9}$$

$$\rho(A) = \rho_{\max}\left(\frac{K_A}{K_A + A}\right)^4 \tag{10}$$

$$\beta_L(Le) = \frac{Le}{\kappa_L + Le} \tag{11}$$

$$\beta_G(Ge) = 1 - \phi_G \frac{Ge}{\kappa_G + Ge} \tag{12}$$

$$\mathscr{M}(L) = \frac{L}{\kappa_M + L} \tag{13}$$

represents the degradation plus dilution rate, the maximum transcription rate per promoter is k_M, and D is the average concentration of promoter copies per bacterium. The function $\mathscr{P}_R(A)$ [defined in Equation (9)] accounts for regulation of transcription initiation by active repressors. The fraction of active repressors is proportional to $\rho(A)$, c.f. Equation (10). Since repressors are inactivated by allolactose (A), ρ is a decreasing function of A. Furthermore the rate of transcription initiation decreases as the concentration of active repressors increases. Concomitantly, $\mathscr{P}_R$ is a decreasing function of ρ, and thus it is an increasing function of A. The function $\mathscr{P}_D(Ge)$ [Equation (9)] denotes the modulation of transcription initiation by external glucose, i.e. through catabolite repression. Production of cyclic AMP (cAMP) molecules is inhibited by extracellular glucose. cAMP further binds the so-called cAMP receptor protein (CRP) to form the CAP complex. Finally, CAP binds a specific site near the *lac* promoter and enhances transcription initiation. The probability of finding a CAP molecule bound to its corresponding site is given by p_c [Equation (8)] and is a decreasing function of Ge. Moreover, $\mathscr{P}_D$ is an increasing function p_c and, therefore, a decreasing function of Ge.

The translation initiation rate of *lacZ* transcripts is k_E, while γ_E is the dilution and degradation rate of *lacZ* polypeptides. Given that the corresponding parameters for *lacY* transcripts and polypeptides attain similar values, that β-galactosidase is a homotetramer made up of four identical *lacZ* polypeptides, and that permease is a *lacY* monomer, it follows that $Q = E$ and $B = E/4$ [Equations (5) and (6)].

Lactose is transported into the bacterium by a catalytic process in which permease proteins play a central role. Thus, the lactose influx rate is assumed to be $k_L \beta_L(Le) Q$, with the function $\beta_L(Le)$ given by Equation (11). Extracellular glucose negatively affects lactose uptake, so-called inducer exclusion, and this is accounted for by $\beta_G(Ge)$ [Equation (12)] which is a decreasing function of Ge. Once inside the cell, lactose is metabolized by β-galactosidase. Approximately half of the lactose molecules are transformed into allolactose, while the rest enter the catalytic pathway that produces galactose. In our model, $\phi_M \mathscr{M}(L)$, with $\mathscr{M}$ defined in Equation (13), denotes the lactose-to-allolactose metabolism rate, which equals the allolactose production rate. Allolactose is further metabolized into galactose by β-galactosidase. From the assumptions that the corresponding metabolism parameters are similar to those of lactose, and that the allolactose production rate is much higher than its degradation plus dilution rate, it follows that $A \approx L$ [Equation (4)].

All of the parameters in the model depicted above are tabulated in Table 2.

TABLE 2. Value of all of the parameters in the equations of Table 1.

μ	$\approx$	$0.02\,\text{min}^{-1}$	K_G	$\approx$	$2.6\,\mu\text{M}$
D	$\approx$	$2\,\text{mpb}$	n_h	$\approx$	1.3
k_M	$\approx$	$0.18\,\text{min}^{-1}$	ξ_2	$\approx$	0.05
k_E	$\approx$	$18.8\,\text{min}^{-1}$	ξ_3	$\approx$	0.01
k_L	$\approx$	$6.0 \times 10^4\,\text{min}^{-1}$	ξ_{123}	$\approx$	163
γ_M	$\approx$	$0.48\,\text{min}^{-1}$	$\rho_{\max}$	$\approx$	1.3
γ_E	$\approx$	$0.03\,\text{min}^{-1}$	K_A	$\approx$	$2.92 \times 10^6\,\text{mpb}$
γ_L	$\approx$	$0.02\,\text{min}^{-1}$	κ_L	$\approx$	$680\,\mu\text{M}$
k_{pc}	$\approx$	30	ϕ_G	$\approx$	0.35
p_p	$\approx$	0.127	κ_G	$\approx$	$1.0\,\mu\text{M}$
ϕ_M	$\approx$	$\in [0, 1.5 \times 10^4]\,\text{min}^{-1}$	κ_M	$\approx$	$7.0 \times 10^5\,\text{mpb}$

Results

The steady-state values for M, E, and L can be calculated by setting $\dot{M} = \dot{Y} = \dot{T} = 0$ in Equations (1)-(3), and solving for M, E, and L. It is straightforward to show after a little algebra that, given Ge and Le, the steady-state values L^* of lactose are the roots of $S(Ge, Le, L^*)$, with this function defined as:

$$S(Ge, Le, L^*) = \frac{Dk_M k_E k_L}{\gamma_M \gamma_E \gamma_L} \left(\beta_L(L_e)\beta_G(G_e) - \frac{\phi_M}{2k_L} \frac{L^*}{L^* + \kappa_L} \right) \mathscr{P}_D(Ge)\mathscr{P}_R(L^*) - L^*. \quad (1)$$

The corresponding steady-state values for M and E can then be calculated as:

$$M^* = \frac{Dk_M}{\gamma_M} \mathscr{P}_D(G_e)\mathscr{P}_R(L^*), \quad (2)$$

$$E^* = \frac{Dk_M k_E}{\gamma_M \gamma_E} \mathscr{P}_D(G_e)\mathscr{P}_R(L^*), \quad (3)$$

It is easy to numerically show that, given the parameter values tabulated in Table 2 and $Ge > 0$, the function $S(L^*)$ may have up to three roots, and therefore that the model has up to three steady states. For very low Le values there is a single stable steady state corresponding to the uninduced state. As Le increases, two more steady states appear via a saddle-node bifurcation. One of these new fixed points is stable, corresponding to the induced state, while the other is a saddle node. With further increases in Le the saddle node and the original stable steady state approach until they eventually collide and annihilate via another saddle-node bifurcation. Afterwards, only the induced stable steady state survives.

Ozbudak et al. [1] carried out a series of clever experiments regarding the bistable behavior of the *lac* operon in *E. coli*. In these experiments, the Le values at which the bifurcations take place were experimentally determined for various extracellular glucose concentrations, using a non-metabolizable inducer (TMG). In the present model, the usage of a non-metabolizable inducer can be modelled by setting $\phi_M = 0$. (Recall that this parameter stands for the maximum rate of lactose-to-allolactose and of lactose-to-galactose metabolism.

Note that given Ge, the values of Le and L^* at the bifurcation points can be found by simultaneously solving the equations: $S(Ge, Le, L^*) = 0$ and $dS(Ge, Le, L^*)/dL^* = 0$. By using this procedure, we numerically calculated the model Le bifurcation values for several Ge concentrations when a non-metabolizable inducer is employed. The results are shown in the bifurcation diagram of Figure 4A. There, they are also compared with the experimental results of Ozbudak et al. [1]. In our opinion, the agreement between experiment and model is good enough to validate the forthcoming results.

Ozbudak et al. further assert that, when their experiments were repeated with the natural inducer, lactose, they were unable identify any bistable behavior. This result, and other studies reporting similar results, have sparked a very interesting debate over the question: "Is bistability a property of the *lac* operon in natural conditions, or it is only an artifact introduced by the use of non-metabolizable inducers?"

To investigate the effects of lactose metabolism we recalculated the bifurcation diagram of Figure 4A with different values of parameter ϕ_M. From [2], we can estimate that $\phi_M \approx 3.6 \times 10^3 \text{min}^{-1}$. However, we used even larger values to better understand how it affects the *lac* operon bistable behavior. Bifurcation diagrams in the *Le vs. Ge* parameter space, respectively calculated with $\phi_M = 0, 3.6\times10^3, 1.5\times10^4$, and $4.0\times10^4\,\text{min}^{-1}$, are shown in Figures 4C to 4F. From these results we conclude that the main effects of lactose metabolism on the *Le vs. Ge* bifurcation diagrams are:

- The border between the uninduced monostable and the bistable regions is located at higher *Le* values as ϕ_M increases.
- At high *Ge* concentrations, the height of the bistable region increases together with ϕ_M.
- Conversely, the bistability region shrinks at low concentrations of external lactose at high ϕ_M values.
- For very high values of ϕ_M (larger than $2.5\times10^4\text{min}^{-1}$), bistability is not present at very low *Ge* concentrations, and only appears (via a cusp catastrophe) after *Ge* exceeds a given threshold.

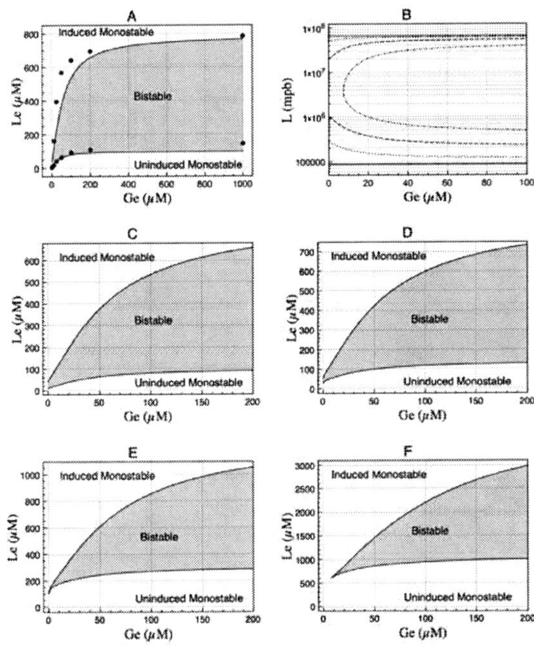

FIGURE 4. A. Bifurcation diagram in the *Le vs. Ge* parameter space, calculated with $\phi_M = 0$. As discussed in the main text, this choice of the parameter ϕ_M simulates induction of the *lac* operon with a non-metabolizable inducer, like TMG. The dots correspond to the experimental results of Ozbudak *et al.* (2004). B. Plots of the critical induced, L_i^* (upper curves), and uninduced, L_u^* (lower curves), lactose steady-state values *vs.* G_e, for different values of the lactose metabolism rate. The solid, dotted, dashed, and dot-dashed lines respectively correspond to $\phi_M = 0\,\text{min}^{-1}$, $\phi_M = 3.6\times10^3\,\text{min}^{-1}$, $\phi_M = 1.5\times10^4\,\text{min}^{-1}$, and $\phi_M = 4.0\times10^4\,\text{min}^{-1}$. Plots C, D, E, and F show bifurcation diagrams calculated with the following lactose metabolism rates: $\phi_M = 0\,\text{min}^{-1}$, $\phi_M = 3.6\times10^3\,\text{min}^{-1}\,\text{min}^{-1}$, $\phi_M = 1.5\times10^4\,\text{min}^{-1}$, $\phi_M = 4.0\times10^4\,\text{min}^{-1}\,\text{min}^{-1}$, respectively.

For a given Ge, let L_u^* denote the L uninduced steady-state value when Le is at the lower border of the bistable region, in the Le *vs.* Ge parameter space. Analogously, let L_i^* denote the L induced stable steady-state value when Le is at the upper border of the bistable region. In what follows, we shall call L_u^* and L_i^*, respectively, the critical uninduced and induced steady-state values of L. From its definition, L_u^* (L_i^*) is the minimal (maximal) value attained by L at the uninduced (induced) state. Therefore, $L_i^* - L_u^*$ is representative of the size of the bistable region in terms of the internal variable L.

In Figure 4B, plots of L_u^* and L_i^* *vs.* Ge are shown for different values of ϕ_M. Notice that both L_u^* and L_i^* are constant when $\phi_M = 0\,\text{min}^{-1}$. This means that when lactose metabolism is not present, the borders of the bistable region in terms of the internal variable L are independent of Ge, in contrast to what happens in terms of the external variable Le, see Figure 4A. This observation allows a better understanding of the effects of lactose metabolism on the model bistable behavior:

- As ϕ_M increases, the size of the bistable region decreases because L_u^* increases and L_i^* decreases.
- When $\phi_M = 0\,\text{min}^{-1}$, L_i^* is three orders of magnitude larger than L_u^*. However, when $\phi_M = 4.0 \times 10^4\,\text{min}^{-1}$ this difference is reduced to two orders of magnitude for large values of Ge.
- The size of the bistable region is shorter in the limit of low external glucose concentrations, and this effect is enhanced with further increments of ϕ_M Indeed, as noticed before, bistability disappears for very low Ge values when $\phi_M > 2.5 \times 10^4\text{min}^{-1}$.

Concluding remarks

We have developed a mathematical model of the *lac* operon that, though not accurate in all detail, nonetheless captures the essential system behavior with respect to its bistable characteristics. The model was validated by simulating the experimental results of Ozbudak et al. [1], who report the bifurcation diagram for the *lac* operon in *E. coli*, when induced with a non-metabolizable inducer.

Since Ozbudak et al. and others report that bistability was not observed when the natural inducer, lactose, was employed, we investigated the influence of lactose metabolism on the system bistable behavior by varying the parameter ϕ_M (the maximum lactose-to-allolactose metabolism rate). From our results, bistability does not disappear because of lactose metabolism, although it is highly modified.

In terms of external lactose, higher Le values are required to enter the bistable region from the monostable uninduced one, as well as to go from to bistable to the monostable induced regions, as ϕ_M increases. If $\phi_M > 2.5 \times 10^4\,\text{min}^{-1}$, bistability is not present in the limit of very low Ge, and it only appears after Ge surpasses a given threshold, via a cusp catastrophe. We suspect that this last situation is the most likely in *wild type* bacteria since, on the one hand, it becomes practically impossible for the operon to switch into the induced state when there is an abundance of glucose but, on the other hand, there is no resistance, caused by bistability, to start activating the *lac* genes, and so employing the energy provided by lactose, when glucose is absent.

Our results also suggest an explanation for the incapability of Ozbudak et al. [1] to observe bistability when lactose is used as an inducer. Namely, it may be that the value of ϕ_M, and so the upper bound of the bistable region, are so high that the external lactose concentration required to jump to the induced state are extremely high and almost impossible to reach.

From a global point of view, the development and persistence of a physiological trait like bistability should be explained by the evolutionary advantages accompanying it. Based on the previous section results, we analyze what possible evolutionary advantages might the *lac* operon in *E. coli* have achieved by having bistability in its regulatory pathway.

E. coli and other bacteria can feed on both lactose and glucose. However, when they grow in a medium rich in both sugars, glucose is finished off before lactose starts being consumed. This phenomenon is known as diauxic growth [3, 4], and represents an optimal thermodynamic solution, given that glucose is a cheaper energy source. In order to take advantage of lactose, the bacteria need to expend some amount of energy in producing the enzymes needed to transport and metabolize it [5].

From energetic considerations, it is important to maintain the lac operon uninduced under conditions of moderate to high external glucose concentrations, even if external lactose is abundant. From Figure 4F, if $\phi_M = 4.0 \times 10^4\,\text{min}^{-1}$, the system remains in the monostable uninduced region for *Le* values as high as $1000\,\mu$M, whenever $Ge > 50\,\mu$M. That is, it bistability in this case makes the induction of the *lac* operon genes practically impossible, as long as $Ge > 50\,\mu$M.

When external glucose is finishing off and there still is a large amount of external lactose, the lac operon must be rapidly induced to make use of lactose; it truly is an expensive energy source, but it is the only one and is abundant. In other words, when *Le* is high, the lac operon expression level should abruptly change from the off to the on state, and this change should be driven by a small variation on the external glucose concentration. From Figure 4F, we can see that bistability can also explain this behavior: if $Le = 1000\,\mu$M, the system goes from the uninduced to the induced monostable regions when *Ge* decreases from 50 to $10\,\mu$M.

According to Kitano [6] robustness, the ability of a system to maintain its optimal functionality across a wide range of operational conditions, is a key to understanding the evolutionary design of virtually all living systems. Robustness, can manifest in two different ways: either the system returns to its current attractor (which ensures an optimal performance), or moves to a new attractor that maintains the system's functions. A transition to a new attractor has to be made robustly, in response to stimuli, so that the system behaves consistently in the face of perturbations. As discussed above, bistability in the *lac* operon fulfills all these characteristics. When the level of external glucose is moderate or high, it keeps the *lac* operon uninduced and allows *E. coli* to make use of the cheaper energy source. However, when *Ge* decreases to very low levels a robust and rapid transition to the induced stated is facilitated, and the bacteria can keep growing by feeding on lactose. In conclusion, the above considerations suggest, to the best of our understanding, that bistability in the *lac* operon is an efficient way to guaranty a robust dynamic performance in the *lac* operon of *E. coli*.

ACKNOWLEDGMENTS

The author thanks the organizers of Cinvestav's Sixth Advanced Summer School in Physics for the invitation to participate as a lecturer. This work was supported by CONACyT (MEXICO), COFAA-IPN (MEXICO), and EDI-IPN (MEXICO)

REFERENCES

1. E. M. Ozbudak, M. Thattai, H. N. Lim, B. I. Shraiman, and A. van Oudenaarden, *Nature* **427**, 737–740 (2004).
2. M. Martínez-Bilbao, R. E. Holdsworth, R. A. Edwards, and R. E. Huber, *J. Biol. Chem.* **266**, 4979–4986 (1991).
3. S. Roseman, and N. D. Meadow, *J. Biol. Chem.* **265**, 2993–2996 (1990).
4. A. Narang, *Biotechnol. Bioeng.* **59**, 116–121 (1998).
5. E. Dekel, and U. Alon, *Nature* **436**, 588–592 (2005).
6. H. Kitano, *Nat. Rev. Genet.* **5**, 826–837 (2004).

Application of FRET Technology to the *In Vivo* Evaluation of Therapeutic Nucleic Acids (ANTs)

María Luisa Benítez-Hess and Luis Marat Alvarez-Salas

Laboratorio de Terapia Génica, Departamento de Genética y Biología Molecular, Cinvestav, México D.F. 07360, Mexico

Abstract. Developing applications for therapeutic nucleic acids (TNAs) (i.e. ribozymes, antisense oligodeoxynucleotides (AS-ODNs), siRNA and aptamers) requires a reporter system designed to rapidly evaluate their in vivo effect. To this end we designed a reporter system based on the fluorescence resonance energy transfer (FRET) engineered to release the FRET effect produced by two green fluorescent protein (GFP) variants linked by a TNA target site. Because the FRET effect occurs instantaneously when two fluorophores are very close to each other (>100nm) stimulating emission of the acceptor fluorophore by the excitation of the donor fluorophore it has been widely use to reveal interactions between molecules. The present system (FRET2) correlates the FRET effect with the in vivo activity of distinct types of TNAs based on a model consisting of RNA from human papillomavirus type 16 (HPV-16) previously shown accessible to TNAs. HPV-16 is the most common papillomavirus associated with cervical cancer, the leading cause of death by cancer in México. The FRET2 system was first tested in vitro and then used in bacteria in which transcription is linked to translation allowing controlled expression and rapid evaluation of the FRET2 protein. To assure accessibility of the target mRNA to TNAs, the FRET2 mRNA was probed by RNaseH assays prior FRET testing. The fluorescence features of the FRET2 system was tested with different FRET-producing GFP donor-acceptor pairs leading to selection of green (donor) and yellow (acceptor) variants of GFP as the most efficient. Modifications in aminoacid composition and linker length of the target sequence did not affect FRET efficiency. In vivo AS-ODN-mediated destruction of the chimerical FRET2 reporter mRNA resulted in the recovery of GFP fluorescent spectrum in a concentration and time dependent manner. Reported anti-HPV ribozymes were also tested with similar results. Therefore, we conclude that the FRET effect can be a useful tool in the development of TNAs.

INTRODUCTION

The green fluorescent protein (GFP) has been widely used as a highly sensitive reporter to monitor gene expression and transfer in cells [1]. When differently colored mutants of GFP, such as EGFP (enhanced green fluorescence protein), ECFP (enhanced cyan fluorescence protein) or YFP (yellow fluorescent protein), are covalently linked together within a few nanometers distance, fluorescence resonance energy transfer (FRET) can be detected [2-4]. FRET is a nonradiative exchange of energy between two fluorescent molecules, one functioning as a donor of energy to an acceptor, which is excited and emits at a longer wavelength [5]. Importantly, in FRET, the energy transfer is highly dependent on the distance and orientation between the donor and the acceptor molecules. FRET is characterized by a reduction of

CP885, *Advanced Summer School in Physics 2006, Frontiers in Contemporary Physics—EAV06*, edited by O. Miranda, M. Carbajal, L. M. Montaño, O. Rosas-Ortiz, and S. A. Tomás Velázquez

fluorescence intensity of the donor fluorophore and re-emission of fluorescence at acceptor fluorophore wavelengths. Although spectral variants of GFP should be expected to produce FRET and therefore suited for studies of protein-protein interactions, the unfavorable location of the fluorophore 15 Å deep inside the GFP molecule has especially impaired this application. However, FRET signaling between spectral variants of GFP can be increased by stabilizing dimer formation and especially by favoring heterodimer formation [6].

The function of gene products can be altered at many levels, including the mutation of gene sequence and the change in steady state levels of mRNA and/or protein by various mechanisms. In the last decade a novel therapeutic approach focused in gene-based information has developed into a promising tool to fight human disease. Therapeuthic nucleic acids (TNAs) refer to the application of DNA and RNA molecules to disrupt the expression or function of disease related genes. Approaches to TNA technology include: 1) blocking of gene transcription by triplex-forming oligodeoxyribonucleotides (TFOs); 2) mRNA translation inhibition by AS-ODNs, antisense RNA (AS-RNA) and catalytic nucleic acids (ribozymes and DNAzymes); 3) translational inhibition by small interfering RNAs (siRNA) and inhibition of protein function by RNA aptamers (fig. 1) [7].

Recent reports suggest that the efficacy of antisense ribozymes, and ODNs exhibits serious dependence on the structure of the target site, thus limiting their convenient use [8-10]. Consequently, considerable efforts have been devoted to prediction of target accessibility [11]. Although computer-assisted prediction of accessibility has become an important tool in antisense design, most algorithms are based on the structural constraints of the target not fully considering the variables present inside the cell. Therefore, extensive trial and error testing is still required to evaluate antisense TNAs *in vivo* performance. This process can be costly and time-consuming, making necessary the use of reporter systems.

Most GFP-based FRET reporter systems focused either on the separation of the two fluorophores by proteolytic activity or conformational changes of the covalent linkage after interaction with co-factors or ligands [12,13]. Although FRET between GFP variants can be difficult to attain, it has been shown that disruption of the covalent bond between two linked fluorophores, such as GFP and YFP, effectively curtails the FRET effect. However, the use of FRET to report on the synthesis of proteins has been largely overlooked in favor of reporter proteins in simple expression systems.

We developed a bacterial FRET reporter system based on GFP variants to monitor the efficiency of antisense TNAs in bacteria. The system consists of two FRET-producing spectral variants of GFP linked together with a target site for antisense TNAs. Interaction of the chimeric mRNA with antisense will cleave the linkage, allowing production of only one GFP variant eliminating the FRET effect. By measuring the residual FRET fluorescence it was possible to estimate the efficiency of the antisense used. Because the FRET effect is relatively independent of concentration, the system can measure subtle differences between antisense molecules regardless of GFP stability.

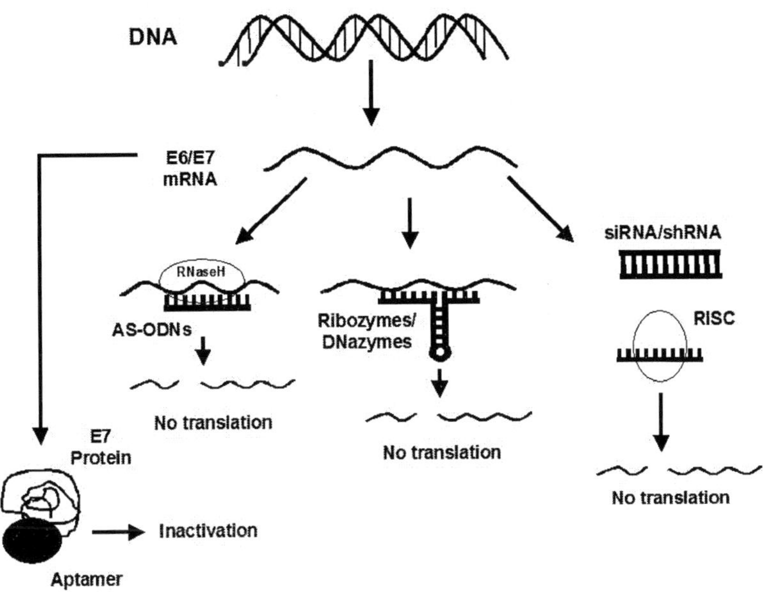

FIGURE 1. Applications of therapeutic nucleic acids (TNAs) to gene silencing.

Application of FRET to TNA Research

To determine the *in vivo* performance of antisense ODNs we designed a bacterial reporter system based on the FRET effect produced by EGFP and several GFP variants. The RNA target sequence (HPV-16 nt 410-445) was placed between the donor and the acceptor. Such configuration will result in the antisense-dependent disruption of the chimeric mRNA measurable by the loss of the FRET effect. As a first step, plasmid constructs were made using the pairs EGFP-EBFP or EGFP-ECFP, with GFP as the acceptor and BFP or CFP as the donors. Additionally, the pair EGFP donor and EYFP acceptor was also produced (Figure 2).

Fluorophore combinations were based on the on the spectral overlap between mutants of GFP. Thus, GFP and YFP emission and excitation spectra overlap considerably better than BFP and GFP.

Because *E. coli* transcription is essentially coupled to translation, we hypothesized that fast interaction of antisense ODNs with the HPV-16 target linkage would interrupt the production of full-length chimeric mRNA.

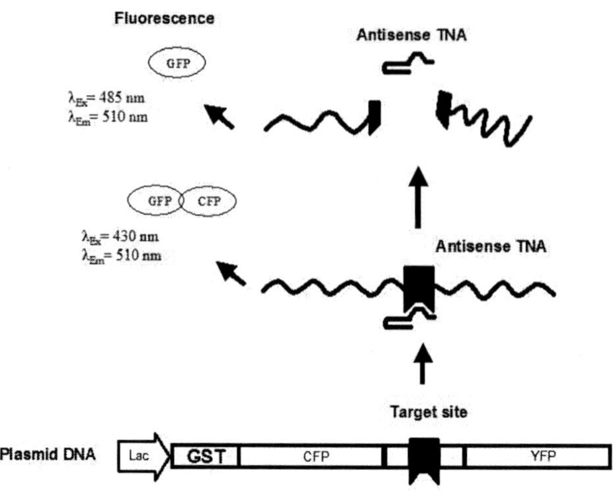

FIGURE 2. Model of FRET application to report TNA *in vivo* activity.

Thus, in the presence of effective antisense ODNs the architecture of FRET2 plasmids will only translate the first fluorophore, eliminating or reducing the average FRET effect in a dose-dependent manner. Efficiency of the antisense hybridization can then be measured by the ratio of fluorescence produced by the donor alone over the fluorescence produced by FRET. Fluorescence of all FRET constructs was quenched on CFP, GFP and YFP excitation wavelengths suggesting that the presence of the two fluorophores resulted in energy transfer. Interaction of the donor and acceptor in the FRET systems was further confirmed using FACS analysis, where the fluorescence intensity of the population of interacting molecules shifted to the middle of that produced by the donor and acceptor alone.

The chimeric mRNA of the FRET plasmid was initially analyzed by M-fold modeling [14]. Secondary structure prediction showed that target sequence for AS-ODNs have a loop structure and a small stem-loop. *In vitro* transcribed FRET reporter mRNA was treated with RNaseH in the presence of AS-ODNs, and the digested products were *in vitro* translated with rabbit reticulocyte lysates. Resulting protein products were analyzed through SDS-PAGE and fluorography. Use of AS-ODN treatment resulted in the translation of the GFP portion of the FRET reporter mRNA. Untreated RNA and AS-HPV ODN produced a protein product of similar size, suggesting that cleavage by RNaseH interrupted FRET reporter mRNA thus impeding translation.

To test the *in vivo* behavior of the FRET system, BL21(DE3) bacteria expressing the FRET reporter were treated with 0.1 or 1μM AS-ODNs, followed by IPTG induction and fluorometric analyses. Bacteria treated with AntiE6 AS-ODN decreased FRET efficiency in a dose-dependent manner. However, bacteria treated with M7 AS-ODN had no significant decrease in FRET efficiency. AS-ODN-treated bacteria were lysed and the translated products were further analyzed by SDS-PAGE. Use of AS-ODN treatments resulted only in the translation of the GFP portion of the FRET reporter mRNA. As with the *in vitro* experiments, treatment with M7 produced a protein product of similar size to the untreated control, suggesting that hybridization of AntiE6 induced cleavage by intracellular RNaseH interrupting translation. Additional fluorographic analysis of the purified GST-protein products showed that treatment with AntiE6 at doses of 0.1 and 1μM produced proteins with similar fluorographic profile to that of GFP. In contrast, treatment with M7 resulted in a protein product identical to the untreated control. Therefore, interruption on the FRET2 mRNA reestablished GFP fluorescence further confirming the presence of FRET effect on the system [15].

ACKNOWLEDGMENTS

This work was partially supported by CONACyT grant 45715-Z.

REFERENCES

1. Misteli, T. and D. L. Spector, *Nat.Biotechnol.* **15**, 961-964 (1997).
2. Heim, R. and R. Y. Tsien, *Curr.Biol.* **6**, 178-182 (1996).
3. Miyawaki, A., J. Llopis, R. Heim, J. M. McCaffery, J. A. Adams, M. Ikura, and R. Y. Tsien, *Nature* **388**, 882-887 (1997).
4. Harpur, A. G., F. S. Wouters, and P. I. Bastiaens, *Nat.Biotechnol.* **19**, 167-169 (2001).
5. dos, R. C. and P. D. Moens, *J.Struct.Biol.* **115**, 175-185 (1995).
6. Jensen, K. K., L. Martini, and T. W. Schwartz, *Biochemistry* **40**, 938- 945 (2001).
7. DiPaolo, J. A. and L. M. Alvarez-Salas, *Expert.Opin.Biol.Ther.* **4**, 1251-1264 (2004).
8. Vickers, T. A., J. R. Wyatt, and S. M. Freier, *Nucleic.Acids Res.* **28**, 1340-1347 (2000).
9. Holen, T., M. Amarzguioui, M. T. Wiiger, E. Babaie, and H. Prydz, *Nucleic.Acids Res.* **30**, 1757-1766 (2002).
10. Miyagishi, M., M. Hayashi, and K. Taira, *Antisense.Nucleic.Acid.Drug Dev.* **13**, 1-7 (2003).
11. Mathews, D. H. and D. H. Turner, *J.Mol.Biol.* **317** , 191-203 (2002).
12. Xu, X., A. L. Gerard , B. C. Huang, D. C. Anderson, D. G. Payan, and Y. Luo, *Nucleic.Acids.Res.* **26**, 2034-2035 (1998).
13. Truong, K., A. Sawano, H. Mizuno, H. Hama, K. I. Tong, T. K. Mal, A. Miyawaki, and M. Ikura, *Nat.Struct.Biol.* **8**, 1069-1073 (2001).
14. Zuker, M. and A. B. Jacobson, *RNA* **4**, 669-679 (1998).
15. Benitez-Hess, M. L., J. A. DiPaolo, and L. M. Alvarez-Salas, *Luminescence.* **19**, 85-93 (2004).

New Radiography Techniques

Gerardo Herrera[*], Claudia C. Díaz[*], Rosa E. Sanmiguel[†]

[*] *Departamento de Física, Centro de Investigación y de Estudios Avanzados del IPN, A. P. 14-740, 07000 México, D. F., Mexico.*
[†] *Centro de Investigación y de Estudios Avanzados del IPN campus Monterrey Cerro de las Mitras 2565, Col. Obispado 64060 Monterrey, Nuevo Leon, Mexico*

Abstract. Improving the quality of radiographic images involves several aspects: the X ray source, the device used to obtain the image, the optical array in the process and the image manipulation. We give a brief and general view of these different aspects and new techniques for radiography.

X RAY SOURCES

There are different sources of X rays. Radioactive sources like Ir-192 that produces gamma rays with an energy of 0.137 and 0.651 MeV, or Cs-137 with emission of 0.66 MeV, and the well known Fe-55 which gives X rays of 5.9 keV, are examples of X ray sources used very often in research and various applications. There are also accelerators of charged particles that generate synchrotron radiation. This kind of radiation is very useful for medical purposes and research. However the most widely sources of X rays used are the traditional X ray tubes.

Since 1895, when Wilhelm Röentgen discovered X rays produced by electrons hitting a target, technological developments have improved the quality of the radiation along the years. One of them is the introduction, in 1912 by W.D. Coolidge, of a separate filament where a current could be induced independently of the target. The anode was then cooled with water. This new array made possible to change current and voltage separately improving thereby the amount of radiation produced.
In conventional X ray tubes an electron cloud is produced around a filament and afterwards is accelerated towards a metallic target by means of an electric potential, whose values usually go from 10 up to 300 kV, depending on the type of the source. The material used as target determines the characteristic energy spectrum of the source: the intensity, i.e. the number of photons, is a function of the energy they have; this function is composed by a continuous spectrum which runs up to the maximum voltage plus the fluorescence lines, originated by the emission of photons as a result of de-excitation process of atomic electrons. Such a X ray source is characterized by parameters like the focal spot size, maximum operation voltage and power, energy spectrum, brilliance and beam quality. In particular coherence, brilliance and focal spot are the important parameters for applications of X ray tubes in medicine.

CP885, *Advanced Summer School in Physics 2006, Frontiers in Contemporary Physics—EAV06,* edited by O. Miranda, M. Carbajal, L. M. Montaño, O. Rosas-Ortiz, and S. A. Tomás Velázquez

Coherence

Figure 1 shows the concept of coherence. Two waves with slightly different wavelengths are in phase at some point. They will be out of phase after a distance L_L, which is called longitudinal coherence length. They will be in phase again at a $2L_L$ distance. By looking at the geometry of the waves one obtains the expression given in Fig. 1(a) for the longitudinal coherence length [1].

The transverse coherence length is illustrated in Fig. 1(b). Two waves travel in slightly different directions. The front of the waves coincides in one point. At some distance, called transverse coherence length L_T, the wave fronts will be out of phase. From the geometry in Fig.1(b) one can get the expression for the transverse longitudinal coherence.

With this in mind, the micro-source X ray tube in our laboratory uses a Cu anode which gives 8 KeV (K_α energy); it corresponds to a wavelength of 1.55 angstroms. Tuned to a focal spot size of 10 μm, and measuring at 1 m distance, the transverse coherence would be L_T = 7.75 μm. It gives an idea of the scales reachable with radiation produced in these conditions. Assuming a perfect crystal with $\Delta\lambda/\lambda \approx 10^{-5}$, L_L is approximately 5 μm for X rays with $\lambda \approx 1$ angstrom.

Brilliance

Brilliance is defined as:

$$brillance = \frac{photons\,/\,seg}{mrad^2\ mm^2\ (source\ area)\,(0.1\%\ bandwidth)}\,. \tag{1}$$

It is a measure of the amount of photons available to produce an image. This quantity depends on the photon energy. The brilliance in synchrotron radiation sources is several orders of magnitude higher than the one obtained with an X ray tube. However it is possible to improve the brilliance of X rays tubes with the help of optical arrays.

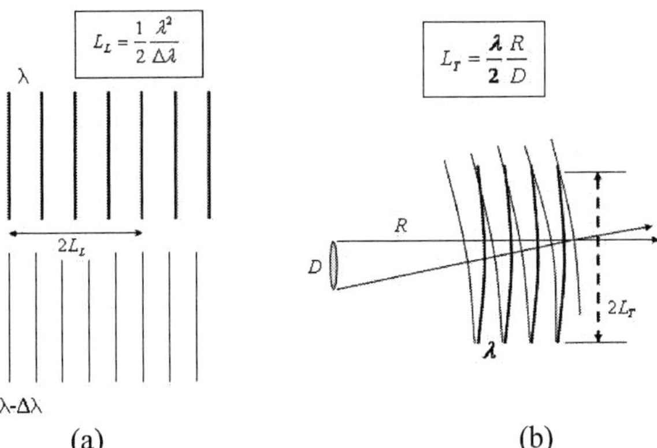

FIGURE 1. (a) Definition of longitudinal coherence length. (b) Transverse coherence length definition.

Focal Spot

The nominal sizes of the focal spots go from 0.1 to 0.3 mm for mammography and from 0.6 to 1.2 mm for radiography in general. By reducing the size of the focal spot the penumbra is reduced as can be seen in Fig. 2(a). The actual size of the focal spot can be determined by several techniques. The more traditional way of measuring the focal spot is making a pinhole image on a piece of film as shown in Fig. 2(b). When the pinhole is midway between the X ray source and the film, the image will be the same size as the source.

Conventional units have a focal spot larger than 0.5 mm. Units with a focal spot ranging from 50 to 500 microns are considered mini-focus, while those with a focal spot smaller that 50 microns are called micro-focus. The micro-source X ray tube in our laboratory running with a cooper target at 30 kV and 6 μA would have focal spot of 8 μm.

DETECTORS

Detectors are based on ionization or excitation of atoms in its sensitive material. Films have been used along the years as detectors in radiography. It is possible to replace the film with a fluorescent screen system. A number of systems like image plates have been developed in this line. Another approach is to replace the films with solid state detectors. These devices work on real time and offer immediate access to digital information and powerful computational tools. By real time imaging we mean the instantaneous conversion of two dimensional X ray intensity distributions into images.

There are many different ways of doing this. X ray photons can give all of its energy to electrons in the semiconductor; such process is called photoelectric absorption.

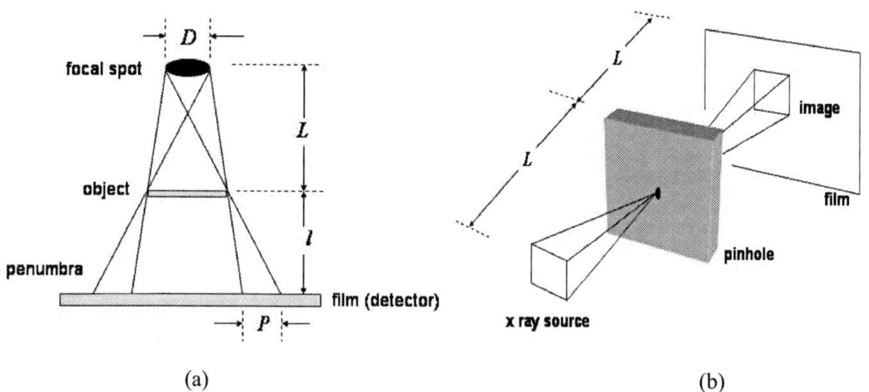

FIGURE 2. (a) The width of the penumbra is given by $P = Dl/L$. The focal spot size can be determined using this expression. When $l = L$, the image produced by a pinhole on a film will be the same size as the source.

If the photon gives some of its energy the process is called Compton scattering. Sometimes the photon scatters without losing energy in a process known as Rayleigh scattering.

The detector we are testing in the laboratory works on base of electrons which have been released from their host atoms by X ray photons. By applying an electric field the electrons can be collected and counted in strips as it is shown in Fig. 3(a). The detector is operated at room temperature with a special, low noise electronics for the read out [2-5]. X rays interact with silicon atoms to create in average 1 electron hole pair for each 3.62 eV of energy deposited in the material.

The use of a collimator (Fig. 3(b)) perpendicular to the strips of the detector converts the system in a 2D device. The advantage of using a slit over an array of two strip detectors is that ghost hits are avoided, increasing in this way the signal to background rate [6-8].

To test the whole digital system we used home-made phantoms which simulate dense small structures as expected in some cases in the body. As it is shown in Fig. 4(a) two different kind of phantoms were made. In Fig. 4(b) it is shown the line profile of a complete analysis for different exposition times using our X-ray tube. It can be observed the variation in number of counts depending on the time it was exposed. Further analyses were made and are reported in references [9-12].

There are major problems that should be solved before these new devices can be used in real life: a) reduction of the focal spot in the X rays source keeping high brilliance, b) significant improvement on spatial resolution in detectors, c) large sensitive area d) good detection efficiency for photons in the energy range 6-25 keV [13,14].

NEW TECHNIQUES

The contrast of an image is the result of different physical processes upon the detector used. Nowadays, the extraction of information is considerably improved, beyond the contrast on the original image, by means of digital image processing.

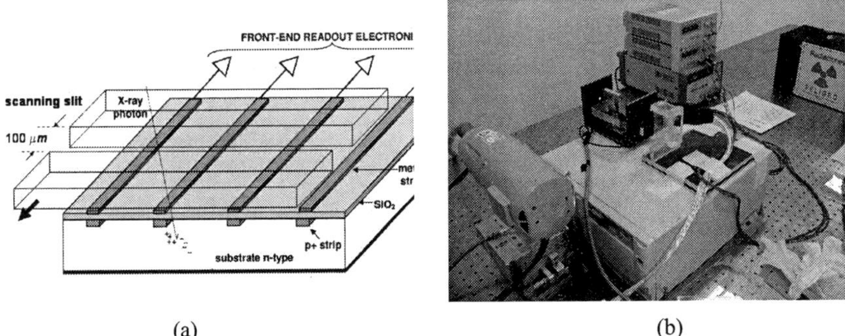

(a) (b)

FIGURE 3. (a) The detector used to obtain the images presented in Fig. 4. It is a silicon wafer with a slit of aluminum in the front. (b) The experimental array in the laboratory showing the X ray tube, the detector and the associated electronics.

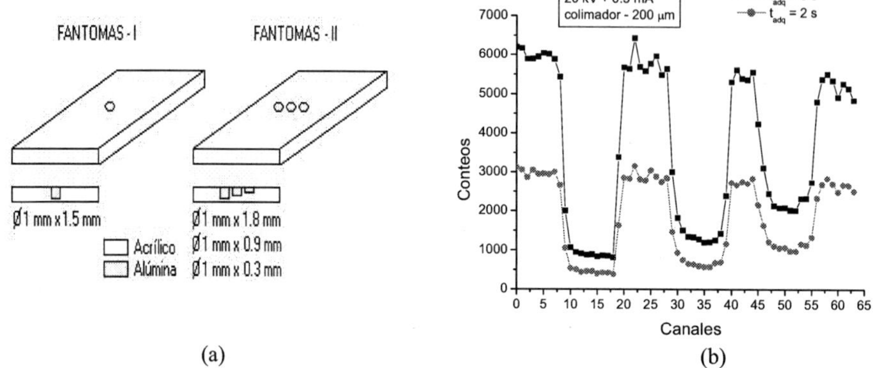

(a) (b)

FIGURE 4. (a) Phantoms prepared to check the experimental set up. On the right we show the measured counts for each channel of the detector for both (low curve) with 2 seconds exposition and (upper curve) with a 5 seconds exposition.

The physical description of the interaction between the X rays and the medium is given in terms of an amplitude and a phase. The index of refraction the material, which is traversed by X is given by

$$n = 1 - d - ib \tag{2}$$

where d is related to the phase shift the photon experiences at media interfaces, and the imaginary term, b, is related to the attenuation by absorption of the photon beam in the media. Traditional X ray imaging bases the contrast on the use of different absorption properties of the bone and soft tissue.

Variations in amplitude, as the photon travels trough a specific medium, show a behavior characterized by several absorption maxima as a function of the photon energy. This is used for imaging of specific tissues. By incorporating in the blood flow an iodine solution as contrast agent, it is possible the identification of veins and arteries by taking X ray images at energies slightly below and above of the iodine absorption peak. A digital subtraction of the images obtained in this way, produces a good quality image. The use of contrast agents is one of the recent techniques, currently in use for imaging of circulatory system, particularly in kidneys and heart.

On the other hand, phase variations are observed in the change of the index of refraction in media, producing a considerably contrast at the borders of the different materials. Information from phase variations is used to construct images using visible light in the so called phase-contrast microscopy.

However, the phase shift observed in the change of refraction index is a function of the photon energy. In the case of X ray photons the refractive index differs from unity by a factor of 10^{-5}. This is the fact that makes extremely difficult the observation of phase shift in X ray images. Different approaches have been developed in order to obtain such images. One of them was proposed in recent years and involves the use of a perfect crystal as an analyzer. The intensity of a photon beam is measured and monitored during the different stages of the imaging setup.

Detectors are sensitive to intensity variations, therefore the acquisition of image with contrast due to phase shift must be based on special arrangements, either applicable only to phase objects, or using an alternative technique which allows the imaging of any object with appreciable absorption. Both ideas have been the objective of research work around the world. Each of them looks for extraction of the information of phase shift using particular physical phenomena. We will not give a review here of the techniques under development. The interested reader can look at ref. [15] for non specialist presentation.

Either absorption or phase contrast images can be obtained by well known sensors like film, image intensifier, or digital sensors. The image recorded by any of these mechanisms has the contrast due exclusively to physical interactions involved in the image arrangement. Further treatment involves the digital processing which allows the use of special algorithms implicated in some techniques.

IMAGE PROCESSING

A digital image is defined in 2D as a function $a(m,n)$ where m defines a row and n a column in a matrix array. The intersection of a row and column defines a pixel. The value assigned to coordinates *(m,n)*, with $\{m = 1,2,...M\}$ and $\{n = 1,2,...N\}$ is a grey level in a monochromatic picture. Each value corresponds to the number of counts associated to the transmitted X-ray photons in the sample (biopsy or phantom).

Digital image processing refers to manipulation of images by means of a computer algorithm. There are three types of computerized processes: low-,mid- and high-level.

Low level processes involve primitive operations such as image preprocessing to reduce noise, contrast enhancement, and image sharpening. With these operations, images can be improved, in the sense of emphasize the region of interest. These operations can be made by means of equalization of the histogram associated to the image or filtering the image using masks in the frequency (or spatial) domain.

The histogram of a digital image with L total possible intensity levels in the range [0, G] is defined as the function:

$$h(r_k) = n_k \tag{3}$$

where r_k is the kth intensity level in the interval [0, G] and n_k is the number of pixels in the image whose intensity level is r_k.

Mid-level processes on images involve segmentation (it means, partitioning an image into regions or objects), description of the objects of interest and recognition of individual objects.

High-level processing involves the cognitive functions normally associated with human vision, that is, detecting an ensemble of recognized objects [16]. It is usually done by training a computer system with several images previously analyzed.

Radiology facilities have now experienced the advantages of digital imaging. Among others, digital image processing has made possible the implementation of techniques like Digital Subtraction Angiography (DSA), Computed Tomography (CT), and Computer-Aided Diagnosis (CAD) Radiology.

Each of the above mentioned techniques involves different stages of image processing and relevance on image reconstruction and contrast. DSA for instance, requires the image processing in order to accomplish the extraction of the physical contrast, by carrying out a subtraction of images at each side of an absorption peak of a specific contrast agent. Research interests currently under development necessarily involve the use of digital image processing, for extraction of the very physical contrast.

Computed Tomography, an application of image processing that involves the acquisition of images by a collection of sensors making possible the reconstruction of 3D images. CT has become a very important tool not only for diagnostic facilities, but for radiotherapy as well. With the obtained information, it is possible to coordinate the application of radiation therapy with modern accelerator equipments. Even for complex profiles and geometries.

Computer-Aided Diagnosis is a new technique for recognition of patterns which are found correlated to real lesions. It does imply the use of the complete area of image processing, as mentioned before. This is not aimed to find special contrast sources, but instead is intended to assist the radiologist by providing additional information besides the pure image contrast or to create a database for statistical or historical studies.

ACKNOWLEDGMENTS

The authors wish to thank to Luis M. Montaño (Departamento de Física-Cinvestav) for his contribution on the microstrips detector section as well as the results shown here. The project has been supported by the World Bank through the Iniciativa Científica del Milenio 2001.

REFERENCES

1. Elements of Modern X Ray Physics, J. A. Nielsen and D. McMorrow, John Wiley & Sons, ISBN 0 471 498572
2. G. Baldazzi, et. al. *Nucl. Instrum. Methods A*, **514**, 206-214 (2003).
3. D. Bollini, et. al., *Nucl. Instrum. Methods A*, **509**, 315-320 (2003).
4. R. Mendez, etl al., *Nucl. Instrum. Methods A*, **509**, 333-339 (2003).
5. D. Bollini, et. al., *Nucl. Instrum. Methods A*, **515**, 458-466 (2003).
6. L. Montaño, "Silicon Detectors Applied to Medical Imaging" in First ICFA Instrumentation School/Workshop, AIP Conference Proceedings 674, Morelia, Mexico 2002, pp. 344-347.

7. L. Montaño, et. al., "Comparison between two Monte Carlo simulation of Angiography Phantom coupled to silicon strips detector" in VIII Mexican Symposium on Medical Physics, AIP Conference Proceedings 724, León Gto. Mexico 2004, pp. 221-225.
8. Proceedings of the International Workshop of Medical Physics, April 8-10 2002, La Habana, Cuba.
9. A. Cabal, et. al. "Feasibility of silicon strip detectors and low noise multichannel readout system for medical digital radiography" in VI Mexican Symposium on Medical Physics, AIP Conference Proceedings 630, Mexico City 2002, pp. 202-208.
10. G. Baldazzi, et. al., "Results about imaging with silicon strips for angiography and mammography" in VII Mexican Symposium on Medical Physics, AIP Conference Proceedings 682, Mexico City 2003, pp. 14-23.
11. G. Baldazzi, et al. "Recent advances on X-ray imaging with a single photon counting system" in 4to. Simposio Internacional de técnicas Nucleares y Conexas, Proceedings of NURT 2003, La Habana, Cuba., pp. 27-31.
12. C. Ceballos, et. al. "Monte Carlo simulation of a silicon strip detector response for angiography application. First approach" in VII Mexican Symposium on Medical Physics, AIP Conference Proceedings 682, Mexico City 2003, pp. 185-191.
13. A. Avila , et. al. , *Med. Phys.* **32,** 3755-3766 (2005).
14. L. Montaño, "Digital mammography: Improvements in Breast Cancer Diagnostic" in Advanced Summer School in Physics 2005, edited by O. Rosas-Ortíz, et. al. , AIP Conference Proceedings 809, Mexico City 2005, pp. 283-289
15. R. Fitzgerald, Phys. Today, 53 (2000)23
16. R. C. González, R. E. Woods, and S. L. Eddins, *Digital Image Processing Using Matlab*, New Jersey, Pearson Prentice Hall, 2004, pp. 76-84, 407-422.

Digital Processing and Segmentation of Breast Microcalcifications Images Obtained by a Si Microstrips Detector: Preliminary Results

Claudia. C. Díaz [1] and Abril A. Angulo[2]

1 Departamento de Física, Centro de Investigación y de Estudios Avanzados del IPN, A. P. 14-740, 07000 México, D. F., Mexico,
2 C.U.C.E.I. Universidad de Guadalajara, Av. Revolución 1500, 1100, Guadalajara, Jal. Mexico.

Abstract. We present the preliminary results of digital processing and segmentation of breast microcalcifications images. They were obtained using a Bede X ray tube with Cu anode, which was fixed at 20 kV and 1 mA. Different biopsies were scanned using a 128 Si microstrips detector. Total scanning resulted in a data matrix, which corresponded with the image of each biopsy. We manipulated the contrast of the images using histograms and filters in the frequency domain in Matlab. Then we intended to investigate about different contour models for the segmentation of microcalcifications boundaries, which were based on the contrast and shape of the image. These algorithms could be applied to mammographic images, which may be obtained by digital mammography or digitizing conventional mammograms.

INTRODUCTION

Today breast cancer is the second cause of women's decease in Mexico. For this reason screening programs are being developed across the country. They have proved to be a method for controlling the disease. Most of mammograms are obtained using a conventional mammographic unit and, in some cases, using the full field digital mammography. Radiologist reads around 100 of these screening mammograms at a sitting, and because only about 0.5% of these cases will have breast cancer, it's difficult to be vigilant to find the subtle indications of malignancy on the mammogram. As a consequence, between 5% and 30% of the diagnosed women have breast cancer. A "second opinion" could be helpful for reducing the chance to miss a possible injury. This "opinion" could be given by a computerized diagnosis, which analyzes the mammograms. Some of the characteristics that could be analyzed are the contrast, the area, the average radius, etc. of microcalcifications [1]. In this case, we studied the contrast of the image using histograms, and made segmentation of the image using different algorithms based on gradients. The goal of segmentation is to separate microcalcifications from the rest of the tissue.

CP885, *Advanced Summer School in Physics 2006, Frontiers in Contemporary Physics—EAV06,*
edited by O. Miranda, M. Carbajal, L. M. Montaño, O. Rosas-Ortiz, and S. A. Tomás Velázquez

DIGITAL IMAGE USING Si MICROSTRIPS DETECTOR

We obtained a digital image of a biopsy, which was supposed to have microcalcifications (see figure 1a), in the following way: x-rays were generated using a Bede X-ray source, with a Cu anode (characteristic energy: 8 keV). It was fixed at 20 kV and 1 mA. These rays passed through a collimator, whose width was of 120 μm and were directed to the biopsy. It absorbed some of these x-rays, and the rest were transmitted. The biopsy was compressed, as shown in figure 1a, to improve the contrast of the image. The transmitted x-rays went to a 128 Si microstrips detector (see figure 1b), which was put on "the edge" configuration. These photons were associated to a relative small production of charge, which was transformed in a digital signal, so we obtained a data vector [2, 3]. Horizontal movement of the biopsy, made by a step motor, allowed to have a data matrix, which was formed by many of those data vectors. The exposure time to radiation was 1 second. The experimental setup is shown in figure 2 [4].

DIGITAL PROCESSING AND SEGMENTATION

The original image obtained is shown in figure 3a. The cluster of dark spots was associated to microcalcifications. Manipulation of the contrast was made in Matlab [5], in order to reduce the effects of the different components of the biopsy, which acts as a camouflaging background to clustered microcalcifications. The contrast was manually varied, analyzing the gray scale histogram associated with the image. Then we used filters for segmentation of the image. We intended to investigate the use of 2D contour model for the segmentation of mass boundaries. First, we analyzed the image of a phantom, which simulates the microcalcifications.
Discontinuities in intensity values were detected using first order derivative, which is the gradient of a function f(x,y), defined as

$$\nabla \mathbf{f} = \begin{bmatrix} G_x \\ G_y \end{bmatrix} = \begin{bmatrix} \dfrac{\partial f}{\partial x} \\ \dfrac{\partial f}{\partial y} \end{bmatrix}$$

The Sobel edge detector uses the masks

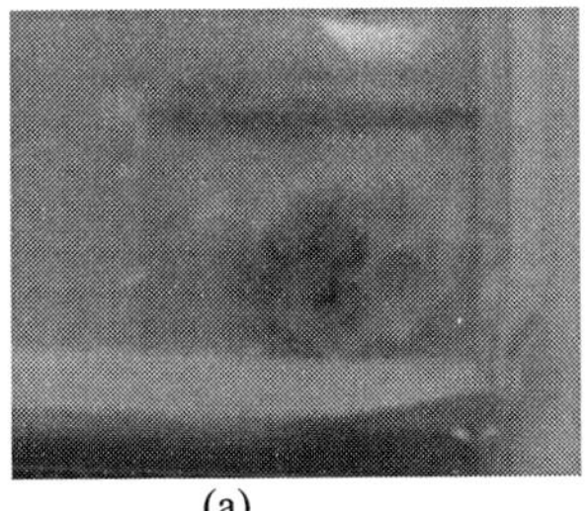

(a)

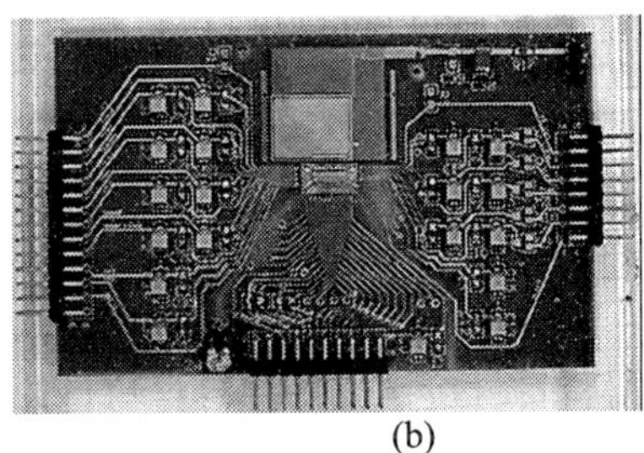

(b)

FIGURE 1. (a) The compression of the biopsy served to distribute de tissues in the volume, and to improve the contrast in the image. (b) 128 Si microstrips detector.

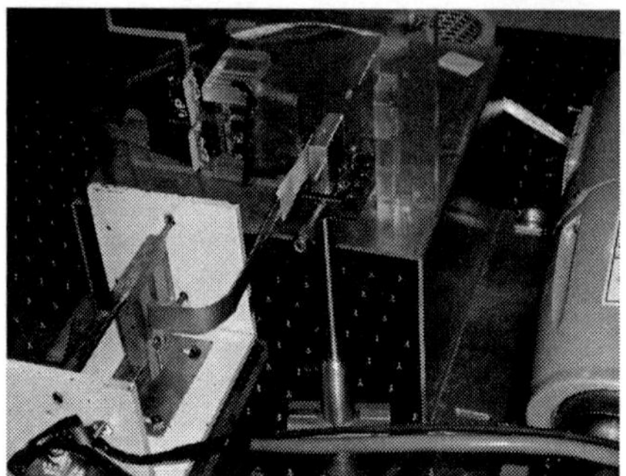

FIGURE 2. Experimental setup to obtain digital images with Si detector.

$$S_x = \begin{bmatrix} -1 & -2 & -1 \\ 0 & 0 & 0 \\ 1 & 2 & 1 \end{bmatrix} \quad S_y = \begin{bmatrix} -1 & 0 & 1 \\ -2 & 0 & 2 \\ -1 & 0 & 1 \end{bmatrix} \tag{1}$$

to approximate digitally the first derivatives G_x and G_y. The gradient at the center point in a neighborhood of the image

$$\begin{bmatrix} z_1 & z_2 & z_3 \\ z_4 & z_5 & z_6 \\ z_7 & z_8 & z_9 \end{bmatrix} \tag{2}$$

was computed as follow by this detector [6]:

$$\begin{aligned} g = \left[G_x^2 + G_y^2\right]^{1/2} = \Big\{&\left[(z_7 + 2z_8 + z_9) - (z_1 + 2z_2 + z_3)\right]^2 \\ &+ \left[(z_3 + 2z_6 + z_9) - (z_1 + 2z_4 + z_7)\right]^2\Big\}^{1/2} \end{aligned} \tag{3}$$

We used the watershed segmentation of the phantom image with these gradients. Figures 3d and 3e show the results. There, marks around the contour of the dark spots separate the microcalcifications from the background. Then, we used this algorithm with the image of the biopsy. The result is shown in figure 3b. In this case, segmentation did not identify all the microcalcifications. The following step was to develop an algorithm for detecting these abnormalities from the background analyzing the shape of the injuries. Figure 3c shows a map which locates them. It was obtained using median filter, modifying gray level histogram and applying the Canny edge detector.

CONCLUSIONS

We studied the segmentation of an image for locating abnormalities. It was done using the available tools in Matlab. This program also offers different commands that

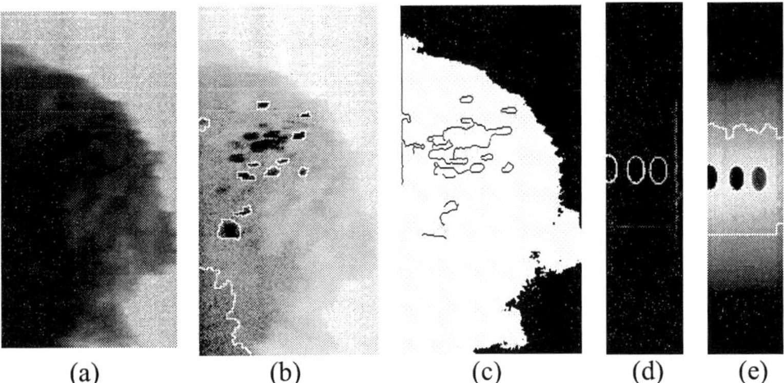

FIGURE 3. (a)Original image .(b) Segmentation of the image using gradients. (c) Map of the location of injuries d) Map of the location of dots in phantoms. (e) Segmentation of dots in phantoms.

detect lineal edges, that may help to classify the shape of the microcalcifications. The following step in this work is designing a program, which will obtain a better contrast of the image and automatically segmenting the image to locate the possible injuries in mammographic images. This algorithm may be applied to images obtained by a digital mammographic unit or digitizing a conventional film.
Also, the X-ray anode will be changed by a Mo anode, which corresponds to the real mammographic conditions.

ACKNOWLEDGMENTS

The authors wish to thank to the organizers of the 2006 Advanced Summer School for the invitation, and to L. M. Montaño (Departamento de Física-Cinvestav) for his reviews and comments to this work.

REFERENCES

1. R. M. Nishikawa, "Detection of Microcalcifications", in *Image Processing Techniques for Tumor Detection*, edited by R. N. Strickland, Tucson Arizona, Marcel Dekker, 2002, pp. 131-153.
2. L. Ramello, "Medical Imaging with Semiconductor Detectors" in Advanced Summer School in Physics 2005, edited by O. Rosas-Ortíz, et. al. ,AIP Conference Proceedings 809, Mexico City 2005, pp. 263-282.
3. A. Cabal, et. al. "Feasibility of silicon strip detectors and low noise multichannel readout system for medical digital radiography" in VI Mexican Symposium on Medical Physics, AIP Conference Proceedings 630, Mexico City 2002, pp. 202-208.
4. L. Montaño, "Digital mammography: Improvements in Breast Cancer Diagnostic" in Advanced Summer School in Physics 2005, edited by O. Rosas-Ortíz, et. al. ,AIP Conference Proceedings 809, Mexico City 2005, pp. 283-289
5. R. C. Gonzâlez, R. E. Woods, and S. L. Eddins, *Digital Image Processing Using Matlab*, New Jersey, Pearson Prentice Hall, 2004, pp. 378-425.
6. B. Sahiner, et. al. *Med. Phys.* **31 (4),** 744-754 (2004).

V. SOLID STATE PHYSICS

TiN surface dynamics: role of surface and bulk mass transport processes

J. Bareño[a], S. Kodambaka[b], S.V. Khare[c], W. Swiech[a], V. Petrova[a], I. Petrov[a], and J.E. Greene[a]

[a]*Department of Materials Science and the Frederick Seitz Materials Research Laboratory University of Illinois, 104 South Goodwin Avenue, Urbana, IL 61801, USA*
[b]*IBM Research Division, T. J. Watson Research Center, Yorktown Heights, New York 10598, USA.*
[c]*Department of Physics and Astronomy, University of Toledo, Toledo, OH 43606, USA*

Abstract. Transition-metal nitrides, such as TiN, have a wide variety of applications as hard, wear-resistant coatings, as diffusion barriers, and as scratch-resistant and anti-reflective coatings in optics. Understanding the surface morphological and microstructural evolution of these materials is crucial for improving the performance of devices. Studies of surface step dynamics enable determination of the rate-limiting mechanisms, corresponding surface mass transport parameters, and step energies. However, most models describing these phenomena are limited in application to simple elemental metal and semiconductor surfaces. Here, we summarize recent progress toward elucidating the interplay of surface and bulk diffusion processes on morphological evolution of compound surfaces. Specifically, we analyze the coarsening/decay kinetics of two- and three-dimensional TiN(111) islands and the effect of surface-terminated dislocations on TiN(111) steps.

Keywords: Microscopic aspects of nucleation and growth, STM, LEEM.
PACS: 68.55.Ac, 68.37.Ef, 68.37.Nq.

I. INTRODUCTION

Growth of nanostructures, and thin films in general, is a complex phenomenon controlled by the interplay of both thermodynamic and kinetic driving forces. Fundamental understanding of the processes governing the formation and stability of nanostructures can be developed via studies of the dynamics of surfaces at the atomic scale. Prior to the invention of scanning tunneling microscopy (STM) [1] and related scanning probe microscopy tools such as atomic force microscopy (AFM) [2], field ion microscopy (FIM) [3] was the only real-space imaging technique available for resolving individual atoms on surfaces and for studying dynamic processes such as surface diffusion of single adatoms and small two-dimensional (2D) islands [4,5]. Despite the restrictions on the type and size of materials that can be used in FIM studies, which requires high electrical fields and sharp single-crystalline tips, detailed investigations of adatom transport mechanisms have been carried out on a wide variety of metal surfaces [6].

The advent of high-speed variable temperature and pressure STM allows studies of surface dynamics at video rates [7-11] for a wider range of materials. Using STM, *in-situ* studies of the diffusion of adspecies (adatoms and advacancies), kinks, steps, and

CP885, *Advanced Summer School in Physics 2006, Frontiers in Contemporary Physics—EAV06,* edited by O. Miranda, M. Carbajal, L. M. Montaño, O. Rosas-Ortiz, and S. A. Tomás Velázquez

2D islands [12-21] are carried out over a wide range of temperatures (20-1500 K). From analyses of the changes in successive images, atomic processes contributing to these phenomena are quantified. The kinetics of nucleation and growth [22,23,24] on crystalline surfaces are also routinely studied in-situ, either in ultra-high vacuum or electrolytic environments. A related tool, the AFM, is also used to study the nucleation and growth kinetics of crystals in a liquid environment [25,26].

Low energy electron microscopy (LEEM) [27] is another surface-sensitive analytical technique (complementary to STM) in which a coherent, low energy electron beam (typically 1-100 eV) illuminates the sample; the electrons undergo diffraction and are captured by an objective lens to form a real-space image of the surface. Using a contrast aperture, diffracted beams corresponding to either specular (0,0) or fractional order reflections are selected to yield bright-field (BF-LEEM) or dark-field (DF-LEEM) images, respectively. In LEEM, single-atom-height steps on the surface can be resolved by geometric phase contrast. Surface lateral resolution is typically of the order of a few tens of Ångstroms. Diffraction and chemical contrast also provide information about the surface. Since LEEM is not a scanning microscope, images are acquired at video rates, thus providing sufficient time resolution for real-time studies of dynamic phenomena such as epitaxy, interface formation, and surface morphological evolution on large lateral length scales (1-10 μm) over a wide range of temperatures. LEEM, best suited for investigating electrically conducting crystalline samples, has been used to study surface diffusion of 2D metal islands [28], surface phase transformations [29], alloy formation, step fluctuation kinetics, interlayer mass transport, and bulk diffusion. Photoelectron emission microscopy (PEEM) is a related technique, in which photoelectrons generated by an incident UV or an X-ray light source (rather than an electron gun) are used for imaging. PEEM has proven to be an effective tool for following chemical reactions on catalytic surfaces [30] and to study surface magnetism with elemental and/or chemical sensitivity [31].

There are several excellent review articles covering different aspects of atomic-scale dynamics of the early stages of thin film growth. Zinke-Allmang and co-workers [32,33] summarized the theoretical and experimental understanding of cluster formation phenomena in general, while focusing on the role of adatom surface diffusion and binding energies on cluster formation kinetics. Tromp and Hannon [34] described methods for quantitative analyses of nucleation and growth processes on surfaces. The reviews by Jeong and Williams [35] and Giesen [36] provide theoretical background and describe experimental and computational techniques for investigating and characterizing the dynamics of metastable structures on surfaces, determining step energetics and mass transport parameters, and hence developing an atomic-scale understanding of the stability of solid/vacuum and solid/liquid interfaces. Recently, Bonzel and co-workers [37,38] reviewed new experimental methods for analyses of temperature-dependent three-dimensional (3D) equilibrium crystal shapes and 2D islands as a means to extract absolute surface, step, and kink formation energies.

All of the above reviews focus on simple elemental metal and semiconductor surfaces. However, similar studies on more complex compound surfaces of technologically-relevant materials such as TiN, GaAs, GaN, ZnO, Al_2O_3, ZrO_2, have not been carried out. TiN, in particular, is widely used as a hard wear-resistant coating on cutting tools, a diffusion-barrier layer in microelectronic devices, a corrosion-resistant coating on mechanical components, and an abrasion-resistant layer on optics and architectural glass. Since the elastic and diffusion-barrier properties of TiN are highly anisotropic, controlling polycrystalline TiN film texture is important for all of these applications. This fact has spurred interest in modeling the growth of polycrystalline TiN as a function of deposition conditions [39]. Such a model, however, requires knowledge of surface, step, and nearest-neighbor interaction energies, all as a function of orientation. Recently, considerable progress has been made toward obtaining absolute orientation-dependent step energies and step stiffnesses as well as the activation barriers for island coarsening on TiN(001) and TiN(111) surfaces [40].

Here, we describe recent progress toward elucidating the interplay of surface and bulk diffusion on morphological evolution of TiN(111) surfaces. Section III is a brief introduction to the phenomenon of Ostwald ripening and the coarsening/decay kinetics of 2D TiN islands on TiN(111) surfaces. Section IV deals with the interplay between surface and bulk diffusion processes and their role on surface step motion. In particular, section IV.A focuses on the effects of step permeability, step-step interactions, and bulk mass transport on the coarsening/decay of 3D island stacks. In section IV.B, we discuss the role of defects such as surface-terminated dislocations on step evolution kinetics.

II. EXPERIMENTAL PROCEDURE

Epitaxial TiN(111) layers, 2000-Å-thick, were grown on polished $Al_2O_3(0001)$ substrates at a temperature T = 1050 K in a load-locked multi-chamber ultra-high vacuum (UHV) system using magnetically-unbalanced dc magnetron sputter deposition [41] following the procedure described in Ref. 42.

The TiN(111) samples were then transferred to one of two UHV multichamber microscope systems, containing either a LEEM or a variable-temperature Omicron STM, and degassed at 1073 K for approximately 2 h. Each of the multichamber microscope systems has a base pressure $< 2\times10^{-10}$ Torr and is equipped with facilities for sample preparation, residual gas analysis, electron-beam evaporation, ion sputtering, Auger electron spectroscopy, and low-energy electron diffraction (LEED). Sample temperatures were measured by optical pyrometry and calibrated using temperature-dependent TiN emissivity data obtained by spectroscopic ellipsometry.

In the following sections, we describe the evolution of ensembles of 2D adatom and vacancy islands on wide terraces, 3D stacks of 2D islands, and spiral surface steps. The adatom and vacancy islands were prepared by first depositing homoepitaxial TiN(111) buffer layers (50-100 Å thick) on the TiN(111) substrates described above at 1023 K by reactive evaporation from Ti rods (99.999% purity) in N_2 (99.999%) at 1×10^{-7} Torr. The buffer layers were annealed in N_2 for 4 h at a temperature T > 1100 K. Partial TiN(111) bilayers (BL)[1] with coverages of 0.1-0.8 BL were then deposited on the buffer layers by reactive-evaporation at room temperature. The samples were annealed *in situ* at T = 1050-1250 K in 1×10^{-7} Torr N_2 for times t = 1-2 h. The overall

[1] The [111] direction in B1-NaCl structure TiN is polar, consisting of alternating layers of Ti and N atoms.

procedure results in surfaces with ≃ 500-Å-wide atomically-smooth terraces, separated by bilayer-height steps with triangular shaped (truncated hexagons) 2D TiN(111) adatom islands for coverages < 0.4 BL and vacancy islands at higher coverages [43].

Samples containing 3D mound structures on large (>1000 Å) terraces were prepared by depositing homoepitaxial TiN(111) overlayers (50-200 Å thick) at 1023 K and a rate of ≃ 0.02 ML/s followed by annealing for 2-3 days in $5x10^{-8}$ Torr N_2 at 1200 K. The deposition/annealing cycles were repeated until 3D mounds, consisting of stacks of 2D TiN adatom islands appeared in defect-free areas and spiral structures were formed in the presence of surface-terminated dislocations [44,45].

Samples investigated by STM were allowed to thermally equilibrate with the tip at each annealing temperature for 2 to 3 h prior to obtaining images at a constant rate of 18 to 44 s/frame. Typical tunneling conditions were 0.4-0.6 nA at -3.5 V. Resolution in the images varied from 2 to 5 Å/pixel. Scan sizes, scan rates, and tunneling parameters were varied to check for tip induced effects. No such effects were observed for the results presented here. The samples studied by LEEM were allowed to thermally stabilize at each temperature for 10 to 15 s prior to acquiring LEEM videos at a rate of 30 frames/s. Typical imaging conditions were 4 μm field of view (corresponding to a resolution ≃ 85 Å/pixel) and electron probe beam energies between 5 and 25 eV. Island boundaries were identified and the island areas determined during both STM and LEEM data analyses using the Image SXM [46] image processing software.

III. OSTWALD RIPENING AND ISLAND DECAY

Ostwald ripening [47] is a phenomenon in which larger islands on a surface grow at the expense of smaller neighboring islands. Figs. 1a and 1b show representative STM images ($1660x1660\ Å^2$) of 2D TiN(111) adatom and vacancy islands acquired at 34 s/frame during annealing at T = 1211 K for times t = 0 and 82 min, respectively. Most of the islands observed at t = 0 (e.g., adatom islands *1-6* and vacancy island *8*) have disappeared by t = 82 min, leaving only the largest island, labeled *7*.

Ostwald ripening is described by the Gibbs-Thomson equation [47] in which the equilibrium free adatom concentration ρ^{eq} associated with an island is related to the equilibrium island curvature κ through the expression

$$\rho^{eq} = \rho_{\infty}^{eq} \exp\left(\frac{\mu}{kT}\right), \tag{1}$$

where ρ_{∞}^{eq} is the equilibrium free adatom concentration associated with a straight step and μ is the island chemical potential. In the specific case of a circular (isotropic) island of radius r, the chemical potential is

$$\mu = \tilde{\beta}\kappa\Omega = \frac{\tilde{\beta}\Omega}{r}, \tag{2}$$

where $\tilde{\beta}$ is the step stiffness, κ is the step curvature, and Ω is the unit molecular area. Smaller islands have higher curvatures and, hence, higher adatom concentrations than larger islands, resulting in mass transfer from smaller to larger islands.

The triangular shapes of the truncated hexagonal TiN(111) islands indicate that they are highly anisotropic, thus Eq. (2) is not applicable. The chemical potential associated with a non-circular (anisotropic) island shape is

$$\mu = \tilde{\beta}(\varphi)\kappa(\theta)\Omega, \tag{3}$$

where κ and $\tilde{\beta}$ are, respectively, functions of the azimuthal angle θ and the step normal φ. For an equilibrium-shaped island, the chemical potential per unit molecular area is independent of step orientation and depends only on island size, [42]

$$\lambda = \tilde{\beta}(\varphi)\kappa(\theta) = \frac{B}{r_{avg}}. \tag{4}$$

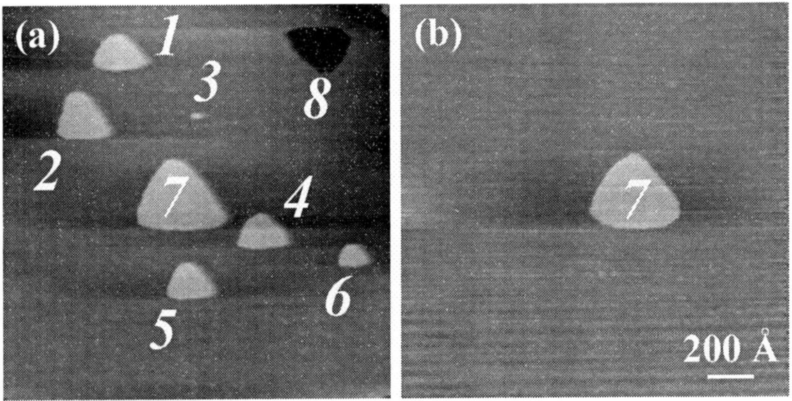

FIGURE 1. Representative STM images (1660x1660 Å^2) of 2D TiN adatom (*1-7*) and vacancy islands (*8*) on an atomically-smooth TiN(111) terrace. The images were acquired at 34 s/frame at times t = (a) 0 and (b) 82 min during annealing at T = 1211 K.

In Eq. (4), $r_{avg} = (A/\pi)^{1/2}$ is the average island radius and B is an orientation-independent constant which sets the energy scale of the equilibrium island chemical potential. Substituting Eq. (4) into Eq. (1), the Gibbs-Thomson equation can be rewritten in terms of the *orientation-independent* parameters r_{avg} and B

$$\rho^{eq} = \rho_{\infty}^{eq} \exp\left(\frac{B\Omega}{r_{avg}kT}\right). \tag{5}$$

Eq. (5) is valid for any equilibrium island shape. For the case of circular islands, the step energy β is constant (orientation-independent), B = β, r_{avg} = r; and Eq. (5) reduces to the expression for the isotropic Gibbs-Thomson equation commonly found in the literature.

In classical mean-field theory, the decay rate dA/dt of an isotropic island exhibiting detachment-limited kinetics, which is the case for TiN(111) [43], is given by

$dA/dt = -K_d L\Omega[\rho^{eq} - \rho_s^{eq}]$ [48], where K_d is a temperature-dependent adatom attachment/detachment rate coefficient, L is the island perimeter, and ρ_s^{eq} is the equilibrium free adatom concentration on the surface. For an anisotropic island, where K_d may depend on step orientation, dA/dt is given by

$$\frac{dA}{dt} = -\Omega\int_0^{2\pi} d\theta\, r(\theta) K_d(\theta)\left[\rho^{eq} - \rho_s^{eq}\right]. \tag{6}$$

$r(\theta)$ in Eq. (6) is the radial distance from the island center to the edge as a function of θ. In the pure detachment-limited island coarsening regime, the case for TiN(111) islands [43], $\rho_s^{eq} = \rho_\infty^{eq}$ and assuming that the exponential term in Eq. (5) can be expanded to the first two terms[2], we obtain

$$\frac{dA}{dt} = -\Omega\rho_\infty^{eq}\int_0^{2\pi} d\theta\, r(\theta) K_d(\theta)\left(\frac{B\Omega}{r_{avg}kT}\right) = -\left(\frac{B\Omega^2\rho_\infty^{eq}}{r_{avg}kT}\right)\int_0^{2\pi} d\theta\, r(\theta) K_d(\theta). \tag{7}$$

We note that Eq. (7), which shows that dA/dt is independent of island size for a given T in the detachment-limited regime, is valid even for anisotropic equilibrium-shaped islands.

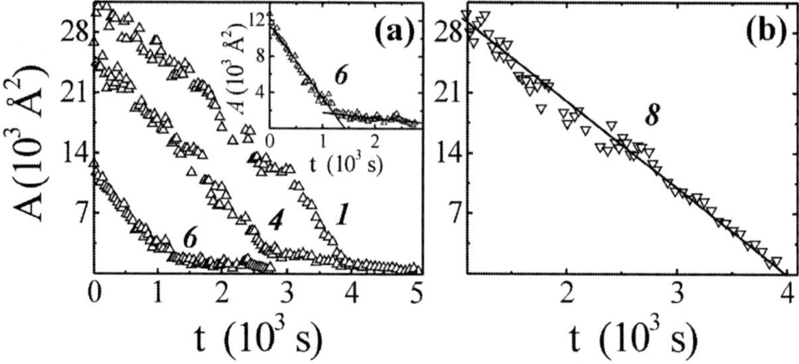

FIGURE 2. (a) Island area A vs. annealing time t for 2D adatom islands *1*, *4*, and *6* at T = 1211 K in Fig. 1a. A discontinuous decrease in decay rates was observed at a critical area A_c. Inset: A vs. t plot for island *6*. (b) A vs. t for vacancy island *8* in Fig. 1a. The solid lines in each case are linear least-squares fits to the data.

Typical results showing the areas A of adatom islands *1*, *4*, and *6* in Fig. 1a vs. t are plotted in Fig. 2a. The areas of all adatom islands were found to decrease linearly with t at a constant decay rate dA/dt, independent of local environment, with an abrupt decrease in slope at A_c = 1600±470 $Å^2$ (220±65 TiN molecules, Ω = 7.2 $Å^2$), irrespective of both T and initial island area. This is illustrated more clearly for island *6* in the inset of Fig. 2a. dA/dt for island *6* decreases from 8.20 Å/s at $A > A_c$ to 1.28 Å/s at $A < A_c$. Measured dA/dt values for all adatom islands with $A > A_c$ were found

[2] For TiN(111), based on the stiffness and curvature values from Ref. [52], the maximum uncertainties introduced due to this assumption are ≃ 0.1% at A > Ac and ≃ 2.5% at A < Ac.

to be a factor of 3 to 20, depending monotonically on T, higher than the same islands with $A < A_c$.

Fig. 2b is a plot of A vs. t for the vacancy island *8* in Fig. 1a. dA/dt remains constant, with no slope transition, over the entire decay process. A vs. t data in Figs. 2a and 2b are typical of STM results obtained from over 80 large ($A > A_c$) and 40 small ($A < A_c$) adatom islands and 5 vacancy islands observed at temperatures in the range 1050-1250 K.

dA/dt, in Eq. (7), is a function of the product of the thermally-activated parameters ρ_∞^{eq} and K_d. Thus, dA/dt can be expressed in the form $dA/dt \propto \exp(E_a/kT)$ with an activation barrier E_a. From an Arrhenius plot of the island decay rates, we obtain an activation energy $E_{ad,small} = 3.3\pm0.4$ eV for small ($A < A_c$) adatom islands and $E_{ad,large} = 2.3\pm0.6$ eV for large ($A > A_c$) adatom islands. The activation energy for decay of vacancy islands is $E_{vac} \lesssim E_{ad,large}$.

In the detachment-limited regime, $E_a = E_f+E_s+E_d$, where E_f, E_s, and E_d are the adatom formation energy, surface diffusion barrier, and attachment/detachment barrier, respectively. Since E_f and E_s are independent of island size, ($E_{ad,small}-E_{ad,large}$) corresponds to a difference in E_d values for small and large adatom islands. It is reasonable to assume, based upon direct observations [49,50], that at relatively low temperatures, which is the case here where T varies from 0.33 to 0.39 of the TiN melting point in K [51], attachment/detachment occurs only at kink sites. Thus K_d, and hence E_d, depend on the kink density (number of kinks per unit step length), which in turn depends on the kink formation energy ε. For TiN(111), $\varepsilon_1 \simeq 0.43$ eV and $\varepsilon_2 \simeq 0.13$ eV, where ε_1 and ε_2 correspond to kink formation energies along S_1 (long steps, "triangle" sides) and S_2 (short steps, rounded corners) respectively [52]. Thus, for large islands, attachment/detachment of diffusing species occurs preferentially along S_2 steps. Islands with $A \leq A_c$ ($A_c = 1600$ Å^2) are bounded by S_2 steps of length ≤ 4 atoms and S_1 steps of length ≤ 12 atoms. Kink formation along 4-atom-long S_2 steps involves removal of an atom to produce a double kink, which is energetically unfavorable. Thus, for islands with $A \leq A_c$, attachment/detachment of diffusing species occurs predominantly along S_1 steps which have higher kink formation energies and, hence, higher E_{ad} values and lower decay rates as we observe. In the case of vacancy islands, E_{vac} and, hence, E_d are independent of island size suggesting different pathways for attachment/detachment at adatom and vacancy islands.

IV. EFFECTS OF BULK DIFFUSION ON SURFACE EVOLUTION KINETICS

In the previous section, we showed that a detachment-limited surface diffusion mechanism, driven by the Gibbs-Thomson effect, is sufficient to explain the coarsening/decay kinetics of both adatom and vacancy islands on TiN(111) surfaces. However, other effects such as bulk diffusion and interlayer mass transport can also affect surface morphological evolution kinetics, and even become dominant at higher temperatures. In the following sections, we describe two step-flow models, developed based upon the Burton-Cabrera-Frank (BCF) theory of crystal growth [53], that account for step-step interactions, step permeability, and both surface and bulk mass transport processes.

IV.A. 3D-Mound Decay: Effects of Bulk Diffusion and Step Permeability

The coarsening/decay behavior of 3D mounds consisting of 2D layer structures in a "wedding cake" configuration has been studied extensively, both experimentally [54-56] and theoretically [57-59]. Experimental studies of the coarsening/decay kinetics of 2D homoepitaxial semiconductor and fcc metal islands stacked in 3D mound geometries indicate that the rate-limiting processes controlling island decay behavior are both qualitatively and quantitatively different from those of isolated islands on terraces [54,55,60-62]. In this section, we describe the coarsening/decay kinetics of concentrically-stacked 2D TiN adatom islands on TiN(111) terraces at elevated temperatures (T = 1550-1700 K). The islands exhibit repulsive step-step interactions and high step permeability rates.

Fig. 3a is a typical BF-LEEM image acquired during annealing a TiN(111) sample at T = 1559 K. We follow the time- and temperature-dependent decay kinetics of several successive layers in the circled region shown at higher magnification in Fig. 3b. This simple configuration allows us to apply the step flow model, developed by Israeli and Kandel [59]. The model is based upon the BCF theory of crystal growth [53] and accounts for the effects of step-step interactions, step permeability, and both surface and bulk mass transport on surface step motion.

The ith island of area A_i in a given 3D mound is characterized by its average radius r_i (i is a running index which increases from the top of the mound to the bottom.) In the absence of net mass change due to deposition, evaporation, and/or bulk diffusion, the adatom concentration fields between the islands are described by the 2D steady-state diffusion equation

$$\nabla^2 \rho_i(r) = 0, \text{ with } r_{i-1} \leq r \leq r_i \, , \tag{8}$$

where $\rho_i(r)$ is the adatom concentration on the ith terrace. A steady-state solution is justified since the time scales associated with equilibration of adatom concentration fluctuations on the terraces are much shorter than those associated with island step motion. We solve Eq. (8) using boundary conditions (given below) which specify adatom fluxes into (or out of) the islands and, hence, determine the rate of change of island radii. Assuming first-order kinetics, the flux boundary conditions at the ith island are

$$\begin{aligned} D_s \frac{\partial \rho_i}{\partial r}\bigg|_{r_i} &= K_d\left(\rho_i\big|_{r_i} - \rho_i^{eq}\right) + p\left(\rho_i\big|_{r_i} - \rho_{i-1}\big|_{r_i}\right), \\ D_s \frac{\partial \rho_{i-1}}{\partial r}\bigg|_{r_i} &= -K_d\left(\rho_{i-1}\big|_{r_i} - \rho_i^{eq}\right) + p\left(\rho_i\big|_{r_i} - \rho_{i-1}\big|_{r_i}\right). \end{aligned} \tag{9}$$

In Eq. (9), D_s is the surface diffusivity, K_d is the attachment/detachment rate, and p is the step permeability. ρ_i^{eq} is the equilibrium adatom concentration in the vicinity of the ith step, which is related to the step chemical potential μ_i and the equilibrium adatom concentration ρ_∞^{eq} at a straight isolated step through the Gibbs-Thomson relation, Eq. (1).

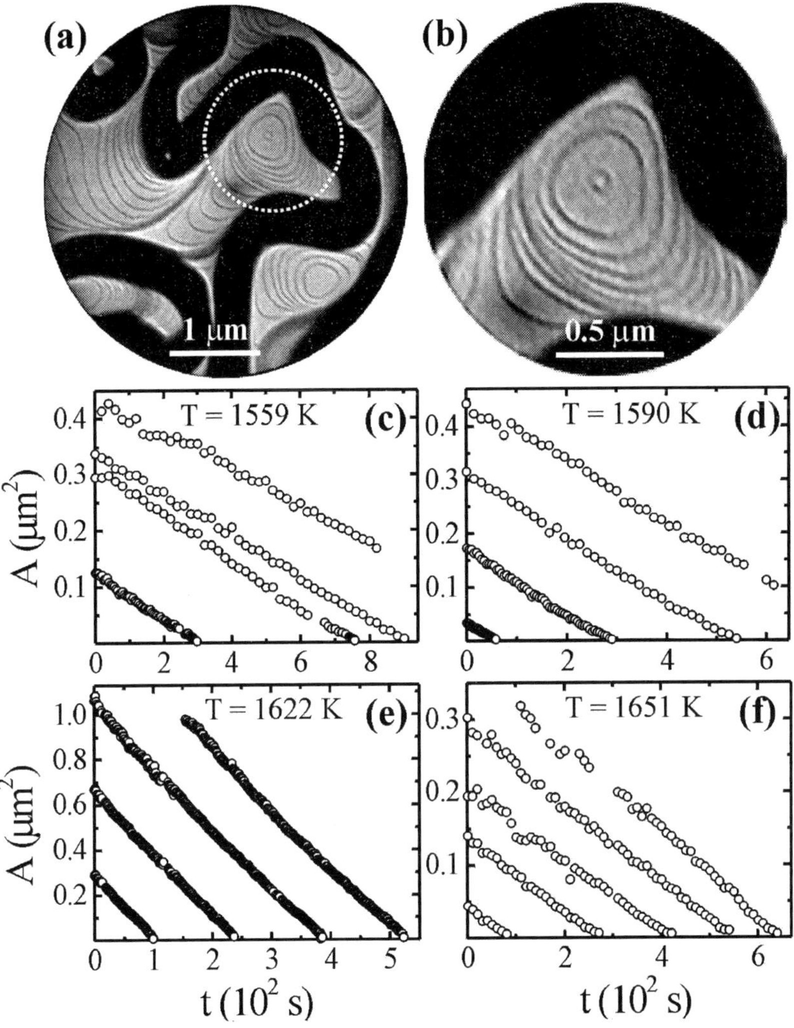

FIGURE 3. (a) Typical BF-LEEM image (field of view = 4.0 μm; E = 13 eV) of a 3D conical stack of 2D homoepitaxial islands on TiN(111) during annealing at T = 1559 K. (b) Higher-resolution image (field of view = 1.7 μm) of the highlighted region in Fig. 3a. (c)-(f): island area A vs. annealing time t plots for four or more successive layers in the region highlighted in Fig. 3b at temperatures T of (c) 1559 K, (d) 1590 K, (e) 1622 K, and (f) 1651 K.

For an island in a stack, μ_i depends on the island curvature and on elastic and entropic repulsive interactions between nearest-neighbor steps, accounting for which yields the relationship[3] [35,54,59]

$$\mu_i = \frac{\Omega B}{r_i} + \Omega G \left(\frac{2r_{i+1}}{(r_{i+1}+r_i)(r_{i+1}-r_i)^3} - \frac{2r_{i-1}}{(r_i+r_{i-1})(r_i-r_{i-1})^3} \right), \tag{10}$$

where *G* is the step-step repulsive interaction strength.

Expanding the exponential term in the Gibbs-Thomson equation to the first two terms, as in Eq. (7), and solving Eq (8) for ρ_i with the boundary conditions specified by Eqs. (9) and (10), yields the rate of change dr_i/dt of each island radius. For the limiting case in which the steps are impermeable, i.e. $p = 0$, and there is no bulk mass transport, we obtain

$$\frac{dr_i}{dt} = \frac{\Omega K_d \rho_\infty^{eq}}{r_i kT} \left[\frac{\mu_i - \mu_{i+1}}{\frac{K_d}{D_s} \ln\left(\frac{r_i}{r_{i+1}}\right) - \left(\frac{1}{r_i} + \frac{1}{r_{i+1}}\right)} - \frac{\mu_{i-1} - \mu_i}{\frac{K_d}{D_s} \ln\left(\frac{r_{i-1}}{r_i}\right) - \left(\frac{1}{r_{i-1}} + \frac{1}{r_i}\right)} \right]. \tag{11}$$

If the steps are permeable ($p > 0$), however, adatoms can hop across without becoming incorporated at the step edges and mass is not conserved locally. This results in coupling of the adatom diffusion fields on all terraces. Consequently, the equations describing the areal rate of change for the ith island are linked with those describing each of the other islands in the stack and we must solve the full equation set.

Adatom transport between the bulk and the surface also leads to local non-conservation of mass. In order to account for this possibility, we follow Ref. [56] and assume that mass exchange with the bulk occurs only near island edges at a rate K_{bulk}. The adatom flux J_i from the ith step to the bulk is then given by

$$J_i = K_{bulk} \left(\frac{\rho_\infty^{eq}}{kT} \right) \mu_i, \tag{12}$$

where we make the assumption that the bulk chemical potential is at equilibrium. Combining Eqs. (11) and (12), we obtain an expression for the velocity dr_i/dt of impermeable steps ($p = 0$) in the presence of bulk diffusion,

$$\frac{dr_i}{dt} = \frac{\Omega K_d \rho_\infty^{eq}}{kT \cdot r_i} \left[\frac{\mu_i - \mu_{i+1}}{\frac{K_d}{D_s} \ln\left(\frac{r_i}{r_{i+1}}\right) - \left(\frac{1}{r_i} + \frac{1}{r_{i+1}}\right)} - \frac{\mu_{i-1} - \mu_i}{\frac{K_d}{D_s} \ln\left(\frac{r_{i-1}}{r_i}\right) - \left(\frac{1}{r_{i-1}} + \frac{1}{r_i}\right)} - K_{bulk} \cdot r_i \cdot \mu_i \right]. \tag{13}$$

The step-flow model outlined above contains several material parameters: Ω, D_s, K_d, K_{bulk}, p, ρ_∞^{eq}, B, and G, of which only Ω (=7.2 Å^2) and B (= 0.23 ± 0.01 eV/Å) are known for TiN(111) [63]. There are four independent variables which control island

[3] In deriving Eq. (10), we assume circular-shaped islands that are concentrically stacked and separated by a distance that is smaller than the island radii.

decay kinetics. The length scale $l = \frac{D_s}{K_d}$ defines the limiting surface mass transport mechanism. In the diffusion-limited island coarsening regime, $l \ll \Delta x$, where Δx is the average terrace width, while $l \gg \Delta x$ in the detachment-limited regime. The ratios $\frac{p}{K_d}$ and $\frac{K_{bulk}}{K_d}$ describe the relative importance of step permeability and surface mass exchange with the bulk, respectively. Finally, the dimensionless quantity $g = \left(\frac{kT}{\Omega}\right)^2 \frac{G}{B^3}$ is a measure of the step-step interaction strength G. ρ_∞^{eq}, the only term not included in these four parameters, affects only the time scale of step motion and can easily be accounted for by rescaling the time unit.

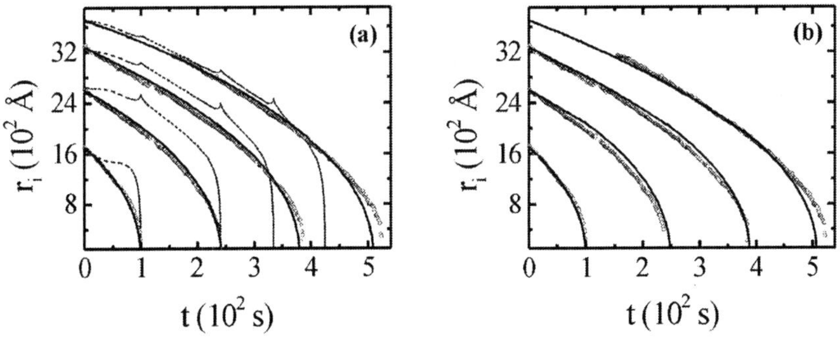

FIGURE 4. (a), (b): Plots of island radii r_i vs. annealing time t at T = 1622 K for the four samples corresponding to the data in Fig. 3e. The dashed and solid lines in Figs. 4a and 4b are calculated curves describing the LEEM data (open circles). The dashed lines in Fig. 4a are obtained assuming that the steps are impermeable and that there is no net bulk mass transport, i.e. $p = K_{bulk} = 0$. The solid lines in Figs. 4a and 4b are obtained assuming permeable steps ($p > 0$) with no bulk mass transport ($K_{bulk} = 0$) and bulk transport ($K_{bulk} > 0$) with impermeable steps ($p = 0$), respectively.

Fig. 4 shows a typical example of the agreement between experimental results at T = 1622 K and model predictions. The open circles in Figs. 4a and 4b are measured TiN(111) island radii as a function of time. The dashed lines in Fig. 4a are best fit calculated curves, obtained under the constraint of local mass conservation, i.e. $p = 0$ and $K_{bulk} = 0$, with $D_s/K_d = 0.1$ μm and $g = 0.0631$, for each island. While this relatively strong step-step interaction tends to minimize the recoil (the spike in lower island radii observed at times corresponding to the complete disappearance of the top island), the quality of the fits are far from satisfactory. Thus, as expected, imposing complete local mass conservation cannot explain the observed results, which require the presence of highly permeable steps and/or bulk mass transport.

The solid lines in Figs. 4a and 4b are calculated best fit solutions obtained by relaxing the local mass conservation constraint. We allow step permeability ($p > 0$) in the absence of bulk mass transport ($K_{bulk} = 0$) in Fig. 4a and bulk transport $K_{bulk} > 0$ with impermeable steps ($p = 0$) in Fig. 4b. For the first case, the best fit parameter values are $D_s/K_d = 100$ μm, $p/K_d = 2000$, and $g = 0.00819$, indicative of detachment-limited decay kinetics with highly permeable steps. In the second case, we obtain $D_s/K_d = 0.5$ μm, $K_{bulk}/K_d = 2.5$, and g = 0.00354. It is important to note that calculated curves obtained with bulk diffusion as the sole mass transport mechanism, e.g. with $D_s = 0$, are not consistent with the experimental results at any annealing temperature, T = 1550-1700 K, suggesting that the observed decay of TiN(111) islands requires the presence of surface mass transport.

TABLE 1.A. Permeable steps ($K_{bulk} = 0$)

T	D_s/K_d (μm)	p/K_d	G	G (eV-Å)
1559 K	100	2000	9.78×10^{-4}	0.034
1590 K	200	2000	2.82×10^{-3}	0.095
1622 K	100	2000	8.19×10^{-3}	0.264
1651 K	200	2000	2.45×10^{-2}	0.7634

TABLE 1.B. Bulk exchange (p=0)

T	D_s/K_d (μm)	p/K_d	G	G (eV-Å)
1559 K	5	2	1.02×10^{-3}	0.036
1590 K	0.5	10	$< 10^{-5}$	$< 3.4\times10^{-4}$
1622 K	0.5	2.5	3.54×10^{-3}	0.114
1651 K	0.5	5.5	2.31×10^{-2}	0.7198

Table 1 summarizes the materials parameters used to obtain the best fit solutions to the experimental data acquired at all four annealing temperatures in the two limits for which mass is not conserved locally: step permeability and bulk mass transport. Excluding bulk diffusion leads to high step permeabilities with D_s/K_d values which are much larger than the average terrace width (~ 1000 Å). This is a signature of detachment-limited kinetics and, as such, is in agreement with previous high-temperature (T = 1000-1250 K) scanning tunneling microscopy (STM) measurements of 2D TiN island coarsening/decay kinetics on TiN(111) terraces [43]. If we include bulk mass transport with impermeable steps, we obtain K_{bulk}/K_d ratios of order unity except at T = 1590 K, where $K_{bulk}/K_d = 10$. The calculated D_s/K_d values in Table 1.B also correspond to detachment-limited kinetics, but they are significantly smaller than the results obtained for permeable steps. This large difference in D_s/K_d values can be understood as follows. Step permeability by itself does not facilitate mass transport. It must be accompanied by fast surface diffusion in order to significantly violate local mass conservation. For bulk diffusion, this is not the case and mass is not conserved locally even with a modest value of D_s/K_d.

Overall, we find that the agreement between the experimental data and the calculated results obtained with non-zero K_{bulk} values is better than that obtained with high p/K_d values. However, the differences are small and we cannot quantitatively distinguish between the two processes. Since step permeability, unlike bulk-diffusion, is a surface process and given that bulk-point defects usually have larger formation and diffusion energies than surface adspecies, the energetics controlling island decay should provide additional insights into the controlling mechanism. To this purpose, we measured island decay rates as a function of temperature for 23 islands at four different temperatures between 1550 and 1700 K [44]. From least-squares analyses of the results, we obtain an activation energy E_d of 2.8 ± 0.3 eV for the decay kinetics of 2D TiN(111) islands stacked in 3D mounds. This is consistent with the previously reported value of 2.3 ± 0.6 eV determined from STM observations of the decay of large 2D TiN(111) islands on atomically-flat terraces [43]. The fact that we obtain an E_d value which is significantly lower than the bulk-mass transport barrier, 4.5±0.2 eV, measured for TiN(111) spiral step growth [64] provides further evidence that the dominant mass transport mechanism is surface, rather than, bulk diffusion. Thus, we attribute the decay of 2D TiN(111) islands in 3D concentric mound structures primarily to the presence of highly permeable steps in the detachment-limited regime.

IV.B. Bulk Diffusion and Surface Spiral Steps

In this section, we focus on the dynamics of surface loops and spiral steps, arising from the surface termination of a screw dislocation. Despite the fact that dislocations terminating on surfaces can strongly influence nanostructure stability, mechanical properties of thin films, chemical reactions, transport phenomena, and other surface processes; most theoretical and experimental studies have focused on dislocation motion in bulk solids under applied stress [65,66] and step formation due to dislocations at surfaces during crystal growth [25,26,53,67] or evaporation [68,69,70]. As a result, very little is known concerning the near-equilibrium dynamics of dislocations at surfaces.

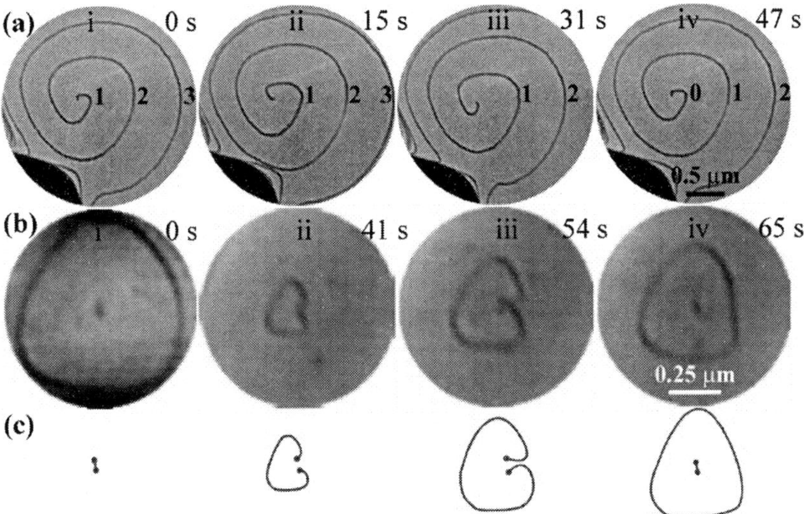

FIGURE 5. LEEM images showing nucleation and growth of bilayer-height surface steps at the cores of dislocations terminating on TiN(111) terraces during annealing in N_2 at temperatures T. The images were acquired as a function of time t. The dark lines in the LEEM images are ⟨110⟩-oriented bilayer-height steps (the [111] direction in B1-NaCl structure TiN is polar, consisting of alternating layers of Ti and N atoms). (a) Field of view ≈ 2.53 μm; T = 1688 K. Spiral steps form due to the pinning of a step edge at a dislocation core. (b) Field of view ≈ 0.93 μm; T = 1653 K. Loops nucleate and grow centered around a step edge pinned at both ends by dislocations of opposite sign. (c) Schematic diagram of the loop generation process observed in (b).

Figures 5a and b illustrate the nucleation and growth of TiN(111) spiral steps and loops at the cores of dislocations terminating on TiN(111) surfaces as observed in-situ by LEEM [71] during annealing in N_2. (We define a loop as a 2D island formed around a surface step segment pinned at both ends by dislocations). Fig. 5a shows the formation of a concentric spiral step structure around a dislocation core, while the images in Fig. 5b capture the nucleation and growth of a closed loop. The spiral and loop steps both exhibit three-fold symmetry with a truncated-hexagonal shape, i.e. the near-equilibrium shape of 2D TiN(111) islands [52], indicative of fast step edge diffusion. Note that these spirals and loops are observed during annealing with no net mass gain or loss and, as we will show, are *not* the growth structures predicted by BCF [53].

Figures 5a i-iv reveal a remarkable feature: the spiral steps rotate around the dislocation core resulting in an increase in the total step length with time t as the spirals undergo a shape-preserving anticlockwise motion with a constant angular velocity ω. That is, the shape and size of the spiral are periodic with time $\tau = 2\pi/\omega$. For the spiral shown in Figs. 5a i-iv, $\tau = 47$ s corresponds to one complete rotation at the annealing temperature T = 1688 K; the spiral shown in i is geometrically identical to the one in iv. During the period τ, the layer labeled 3 disappears from the field of view in Fig. 5a iii and a new layer, labeled 0, is formed (Fig. 5a iv) as the dislocation core climbs by a unit step height, $a_{\perp} = 2.4$ Å, normal to the surface.

Figs. 5b i-iii show nucleation and growth of a loop originating at a defect. Upon detaching from the defect, the expanding loop (Fig. 5b iv) regains the equilibrium shape of 2D TiN(111) islands [52]. This process, like the nucleation and growth of spiral steps, is also periodic. For the loops shown in Figs. 5b i-iv, $\tau = 85 \pm 5$ s corresponds to the loop nucleation period at T = 1653 K.

Analogous to the growth spirals and loops observed in bulk solids due to the operation of Frank-Read [72] and/or Bardeen-Herring [73] sources under applied stress [74], and on surfaces during crystal growth as predicted by the BCF theory [53], the features observed in Figs. 5a and b are due to surface steps pinned by a single dislocation (in the case of a spiral) and a pair of oppositely signed dislocations (for a loop), respectively. For a single dislocation, the steps emanating from the cores wind into expanding spirals. With a pair of oppositely signed dislocations, as illustrated in the schematic diagrams of Fig. 5c, the expanding spirals originating from the cores rotate in opposing directions, eventually coming into contact to form a loop which increases in size.

Our model for this phenomenon is based upon two assumptions. (1) A non-equilibrium concentration of point defects exists in the bulk. This is reasonable based upon the fact that TiN_x is known to have a very wide single phase region extending from x = 0.6 to x = 1.2 and can sustain both high anion (N) and cation (Ti) vacancy concentrations [75]. Given that we observe steps emanating from the grooves (the thick dark lines visible in the LEEM images in Fig. 5) annihilating the spiral steps, we conclude that spirals grow inward, normal to the surface. Note, however, that there is no net mass gain or loss. The surface point defect concentration C_s^{eq} is at thermal equilibrium since a N-terminated surface is energetically favourable for TiN(111) [76]. (2) Dislocation cores emit/absorb point defects at a thermally activated time-independent rate R; a plausible assumption given that dislocation cores act as sources/sinks for point defects [77,78].

Consider, for simplicity, a circular loop of radius r_{loop} centered around a core region of finite radius r_{core}. The surface point defect concentration C(r) at any position r, with $r_{core} < r < r_{loop}$, is given by the diffusion equation $\partial C(r)/\partial t = D_s \nabla^2 C(r)$, where D_s is the surface diffusivity. Within the quasistatic approximation, i.e. surface diffusion is much faster than the step edge velocity, $\partial C(r)/\partial t = 0$. The general solution of the resulting Laplace equation $\nabla^2 C(r) = 0$ is $C(r) = C_1 \ln(r) + C_2$, where the constants C_1 and C_2 are determined by the boundary conditions at the core and the loop. From assumption (2) above, the flux $-D_s \nabla C(r)$ of point defects at the core is

$$-D_s \nabla C(r)\big|_{r = r_{core}} = \frac{R}{2\pi r_{core}}, \qquad (14)$$

while at the loop:

$$-D_s \nabla C(r)\big|_{r = r_{loop}} = k_s[C(r_{loop}) - C_{loop}^{eq}]. \qquad (15)$$

In Eq. (15), k_s is the rate of attachment/detachment at the step and C_{loop}^{eq} = $C_s^{eq} \exp\left(\mu_{loop}/k_B T\right)$, from the Gibbs-Thomson relation (1), is the equilibrium point defect concentration due to the curvature-dependent chemical potential μ_{loop} associated with the loop. Solving for C_1 and C_2 yields the normal component of the loop velocity dr_{loop}/dt as

$$\frac{dr_{loop}}{dt} = \Omega k_s[C(r_{loop}) - C_{loop}^{eq}] = \frac{\Omega}{2\pi r_{loop}} R. \qquad (16)$$

In deriving the above formalism, we have neglected the curvature-driven flux from the loop to the local environment. In the detachment-limited regime, the contribution to dr_{loop}/dt due to this flux, which is proportional to $-1/r_{loop}$, is small and hence will have little effect on the growth velocity dr_{loop}/dt in Eq. (16). The important point is that the form of Eq. (16) is qualitatively different from Eq. (7), the parallel equation describing the 2D island decay (Ostwald ripening) process. We note that Eq. (16) is also valid, without loss of generality, for non-circular loops and spiral steps far from the core. Clearly, from Eq. (16), the loop (and spiral) growth rate $dA/dt = \Omega R$ is time-invariant consistent with the constant slopes obtained in our measurements of A vs. t in Fig. 6.

In formulating Eq. (16), we assume an equilibrium defect concentration C_{loop}^{eq} associated with the loop. This is justified for the first loop, since steps near the core and far from the boundary maintain a truncated-hexagonal shape with 3-fold symmetry, the equilibrium shape of TiN(111) islands. (This observation also suggests that localized growth flux at the cores may have negligible effect on step shapes.) However, far from the cores we find: (i) steps with non-equilibrium shapes that vary with the spiral geometry and (ii) step bunching. We attribute these observations to the presence of grooves bounding the spirals. Hence, modelling growth kinetics of multiple loop/spiral steps requires a better understanding of the effects of the geometric constraints imposed by physical boundaries on step chemical potentials

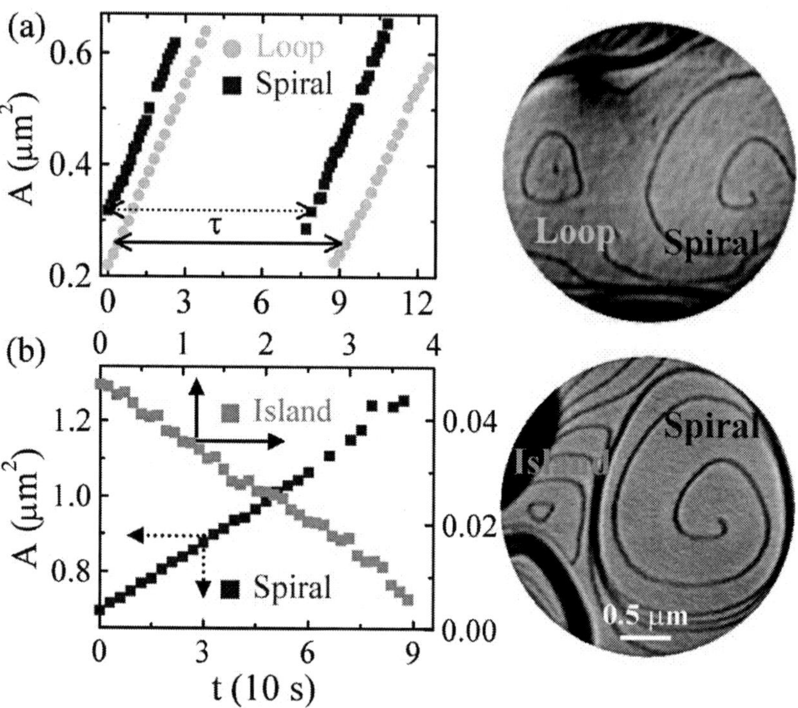

FIGURE 6. Area vs. annealing time for 2D TiN loops, spirals, and islands on TiN(111). (a) Time-dependent areas A of 2D TiN loops and spirals on a TiN(111) surface (shown in the image to the right) during annealing at T = 1653 K. (We only measure areas of loops that formed upon detaching from the central core and have attained near-equilibrium truncated-hexagonal shapes as in Fig. 5b iv). Spiral areas are calculated as $A(t)=\frac{1}{2}\int_{\pi}^{3\pi}\left[r(\theta,t)\right]^2 d\theta$. The spiral shape function r(θ,t) is defined for θ in the range 0 to ∞, and $\pi \le \theta \le 3\pi$ corresponds to an outwardly moving spiral step segment, far from the core, which area increases linearly with time. τ is the average time required to generate successive spirals and loops that have the same area. (b) The time-dependent areas A of spirals and 2D TiN islands on TiN(111) terraces (shown in the image to the right) during annealing at T = 1694 K. Note that the left and bottom axes of the plot correspond to that of the spiral while the right and top axes correspond to the island. The spiral area increases linearly with time while the island area decreases. The fields of view in both images are ≈ 2.75 μm.

Finally, our model assumes a constant rate R for the emission/absorption of point defects at dislocation cores. R depends on two driving forces: (1) minimization of dislocation line energy and (2) equilibration of bulk point defects. In the first case, spiral and loop growth will continue to occur until the disappearance of the dislocation core. In the second case, however, R is a function of the concentration gradient between the surface and the bulk, which decreases with time. Hence, this driving force terminates upon equilibration of the bulk point defect concentration.

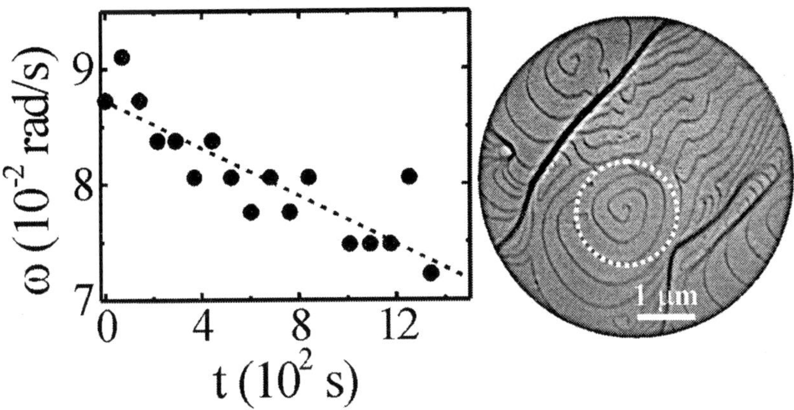

FIGURE 7. ω vs. t plot for successive spiral steps in the region highlighted in the associated LEEM image (field of view = 5.6 μm). The data were acquired during annealing a TiN(111) sample at T = 1727 K with p_{N_2} = 5x10^{-8} Torr. The dashed line is a linear least-squares fit to the measured data.

In order to test the validity of our model, we studied the long time behavior of spiral growth rates. Fig. 7 is a typical plot of ω vs. t values (solid circles) obtained from 17 successive spirals generated from the same dislocation core (highlighted in the associated LEEM image) while annealing a TiN(111) sample at T = 1727 K with p_{N_2} held constant at 5x10^{-8} Torr. We find that ω decreases monotonically from ≃ 0.09 rad/s to ≃ 0.07 rad/s, i.e. an ≃22% reduction in ω, within 1340 s at 1727 K. If we assume a linear relationship, the ω(t) data can be fit using least-squares analyses (dashed line in Fig. 7). The observed decrease in ω with time can be attributed to: (1) the effect of already existing steps on the nucleation of new steps (a "back-force" effect [79]) and/or (2) equilibration of the point-defect concentration in the bulk. In the first case, the geometry of the spiral structures, while in the second case, the N_2 pressure, could affect the TiN(111) spiral step growth kinetics. Since the spirals in our experiments are situated in very similar surface geometries, the effect of local environment on spiral growth kinetics cannot be determined. Thus, we focused on the effects of p_{N_2}.

We measured ω while varying p_{N_2} between zero and 5x10^{-7} Torr at constant T. Fig. 8 contains typical plots of ω (solid circles) and dA/dt (open squares) as a function of t at T = 1670 K. p_{N_2}, set initially to 5x10^{-8} Torr, is suddenly reduced to zero (≤ 5x10^{-9} Torr) at t = 390 s and then varied between zero and 5x10^{-7} Torr. ω and dA/dt, while exhibiting an overall decrease with time, do not vary systematically with p_{N_2}; suggesting that molecular N_2 does not significantly change the surface and bulk compositions of N-terminated TiN(111).

From an Arrhenius plot of ω(T) data, determined from separate sets of LEEM images acquired at four different temperatures T, we find an activation energy E_d = 4.5±0.2 eV; compared to 2.3±0.6 eV for the decay of 2D TiN islands on TiN(111).

The fact that we measure a higher activation barrier E_d for spiral growth than for island decay (and E_d is significantly lower than the desorption energies for TiN and Ti adspecies [76]) provides strong evidence that the dominant mass transport is along dislocation lines rather than surface diffusion or evaporation. Hence, E_d corresponds to the activation barrier for dislocation motion which is generally associated with point defect formation and migration along the dislocation (also referred to as "pipe diffusion") [80]. This is physically reasonable since bulk point defect migration, expected to influence surface dynamics at sufficiently high temperatures [56], has a smaller activation barrier along dislocation lines [80] than in dislocation-free areas.

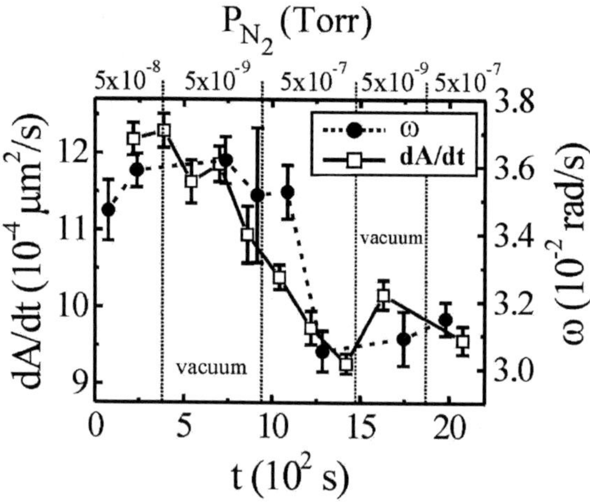

FIGURE 8. Plots of dA/dt (open squares) and ω (solid circles) vs. t for 2D TiN(111) islands during annealing at T = 1670 K while varying p_{N_2} between zero and $5x10^{-7}$ Torr as shown.

V. CONCLUSIONS

In this chapter, we discuss recent progress toward developing a fundamental understanding of morphological evolution on surfaces of compounds such as TiN. In situ high-temperature STM and LEEM studies provide insights into the effects of surface- and bulk-mass transport processes on surface step motion. Specifically, we focused on the effects of step permeability, step-step interactions, bulk diffusion, and surface-terminated dislocations on the coarsening/decay kinetics of 3D island stacks; and the dynamics of surface steps.

This report is intended to complement and extend the existing literature on both isotropic and anisotropic island dynamics. With the recent surge of interest in synthesis and characterization of nanostructures, there is growing need for understanding the atomic-scale mechanisms governing the formation and stability of nanostructures. Our studies are an effort to extend the existing formalism describing 2D island dynamics to the analyses of surface defects and anisotropic 3D nanostructures.

ACKNOWLEDGEMENTS

The authors gratefully acknowledge the financial support of the U.S. Department of Energy (DOE), Division of Materials Science, under Contract No. DEFG02-91-ER45439 through the University of Illinois Frederick Seitz Materials Research Laboratory (FS-MRL). We also appreciate the use of the facilities in the Center for Microanalysis of Materials, partially supported by DOE, at the FS-MRL.

REFERENCES

1. G. Binnig, H. Rohrer, C. Gerber, *Appl. Phys. Lett.* **40**, 178 (1982).
2. G. Binnig, C.F. Quate, C. Gerber, *Phys. Rev. Lett.* **56**, 930 (1986).
3. E.W. Müller, *Z. Phys.* **131**, 136 (1951).
4. S.C. Wang and G. Ehrlich, *Phys. Rev. Lett.* **79**, 4234 (1997).
5. S. C. Wang, U. Kürpick, and G. Ehrlich, *Phys. Rev. Lett.* **81**, 4923 (1998).
6. G. Ehrlich, *Surf. Sci.* **331-333**, 865 (1995) and references therein.
7. F. Besenbacher, P.T. Sprunger, L. Ruan, L. Olesen, I. Stensgaard and E. Lægsgaard, *Topics in Catalysis* **1**, 325 (1994).
8. L. Petersen, M. Schunack, B. Schaefer, T.R. Linderoth, P.B. Rasmussen, P.T. Sprunger, E. Lægsgaard, I. Stensgaard and F. Besenbacher, *Rev. Sci. Instr.* **72**, 1438 (2001).
9. E. Lægsgaard, L. Österlund, P. Thostrup, P.B. Rasmussen, I. Stensgaard, and F. Besenbacher, *Rev. Sci. Instr.* **72**, 3537 (2001).
10. K. Morgenstern, *Phys. Stat. Sol. (b)* **242**, 773 (2005).
11. M.J. Rost, L. Crama, P. Schakel, E. van Tol, G.B.E.M. van Velzen-Williams, C.F. Overgauw, H. ter Horst, H. Dekker, B. Okhuijsen, M. Seynen, A. Vijftigschild, P. Han, A. J. Katan, K. Schoots, R. Schumm, W. van Loo, T. H. Oosterkamp, and J.W.M. Frenken, *Rev. Sci. Instr.* **76**, 053710 (2005).
12. J. de la Figuera, J. E. Prieto, C. Ocal, and R. Miranda, *Solid State Communications* **89**, 815 (1994).
13. J.-M. Wen, S.-L. Chang, J.W. Burnett, J.W. Evans, and P.A. Thiel, *Phys. Rev. Lett.* **73**, 2591 (1994).
14. K. Morgenstern, G. Rosenfeld, B. Poelsema, and G. Comsa, *Phys. Rev. Lett.* **74**, 2058 (1995).
15. J.-M. Wen, J.W. Evans, M.C. Bartelt, J.W. Burnett, and P.A. Thiel, *Phys. Rev. Lett.* **76**, 652 (1996).
16. C.R. Stoldt, C.J. Jenks, P.A. Thiel, A.M. Cadilhe, and J.W. Evans, *J. Chem. Phys.* **111**, 5157 (1999).
17. W.W. Pai, A.K. Swan, Z. Zhang, and J.F. Wendelken, *Phys. Rev. Lett.* **79**, 3210 (1997).
18. T. R. Linderoth, S. Horch, L. Petersen, S. Helveg, E. Lægsgaard, I. Stensgaard, and F. Besenbacher, *Phys. Rev. Lett.* **82**, 1494 (1999).
19. K. Morgenstern, E. Lægsgaard, and F. Besenbacher, *Phys. Rev. Lett.* **86**, 5739 (2001).
20. K. Morgenstern, E. Lægsgaard, and F. Besenbacher, *Phys. Rev. B* **66**, 115408 (2002).
21. D. Schlößer, K. Morgenstern, L.K. Verheij, G. Rosenfeld, F. Besenbacher, and G. Comsa, *Surf. Sci.* **465**, 19 (2000).
22. Y. W. Mo, J. Kleiner, M. B. Webb, and M. G. Lagally, *Phys. Rev. Lett.* **66**, 1998 (1991).
23. H. Brune, H. Röder, C. Boragno, and K. Kern, *Phys. Rev. Lett.* **73**, 1955 (1994).
24. H. Brune, H. Röder, K. Bromann, and K. Kern, *Thin Solid Films* **264**, 230 (1995).
25. H.H. Teng, P.M. Dove, C.A. Orme, and J.J. De Yoreo, *Science* **282**, 724 (1998).
26. C.M. Pina, U. Becker, P. Risthaus, D. Bosbach, and A. Putnis, *Nature* **395**, 483 (1998).
27. E. Bauer, *Rep. Prog. Phys.* **57**, 895 (1994).
28. A.K. Schmidt, N.C. Bartelt, and R.Q. Hwang, *Science* **290**, 1561 (2000).
29. J.B. Hannon and R.M. Tromp, *Annual Review of Materials Research*, **33**, 263 (2003).
30. G. Ertl and H.H. Rotermund, *Current Opinion in Solid State and Materials Science* **1**, 617 (1996).
31. C.M. Schneider and G. Schönhense, *Rep. Prog. Phys.* **65**, 1785 (2002).
32. M. Zinke-Allmang, L.C. Feldman, and M.H. Grabow, *Surf. Sci. Rep.* **16**, 377 (1992).
33. M. Zinke-Allmang, *Thin Solid Films* **346**, 1 (1999).
34. R. M. Tromp and J. B. Hannon, *Surf. Rev. Lett.* **9**, 1565 (2002).
35. H.-C. Jeong and E.D. Williams, *Surf. Sci. Rep.* **34**, 171 (1999).
36. M. Giesen, *Prog. Surf. Sci.* **68**, 1 (2001) and references therein.
37. M. Nowicki, A. Emundts, and H. P. Bonzel, *Prog. Surf. Sci.* **74**, 123 (2003).
38. H. P. Bonzel, *Phys. Rep.* **385**, 1 (2003).

39. F.H. Baumann, D.L. Chopp, T. Díaz de la Rubia, G. H. Gilmer, J.E. Greene, H. Huang, S. Kodambaka, P. O'Sullivan, and I. Petrov, *MRS Bull.* **26**, 182 (2001).
40. S. Kodambaka, S.V. Khare, I. Petrov, J.E. Greene. *Surface Science Reports* **60**, 55 (2006).
41. I. Petrov, F. Adibi, J.E. Greene, W.D. Sproul, and W.-D. Münz, *J. Vac. Sci. Technol. A* **10**, 3283 (1992).
42. S. Kodambaka, S.V. Khare, V. Petrova, A. Vailionis, I. Petrov, and J.E. Greene, *Surf. Sci.* **513**, 468 (2002).
43. S. Kodambaka, V. Petrova, S.V. Khare, D. Gall, A. Rockett, I. Petrov, J.E. Greene, *Phys. Rev. Lett.* **89**, 176102 (2002).
44. S. Kodambaka, N. Israeli, J. Bareño, W. Święch, K. Ohmori, I. Petrov, and J.E. Greene, *Surf. Sci.* **560**, 53 (2004).
45. S. Kodambaka, J. Bareño, S.V. Khare, W. Swiech, I. Petrov, J.E. , Greene, *J. Appl. Phys.* **98**, 34901 (2005).
46. Image SXM, developed by Prof. Steve Barrett, Liverpool, UK (http://www.liv.ac.uk/~sdb/ImageSXM/).
47. W.W. Mullins, *Interface Sci.* **9**, 9 (2001) and references therein.
48. J.G. McLean, B. Krishnamachari, D.R. Peale, E. Chason, J.P. Sethna, and B.H. Cooper. *Phys. Rev. B* **55**, 1811 (1997).
49. L. Kuipers, M.S. Hoogeman, J.W.M. Frenken, and H. van Beijeren, *Phys. Rev. B* **52**, 11387 (1995).
50. C. Pearson, B. Borovsky, M. Krueger, R. Curtis, and E. Ganz. *Phys. Rev. Lett.* **74**, 2710 (1995).
51. *CRC Handbook of Chemistry and Physics*, 82nd edition 2001-2002, pp. 4.
52. S. Kodambaka, V. Petrova, S.V. Khare, D.D. Johnson, I. Petrov, and J.E. Greene, *Phys. Rev. Lett.* **88**, 146101 (2002).
53. W.K. Burton, N. Cabrera, and F.C. Frank, *Philos. Trans. R. Soc. London A* **243**, 299 (1951).
54. S. Tanaka, N.C. Bartelt, C.C. Umbach, R.M. Tromp, and J.M. Blakely, *Phys. Rev. Lett.* **78**, 3342 (1997).
55. K. Thürmer, J.E. Reutt-Robey, E.D. Williams, M. Uwaha, A. Emundts, and H.P. Bonzel, *Phys. Rev. Lett.* **87**, 186102 (2001).
56. K.F. McCarty, J.A. Nobel, and N.C. Bartelt, *Nature* **412**, 622 (2001).
57. A. Rettori and J. Villain, *J. Phys. (France)* **49**, 257 (1988).
58. N. Israeli and D. Kandel, *Phys. Rev. Lett.* **80**, 3300 (1998).
59. N. Israeli and D. Kandel, *Phys. Rev. B* **60**, 5946 (1999).
60. M. Giesen, G. Schulze Icking-Konert, and H. Ibach, *Phys. Rev. Lett.* **80**, 552 (1998).
61. K. Thürmer, J.E. Reutt-Robey, E.D. Williams, M. Uwaha, A. Emundts, and H.P. Bonzel, *Phys. Rev. Lett.* **87**, 186102 (2001).
62. K. Morgenstern, G. Rosenfeld, G. Comsa, M.R. Sørensen, B. Hammer, E. Lægsgaard, and F. Besenbacher, *Phys. Rev. B* **63**, 045412 (2001).
63. S. Kodambaka, S. V. Khare, V. Petrova, D.D. Johnson, I. Petrov, and J.E. Greene, *Phys. Rev. B* **67**, 035409 (2003).
64. S. Kodambaka, S.V. Khare, W. Święch, K. Ohmori, I. Petrov, and J.E. Greene, *Nature* **429**, 49 (2004).
65. V. Bulatov, F.F. Abraham, L. Kubin, B. Devincre, and S. Yip, *Nature* **391**, 669 (1998).
66. B.W. Lagow, I.M. Robertson, M. Jouiad, D.H. Lassila, T.C. Lee, and H.K. Birnbaum, *Material Science & Engineering A*. **309-310**, 445-450 (2001).
67. A.R. Verma and S. Amelinckx, *Nature* **167**, 939 (1951).
68. N. Cabrera and M.M. Levine, *Phil. Mag.* **1**, 450 (1956).
69. T. Surek, G.M. Pound, and J.P. Hirth, *Surf. Sci.* **41**, 77 (1974).
70. S.-J. Tang, S. Kodambaka, W. Święch, I. Petrov, C.P. Flynn, and T.C. Chiang, *Phys. Rev. Lett.* **96**, 126106 (2006).
71. R.M. Tromp and M.C. Reuter, *Ultramicroscopy* **36**, 99 (1991).
72. F.C. Frank and W.T. Read, *Phys. Rev.* **79**, 722 (1950).
73. J. Bardeen and C. Herring in *Imperfections in Nearly Perfect Crystals,* edited by W. Shockley, J.H. Hollomon, R. Maurer, and F. Seitz, New York: Wiley, 1952.
74. D. Hull and D.J. Bacon, in *Introduction to Dislocations,* Oxford: Pergamon Press, 1984.
75. J.E. Sundgren, B.O. Johansson, A. Rockett, S.A. Barnett, and J.E. Greene, in *Physics and Chemistry of Protective Coatings,* edited by J.E. Greene, W.D. Sproul, and J.A. Thornton, American Institute of Physics, New York, 1986.
76. D. Gall, S. Kodambaka, M.A. Wall, I. Petrov, and J.E. Greene, *J. Appl. Phys.* **93**, 9086 (2003).
77. J. Weertman and J.R. Weertman, in *Elementary Dislocation Theory*, edited by M.E. Fine, J. Weertman, and J.R. Weertman, New York: Macmillan Company, 1964.
78. J.F. Justo, M. de Koning, W. Cai, and V.V. Bulatov, *Material Science & Engineering A* **309-310**, 129 (2001).
79. T. Surek, J.P. Hirth, and G.M. Pound, *Journal of Crystal Growth* **18**, 20 (1973).
80. V.L. Indenbom and Z.K. Saralidze, in *Elastic Strain Fields and Dislocation Mobility,* edited by V. Indenbom and J. Lothe, Amsterdam: Elsevier Science,1992.

Electrodynamic Properties of Nanoscopically Inhomogeneous Materials

D. E. Aspnes

Department of Physics, NC State University, Raleigh, NC 27695-8202, USA

Abstract. Electromagnetic properties of nanoscopically inhomogeneous materials are discussed, starting from the fundamental treatment of the dielectric response. Topics covered include effective-medium theory, field enhancement through engineered nanostructures, plasmons, and negative-index metamaterials.

BACKGROUND AND INTRODUCTION

The interaction of light with materials has been a topic of investigation since the development of electromagnetic theory. The renewed interest today is a result of advancing technologies that can create a wide variety of engineered nanostructures. This has stimulated new applications that require, and improvements in diagnostic tools that enable, a better understanding of the properties of materials and structures with dimensions on nanoscopic scales. Applications include the development of materials with new catalytic properties; the use of highly localized interface plasmons for transmitting information on length scales much smaller than the wavelength λ of light for telecommunications; the analysis of scatterometry data to determine critical dimensions of structures in integrated-circuits technology, thereby replacing electron microscopy in applications where damage to materials during fabrication cannot be tolerated; the use of inhomogeneities to produce locally high fields from localization and resonance effects; and finally the possibility of creating novel physical phenomena, for example those associated with "left-handed" (negative-index) materials, by tailoring micro- and nanostructures.

Clearly, on the atomic scale everything is inhomogeneous, so some comments about relative dimensions are a reasonable place to start. First, it is important to recognize that physics always takes place in real time and on the atomic scale, so a fundamental understanding of any physical phenomenon must begin here. However, the probes that we use, and particularly optical probes, are not sensitive to atomic-scale dimensions but rather respond to properties averaged over longer length scales. This is actually an advantage, because the detailed knowledge of the behavior of every electron and atom in a solid, for example their individual responses to an electromagnetic field, would provide far more information that we need, or could use. In principle the connection between the atomic scale and our observations is made by (1) solving the atomic-scale response problem exactly, then (2) averaging the responses to obtain the observables of interest.

Regarding the interaction of light with materials, the first step involves solving a force equation to determine the dipole (charge times displacement) for each charge that results

CP885, *Advanced Summer School in Physics 2006, Frontiers in Contemporary Physics—EAV06,* edited by O. Miranda, M. Carbajal, L. M. Montaño, O. Rosas-Ortiz, and S. A. Tomás Velázquez

from the incoming electric field, while the second step involves obtaining the dielectric response function ε by calculating the resulting macroscopic, or average, dipole density **P**. The formal definition of ε is given by

$$\mathbf{D} = \varepsilon \mathbf{E} = \mathbf{E} + 4\pi \mathbf{P}, \tag{1}$$

where **D** and **E** are the displacement and electric fields, respectively, averaged over macroscopic length scales. Further details will be given below. The quantity ε is then used in the macroscopic Maxwell equations to describe the interaction of light with materials. In the following we will assume that λ is substantially larger than both atomic and nanoscopic scales, so averages yield macroscopic parameters that can be used in the macroscopic equations.

The first derivation that directly related atomic-scale properties to macroscopic properties (in this case, the refractive index n = $\sqrt{\varepsilon}$) was reported by Clausius in 1879 [1], shortly after Maxwell published his famous treatise on electromagnetic theory. Clausius' historically significant result is the Clausius-Mossotti (CM) relation

$$\frac{\varepsilon - 1}{\varepsilon + 2} = \frac{4\pi}{3} n\alpha, \tag{2}$$

where α is the polarizability at a given point of a simple cubic lattice and (in this equation) n is the number density of such points. Mossotti's name also appears because in his derivation Clausius used to advantage the "cavity" approach that Mossotti [2] introduced to establish the connection between molecules and an effective medium. An atom-based treatment was also used by Ewald [3] in 1912 and Oseen [4] in 1915 to describe reflectance on the atomic scale, which for linear optics is summarized as the Ewald-Oseen extinction theorem. Unfortunately, this derivation was never incorporated into textbooks and is now largely forgotten.

With this background we can discuss the treatment of inhomogeneities on length scales that are larger than atomic. If the homogeneous regions are much larger than λ, we describe the interaction by geometric optics. Rigorously speaking, this approximation also requires us to make some assumptions about the incoming radiation, specifically that the regions are large compared to the coherence length of the source, but we do not consider this case further. If the sizes are of the order of λ, then scattering theory is required. Scattering calculations involve significant amounts of computation, but these are now manageable with high-speed computers. However, scattering is a complete topic in itself, so this case will also not be considered further. If the sizes are much smaller than λ but the individual regions are large enough to retain the dielectric properties of bulk material, then the appropriate approach is effective-medium theory. This is the general topic covered in this paper. If the sizes become sufficiently small, then the response of the regions must also be modified, including for example the effects of quantum confinement. For metals this usually results in increased scattering and therefore broadening of structure in optical spectra, whereas for semiconductors shifts of energy levels will occur.

In the nanoscale inhomogeneity regime, the first lasting contribution was that of Garnett [5, 6], who established the mathematical framework for dealing with *cermet* nanostructures. A cermet nanostructure consists of inclusions embedded in a host dielectric, where the dielectric response of the inclusions is different from that of the host. The defining characteristic of the host phase is continuity, so the inclusions are necessarily isolated from each other. Garnett was interested specifically in explaining the apparently anomalous red color of glass containing relatively sparse and therefore noninteracting nanoscopically sized particles of Au, but the theory that he developed had much wider applicability and is now known as the Maxwell Garnett (MG) effective-medium theory. In effect, this is a generalization of the CM relation to configurations where the host is not empty space, but a dielectric in its own right. What Garnett also discovered is the type of bulk plasmon that is characteristic of nanoscopically inhomogeneous material. This plasmon is antisymmetric and hence couples to electromagnetic radiation very efficiently, as will be discussed more completely below.

A more general cermet theory, applicable even when the volume fraction of the host is vanishingly small, was developed by Sen, Scala, and Cohen (SSC) [7] in the context of describing the percolation properties of sedimentary rocks. Their purpose was to develop more efficient methods of extracting oil, but the formalism is applicable to dielectrically inhomogeneous materials as well. In the original application the host material is water and the inclusions are individual grains of rock that contact each other at points of measure zero. In the electromagnetic context an analogous configuration occurs in polycrystalline metal films where each metal grain is completely covered by a thin oxide layer. Since the oxide isolates the grains it functions as the host material, but since it is very thin the grains are necessarily highly interacting. Hence MG theory is not applicable. The SSC formalism was shown to provide an accurate description of the dielectric properties of evaporated Rh films [8], which has this nanostructure.

The other commonly encountered nanostructure is the *aggregate* configuration, where all constituents are mixed together at random. In this case the individual regions of each subspecies can contact other regions of the same subspecies, so a host phase does not exist. Most deposited thin films fall in this category, along with surface roughness on the nanoscopic length scale. Again, MG theory cannot be used. The appropriate effective-medium theory for aggregate nanostructures was developed in 1935 by Bruggeman [9] as a mean-field extension of the MG approach. Bruggeman's model is now known as the Bruggeman effective-medium approximation, or simply the EMA.

Limit theorems are another important aspect of the theory of the dielectric response of nanoscopically inhomogeneous materials. The first such were developed in 1910 by Wiener [10]. Wiener showed that no matter how clever one might be in creating a nanoscopically inhomogeneous two-phase composite, ε could not assume arbitrary values but had to lie within well-defined boundaries, the Wiener limits. For a two-phase composite the allowed range of ε can be calculated even if nothing is known about its composition or nanostructure. A more stringent set of limits was obtained by Hashin and Shtrikman in 1962 [11], which pertain to a two-component composites where the relative volume fraction of the two constituents is known. A further refinement was obtained by Bergman and Milton in 1980 [12, 13], which applies to configurations where the relative volume fraction is known and in addition the composite is known to be macroscopically isotropic in two or three dimensions.

Thus our understanding of the optical properties of nanostructurally inhomogeneous materials was rather thoroughly established well before these materials began to receive their present level of attention. Unfortunately, much of this work is unfamiliar to present workers in the nano field, which has led to confusion, particularly regarding plasmons. Here, I discuss basic principles and methods of evaluating the dielectric properties of nanoscopically inhomogeneous materials, then discuss in more detail effective-medium theory, plasmons, and negative-index structures. The unusual properties of the latter structures have recently come under intensive investigation after it was discovered that they could be realized in the microwave region of the spectrum [14].

FUNDAMENTALS

The fundamental equation for describing the interaction of fields and materials is the Lorentz force law, which for a point charge q is

$$\mathbf{F} = q\mathbf{E} + \frac{\mathbf{v}}{c} \times \mathbf{B}. \tag{3}$$

where $\mathbf{B}$ is the magnetic flux density. A crucial point, which is obscured with SI units but highlighted with the cgs units used above, is the prefactor $\mathbf{v}/c$ in the magnetic term. That this prefactor is necessary follows from the fact that $\mathbf{E}$ and $\mathbf{B}$ are both elements of the second-rank field tensor of special relativity, and hence must have the same units. An immediate consequence is that magnetic interactions are much weaker than electric interactions for particles moving at laboratory speeds. This explains the focus on $\mathbf{E}$ in treatments of the interaction of plane waves with materials. The fact that there is no magnetic charge is a second reason, because the lowest nonvanishing magnetic interactions involve dipoles that typically respond orders of magnitude too slowly to follow optical frequencies. However, for negative-refractive-index behavior, magnetic effects are required. At present these can be simulated only by engineered structures, as will be discussed below.

More generally, it can be shown that magnetic fields are a purely relativistic phenomenon, since all of electrodynamics follows from Coulomb's Law of electrostatics and special relativity. It is sometimes surprising to realize that relativity is not some exotic phenomenon that applies only to e.g. muons moving near the speed of light, but is in fact part of our everyday existence. For example if in our universe c were twice as large, electric motors would develop only half as much torque at a given current.

We now fill in the details regarding the origin of ε, considering first homogeneous materials. As mentioned in the Introduction, the calculation of ε is a two-step process, requiring first the solution of Newton's Second Law of Motion $\mathbf{F} = m\mathbf{a}$ on the atomic scale followed by spatially averaging over the resulting atomic-scale electric field $\mathbf{e}(\mathbf{r})$ and dipoles $q\Delta\mathbf{r}$. Since λ is much larger than atomic dimensions, the force problem can be treated by electrostatics even though optical frequencies are in the 10^{15} Hz range. Accordingly, we suppose that an electric field $\mathbf{E}_0$ is applied to a material consisting of charges q that are originally in equilibrium positions. Since the field exerts a force $\mathbf{F} = q\mathbf{E}$ on the charges q, they will move by displacements $\Delta\mathbf{r}$ to new positions to achieve a new equilibrium. The calculation is intrinsically self-consistent, because the local fields at the charge sites will be modified as a result of the charge displacements themselves. Hence iteration may be necessary to arrive at a suitably accurate solution. The results of the solution are the new atomic-scale electric field $\mathbf{e}'(\mathbf{r})$ and dipoles $q\Delta\mathbf{r}$.

The second step is to average the atomic-scale solution to calculate the macroscopic observables. The macroscopic electric field $\langle \mathbf{E} \rangle$ is defined for example by

$$\langle \mathbf{E}(\mathbf{r}) \rangle = \int d^3r' W(\mathbf{r}-\mathbf{r}')\mathbf{e}(\mathbf{r}'), \tag{4}$$

where $W(\mathbf{r}-\mathbf{r}')$ is a weighting function that is positive definite, slowly varying on the atomic scale but varying sufficiently rapidly on the laboratory scale to follow laboratory-scale spatial dependences, and integrates to a value of 1 over all space. Surprisingly, W need not be specified any more completely than this to allow macroscopic quantities to be calculated from their microscopic counterparts.

If we now take the divergence of the above equation we obtain

$$\begin{aligned} \nabla_{\mathbf{r}} \cdot \langle \mathbf{E}(\mathbf{r}) \rangle &= \nabla_{\mathbf{r}} \cdot \int d^3r' W(\mathbf{r}-\mathbf{r}')\mathbf{E}(\mathbf{r}') \\ &= \nabla_{\mathbf{r}} \cdot \int d^3r'' W(\mathbf{r}-\mathbf{r}'')\mathbf{E}(\mathbf{r}'') \\ &= \int d^3r'' W(\mathbf{r}'')\nabla_{\mathbf{r}} \cdot \mathbf{E}(\mathbf{r}-\mathbf{r}'') \\ &= 4\pi \int d^3r'' W(\mathbf{r}'')\rho(\mathbf{r}-\mathbf{r}'') \\ &= \langle \rho(\mathbf{r}) \rangle. \end{aligned} \tag{5}$$

This does not seem like progress, but we now note that the macroscopic average $\langle \rho \rangle$ is not the average $\langle \rho_0 \rangle$ that existed before the field was applied, which we presumably know, but includes the displacements caused by the applied field. To isolate these field-induced changes we do the equivalent of perturbation theory in quantum mechanics, writing

$$\rho(\mathbf{r}') = \rho_0(\mathbf{r}') + [\rho(\mathbf{r}') - \rho_0(\mathbf{r}')]. \tag{6}$$

Inserting this expression in the divergence equation yields

$$\nabla \cdot \langle \mathbf{E}(\mathbf{r}) \rangle = 4\pi \langle \rho_0 \rangle = +4\pi \int d^3r' W(\mathbf{r}-\mathbf{r}')[\rho(\mathbf{r}') - \rho_0(\mathbf{r}')]. \tag{7}$$

We now consider a discrete-charge model, writing

$$\rho_0(\mathbf{r}') = \sum_i q_i \delta(\mathbf{r}' - \mathbf{r}_i - \Delta\mathbf{r}_i), \tag{8}$$

for each charge q_i located at the unperturbed lattice point $\mathbf{r}_i$. We obviously obtain ρ_0 by setting $\Delta\mathbf{r}_i = 0$. The resulting integral is easily evaluated with the result

$$\nabla \cdot \langle \mathbf{E}(\mathbf{r}) \rangle = 4\pi \langle \rho_0 \rangle = +4\pi \sum_i q_i [W(\mathbf{r}-\mathbf{r}_i - \Delta\mathbf{r}_i) - W(\mathbf{r}-\mathbf{r}_i)]. \tag{9}$$

By performing a Taylor-series expansion of $W(\mathbf{r}-\mathbf{r}'-\Delta\mathbf{r}')$ we can convert the sum to the divergence of an integral:

$$4\pi \sum_i q_i [W(\mathbf{r}-\mathbf{r}_i - \Delta\mathbf{r}_i) - W(\mathbf{r}-\mathbf{r}_i)] = -4\pi \nabla_{\mathbf{r}} \cdot \int d^3r' W(\mathbf{r}-\mathbf{r}')[n(\mathbf{r}')q\Delta\mathbf{r}'] \tag{10}$$

where $n(\mathbf{r}')$ is the number density of lattice sites. But $q\Delta\mathbf{r}'$ is just the dipole at the lattice site, hence we have

$$4\pi \sum_i q_i [W(\mathbf{r}-\mathbf{r}_i-\Delta\mathbf{r}_i) - W(\mathbf{r}-\mathbf{r}_i)] = -4\pi\nabla_{\mathbf{r}} \cdot \int d^3r' W(\mathbf{r}-\mathbf{r}')[n(\mathbf{r}')p(\mathbf{r}')] \quad (11)$$
$$= \nabla_{\mathbf{r}} \cdot \langle \mathbf{P}(\mathbf{r}) \rangle,$$

where $\mathbf{P}(\mathbf{r})$ is the required macroscopic dipole density. We now combine the divergences, obtaining finally

$$\nabla \cdot [\langle \mathbf{E}(\mathbf{r}) \rangle + 4\pi \langle \mathbf{P}(\mathbf{r}) \rangle] = 4\pi \langle \rho_0 \rangle. \quad (12)$$

We define the quantity in brackets to be the displacement field $\mathbf{D}$, writing

$$\mathbf{D} = \varepsilon \langle \mathbf{E}(\mathbf{r}) \rangle. \quad (13)$$

What we have accomplished is the following. First, we have identified the macroscopic equivalent of the microscopic continuity equation on normal $\mathbf{E}$, which now applied to normal $\mathbf{D}$. Second, we have developed specific recipes for calculating $\langle \mathbf{E} \rangle$ and bold $\langle \rho_0 \rangle$, and in particular $\langle \mathbf{P} \rangle$, and therefore ε in terms of the atomic-scale properties of the system. Third, if we generalize the force equation to include anisotropic and anharmonic terms we can extend our calculation not only to anisotropic materials but to nonlinear optics as well. Curiously, this is one example where the calculation of nonlinear optical properties is easier than that of linear properties, because nonlinear responses are weak enough that self-consistency is not required [15]. As a comment, a configuration that can be solved exactly following the above procedure the simple cubic lattice of polarizable points. The result is the CM relation.

The most elementary form of ε is obtained by applying the above procedure to the Lorentz oscillator, which is described by the force equation

$$\mathbf{F} = m\mathbf{a} = q\mathbf{E}e^{-i\omega t} - b\mathbf{v} - \kappa\Delta\mathbf{r}, \quad (14)$$

where m and q are the mass and charge, respectively, of the particle, b is a damping constant representing energy loss, and κ is a Hookes-Law restoring force constant. Assuming that the displacement $\Delta\mathbf{r}$ has the form

$$\Delta\mathbf{r} = \mathbf{r}e^{-i\omega t}, \quad (15)$$

a direct substitution yields

$$\Delta\mathbf{r} = \frac{q\mathbf{E}}{\kappa - m\omega^2 - i\omega b}. \quad (16)$$

Substituting this result into the defining equation for ε yields

$$\varepsilon = 1 + \frac{4\pi n q^2}{\kappa - m\omega^2 - i\omega b} \quad (17)$$
$$= 1 + \frac{\omega_p^2}{\omega_0^2 - \omega^2 - i\omega\Gamma},$$

where n is the volume density of point charges q, and the parameters in the last equation are the square of the *plasma frequency* $\omega_p^2 = 4\pi nq^2/m$, the square of the resonant frequency $\omega_0^2 = \kappa/m$, and the *broadening parameter* $\Gamma = b/m$. The above expression contains all the physics of dielectric responses, as seen in its general properties: ε is greater than 1 at $\omega = 0$, increases as ω increases from 0 up to near ω_0, approaches 1 as $\omega \to \infty$, and exhibits resonance behavior with a Lorentzian absorption lineshape with width Γ and centered about ω_0. The above expression also satisfies linearity, causality (the poles are both located in the lower half of the complex ω plane), and the Kramers-Kronig relations. Because Maxwell's Equations are linear, more Lorentz-oscillator terms can be added to ε to represent more complicated structures, for example the dielectric functions of semiconductors. This is the basis for the *spectral representation* of the dielectric functions of materials.

ASPECTS OF NANOSCOPIC INHOMOGENEITIES

Effective-medium theories

The above concepts, developed for inhomogeneities on the atomic scale, apply to nanoscopically inhomogeneous materials as well. We assume that the characteristic dimensions of the inhomogeneous regions are small compared to λ yet large enough for the regions to retain their own dielectric identities. The prototypical configuration for effective-medium theory is an isolated sphere of radius a and dielectric function ε_a located in a host medium of dielectric function ε_h that is effectively infinite in extent. If a uniform field $\mathbf{E}_0 = E_0\hat{\mathbf{z}}$ is applied to this configuration, a straightforward electrostatics calculation shows that the field in the spherical inclusion is

$$\mathbf{E}_{in} = \frac{3\varepsilon_h}{\varepsilon_a + 2\varepsilon_h} E_0\hat{\mathbf{z}} \tag{18}$$

and that outside the inclusion is

$$\mathbf{E}_{out} = E_0\hat{\mathbf{z}} - \frac{\varepsilon_a - \varepsilon_h}{\varepsilon_a + 2\varepsilon_h} E_0 \frac{3(\mathbf{r}\cdot\hat{\mathbf{z}})\mathbf{r} - r^2\hat{\mathbf{z}}}{r^5}. \tag{19}$$

Averaging both **E** and **D** over inclusion and host materials and using the definition of ε yields the Maxwell Garnett expression:

$$\frac{\varepsilon - \varepsilon_h}{\varepsilon_a + 2\varepsilon_h} = f_a \frac{\varepsilon_a - \varepsilon_h}{\varepsilon_a + 2\varepsilon_h}, \tag{20}$$

where f_a is the volume fraction of material a. The result can clearly be generalized to more than one type of inclusion, for example a second type with dielectric function ε_b:

$$\frac{\varepsilon - \varepsilon_h}{\varepsilon_a + 2\varepsilon_h} = f_a \frac{\varepsilon_a - \varepsilon_h}{\varepsilon_a + 2\varepsilon_h} + f_b \frac{\varepsilon_b - \varepsilon_h}{\varepsilon_b + 2\varepsilon_h}. \tag{21}$$

Suppose now that the volume fractions $f_a + f_b = 1$, that f_a and f_b are comparable, and that the regions are thoroughly mixed on a random basis (aggregate nanostructure).

Bruggeman argued that the response of this configuration is best approximated by a model where each grain of either species is considered to be an inclusion in the effective medium itself, in which case $\varepsilon_h = \varepsilon$ and the left-hand side of Eq. (20) vanishes. The result is an expression that does not favor one species over the other, but can still be solved for ε. This mean-field approach is the EMA. Obviously, for more complex materials additional terms can also be added.

The spherical geometry is by no means unique, and if the inclusions are nonspherical the *screening factor* 2 in the denominators of the above equations will take on different values. These values can be obtained by repeating the electrostatic solution for different shapes. For example for two-dimensional symmetry (cylinders of circular cross section) the value is 1, whereas for laminar structures the value is 0 if the field is perpendicular to the laminations or ∞ if the field is parallel to the laminations. In the former case the configuration is equivalent to capacitors in series, and in the latter to capacitors in parallel.

Limit Theorems

Limit theorems establish constraints on possible values of ε for two-component composites, and therefore give some measure of reliability concerning parameters used to describe these materials even if nothing is known about their compositions or nanostructures. The extreme limiting cases occur for the laminar nanostructure mentioned above, which can be oriented such that all internal boundaries are either parallel or perpendicular to the applied field. By the continuity conditions on normal **E** and **D**, if all interfaces are parallel to the applied field **E**, no screening occurs and **E** is uniform throughout. On the other hand, if all boundaries are perpendicular to the applied field then normal **D** is uniform throughout and the screening charge is maximized. Since there can never be less screening than no screening nor more screening charge than the maximum possible, it follows that these two orientations of the *same* nanostructure define absolute limits on possible values of ε for any composite regardless of relative fractions of the constituents and/or their nanostructures. These are the *Wiener limits*.

For a two-phase composite with dielectric functions ε_a and ε_b the former boundary is easily constructed graphically as the straight line joining the points ε_a and ε_b in the complex ε plane. Because the transformation is conformal, the other boundary, which is also defined by the line $0 \leq f_a = 1 - f_b \leq 1$, is transformed into a circle. The circle is also easily constructed graphically because it passes through ε_a, ε_b, and 0. The fact that 0 also lies on the circle is easily seen by letting either f_a or f_b approach ∞. This can be done mathematically, although values of f_a and f_b greater than 1 obviously cannot be realized physically.

The more restrictive HS and BM limit theorems pertain when other information is available. The HS limits apply when f_a and f_b are known. Here, the allowed values of ε are those contained between the circles that pass through the points ε_a or ε_b and the values of the two Wiener limits at the known value of f_a. The BM limits apply if in addition the configuration is isotropic in two or three dimensions. Here, the allowed region now lies between the circles defined by the previous two HS points and those obtained by evaluating the MG expression first with a then b as the host material.

All the above results can be summarized in a single equation [16]

$$\varepsilon = \frac{q\varepsilon_a\varepsilon_b + (1-q)\varepsilon_b(f_a\varepsilon_a + f_b\varepsilon_b)}{(1-q)\varepsilon_h + q(f_a\varepsilon_a + f_b\varepsilon_b)} \tag{22}$$

where $f_a + f_b = 1$, $0 \leq q \leq 1$ is the effective screening parameter, and ε_h is the host dielectric function. The Wiener limits are obtained by setting $q = 0$ and 1 then evaluating the expression as f_a increases from 0 to 1. The HS limits are obtained by fixing f_a at its known value, setting $\varepsilon_h = \varepsilon_a$ or ε_b, and letting q increase from 0 to 1. The BM limits are obtained by fixing f_a at its known value, setting $q = 1/2$ or 1/3 for two- or three-dimensional isotropy, respectively, setting $\varepsilon_h = x\varepsilon_a + (1-x)\varepsilon_b$ and $\varepsilon_h^{-1} = x\varepsilon_a^{-1} + (1-x)\varepsilon_b^{-1}$, then letting x increase from 0 to 1. The BM limits are therefore obtained by imposing the Wiener limits on ε_h. Consistent with conformal mapping, all boundaries defined above are circular arcs in the complex ε plane.

Effective-medium theories can be applied to describe a wide range of physical phenomena. Most evaporated films are polycrystalline, and hence composites of the film material and empty space. In particular, the boundaries between grains in close-packed fcc and hcp polycrystalline materials such as Au typically correspond to about 1/2 monolayer of voids. Depending on preparation method, the void fraction may be higher. Au is a good test case, for the EMA with appropriate values of void fractions was shown to convert any of the relatively wide differences among reported ε spectra for Au into any other [17]. Other applications of the EMA include representation of nanoscopically inhomogeneous surface regions of deposited materials.

Generation of high internal fields

Having outlined the general treatment of the dielectric response of inhomogeneous materials, we now consider implications of the solutions, beginning with the generation of high internal fields. Such phenomena depend entirely on nanostructure, because the overall average electric field $\langle \mathbf{E} \rangle$ within a medium is completely determined by the surface potential and hence is fixed. This is easily shown by considering the definition of the average field $\langle \mathbf{E} \rangle$:

$$\begin{aligned} \langle \mathbf{E} \rangle &= \frac{1}{V}\int_V d^3r'\mathbf{E}(\mathbf{r}') \qquad (23) \\ &= -\frac{1}{V}\int_V d^3r'\nabla\phi(\mathbf{r}') \\ &= -\frac{1}{V}\int_S d^2r'\hat{\mathbf{n}}\phi_S(\mathbf{r}'), \qquad (24) \end{aligned}$$

where S is the surface containing the volume V, and $\hat{\mathbf{n}}$ is the unit outward-pointing normal vector to S at the point $\mathbf{r}'$. Since the surface potential ϕ_S is assumed to be fixed, the average field inside a given material cannot depend on internal details.

However, nanostructure engineering can be used to create inhomogeneities that generate screening charge, and therefore open a pathway to high internal fields. This does not violate the constraint on the overall average field, because internal fields can be either positive or negative. Thus large local fields can be generated even though the overall average remains the same. To show this we again consider the solution of the electrostatic problem of a spherical inclusion of dielectric function ε_a in the host medium of dielectric function ε_h, Eqs. (18) and (19). Here, the maximum field occurs at the host-medium side of the spherical boundary and has the value

$$\mathbf{E}_{max} = \frac{3\varepsilon_a}{\varepsilon_a + 2\varepsilon_h}\mathbf{E}. \tag{25}$$

The possibility of high internal fields follows from the structure of Eq. (25). If the inclusion ε_a is a metal and the host ε_h a dielectric, then below the plasma edge ε_a will be essentially negative whereas ε_h is typically positive. At some wavelength the real part of ε_a will be equal to the negative of the real part of $2\varepsilon_h$, so depending on the residual imaginary parts of ε_a and ε_h very high internal fields are possible. As expected, at this resonance condition the fields inside and outside the inclusion are essentially 180° out of phase. For Au nanospheres in glass this resonance occurs in the red part of the optical spectrum, leading to the colors that initiated the study by Garnett. Note that the resonance depends only on λ; as long as the characteristic dimensions of the spherical inclusions are significantly less than λ but not so small that quantum effects are important, the actual sizes of the inclusions are irrelevant. The mathematical structure of the EMA leads to a weaker singularity. This shows, not surprisingly, that more definitive resonances are obtained with cermet rather than aggregate nanostructures.

The limit theorems can be applied to estimate the magnitudes of internal fields, and more important for nanostructural engineering, to identify the nanostructures for which maximum values are obtained [16]. We begin by defining a field enhancement factor $\gamma_i = E_i/E_0$ for each medium i, where E_i is the average field within the medium $i = a$ or b, and E_0 is the applied external field, uniform in the absence of inhomogeneities. With this definition it can be shown that

$$\begin{aligned} \gamma_a &= \frac{\varepsilon_b - \varepsilon}{f_b(\varepsilon_b - \varepsilon_a)}; \\ \gamma_b &= \frac{\varepsilon - \varepsilon_a}{f_a(\varepsilon_b - \varepsilon_a)}. \end{aligned} \tag{26}$$

Using the previous expression it follows that

$$\gamma_{a,b} = \frac{q\varepsilon_{a,b} + (1-q)\varepsilon_h}{(1-q)\varepsilon_h + q(f_a\varepsilon_b + f_b\varepsilon_a)}. \tag{27}$$

The resonance structure of this equation is obvious. By the principles of conformal mapping, the boundaries obtained as q or x vary from 0 to 1 are circles in the complex ε plane. Again, the Wiener limits are particularly simple. For $q = 0$ no screening occurs and the enhancement factors are identically 1 in both materials over the entire range of possible compositions. For $q = 1$ the allowed values of γ_a lie on a circular arc passing through 0, 1, and $\varepsilon_a/\varepsilon_b$ while for γ_b the arc passes through 0, 1, and $\varepsilon_b/\varepsilon_a$.

These expressions permit an enlightened choice of composition and nanostructure to achieve the highest possible internal fields in a two-phase composite. Referring to the Au film with voids used as an example in ref. [16], a maximum field enhancement of approximately 8.3 occurs in the void regions for the laminar nanostructure with an Au volume fraction of 0.84. The fact that the maximum enhancement occurs with this nanostructure is not surprising, because as we have seen it maximizes the screening charge. If the Au fraction is fixed at 0.80, which is approximately the value of f_a that would be obtained for Au films evaporated at liquid-nitrogen temperature, the HS limits show that the maximum enhancement in the voids drops to about 6.7 and occurs nearly, but not quite, for the laminar-nanostructure limit. If the nanostructure is constrained to be macroscopically isotropic in two dimensions the BM limits apply, and these show that the maximum enhancement factor is reduced to 4.0. Thus there is considerable variation of the maximum average fields that can be obtained depending on nanostructure.

Plasmons

The resonances discussed in the previous section are actually *plasmons*, which are solutions of Maxwell's equations that are localized and occur in nanoscopically inhomogeneous materials as a result of the polarization charge that develops at boundaries when an external field is applied. The plasmon associated with a spherical inclusion can be identified by writing Eqs. (18) and (19) in terms of the corresponding scalar potentials:

$$\begin{aligned} \phi_{in} &= -\frac{3\varepsilon_h}{\varepsilon_a + 2\varepsilon_h} E_0 z \\ &= -E_0 z - \frac{\varepsilon_h - \varepsilon_a}{\varepsilon_a + 2\varepsilon_h} E_0 r cos\theta; \\ \phi_{out} &= -E_0 z - \frac{\varepsilon_h - \varepsilon_a}{\varepsilon_a + \varepsilon_h} E_0 \frac{a^3}{r^2} cos\theta; \end{aligned} \tag{28}$$

where a is the radius of the inclusion. By extracting the part associated with the uniform field we have highlighted the part associated with the inhomogeneity, which is the plasmon. Its structure is seen to be self-contained, i.e., independent of the applied field. It is also localized, decaying as a dipole function for $r > a$. It also satisfies the boundary condition of continuity at the interface, as required by the continuity condition on tangential **E**.

The charge associated with the plasmon is the polarization charge at the surface of the inclusion. This represents a *separation* of charge, rather than extra charge introduced into the system, and hence satisfies the macroscopic equation

$$\nabla \cdot \mathbf{D} = 0. \tag{29}$$

This can be considered the most fundamental defining property of a plasmon: it is a charge separation and therefore has no net charge. The charge separation associated with the spherical inclusion is antisymmetric, so this plasmon couples efficiently to the electric field of the plane wave. Because the plasmon is already a part of the expression for the effective dielectric function $\langle\varepsilon\rangle$ of the nanoscopically inhomogeneous medium, it is already built into the Fresnel equations describing reflectance. Its effect on transmittance and reflectance is therefore described without additional equations.

From the form of Eq. (29) it is clear that the plasmon attains its maximum amplitude at the wavelength where resonance occurs, that is, where $Re(\varepsilon_a) = -2Re(\varepsilon_b)$. If the denominator vanishes completely, the plasmon can in principle exist independent of the exciting field, although this would be an unusual event. Because the screening factor depends on the shape of the inclusion, it follows that the resonance condition, and therefore the wavelength at which the resonance occurs, can be tailored by modifying the shape of the inclusions. This opens the possibility of engineering inclusion shapes to generate plasmon resonances at different wavelengths.

For perspective, the plasmon that is typically discussed in textbooks is the excitation that occurs if a point charge q is suddenly inserted in a metal. This insertion would result in a "breathing" mode, where the electrons oscillate radially about q until the energy is dissipated and equilibrium is reached. This particular excitation is spherically symmetric, and hence (as is well known) cannot couple to the electric field of a plane wave, since the field would act to move all the electrons in the same direction. It is apparent that the point charge q is not essential; its function can be performed by any local fluctuation away from charge neutrality. Even though bulk plasmons of this nature cannot be excited directly, the change in refractive index as ε goes from positive to negative values has a large effect on the optical properties of materials.

The condition for the existence of bulk plasmons analogous to Eq. (29) above, is usually obtained starting with the dispersion equation for a plane wave propagating in a dielectric medium. The dispersion equation,

$$\mathbf{k}(\mathbf{k}\cdot\mathbf{E}) - \mathbf{k}^2\mathbf{E} - \varepsilon\mathbf{E} = 0, \tag{30}$$

is easily derived by taking the curl of Faraday's Law, substituting Ampère's Law, then applying the result to a plane wave propagating in a medium of dielectric function ε. If the medium is isotropic and the plane wave has the space and time dependence

$$\mathbf{E}(\mathbf{r},t) \sim e^{ikz-i\omega t}, \tag{31}$$

the dispersion equation reduces to

$$\begin{bmatrix} \varepsilon(\omega)-n^2 & 0 & 0 \\ 0 & \varepsilon(\omega)-n^2 & 0 \\ 0 & 0 & \varepsilon(\omega) \end{bmatrix} \begin{bmatrix} E_x \\ E_y \\ E_z \end{bmatrix} = 0 \tag{32}$$

where $n = ck/\omega$. The first two rows lead to the transverse-wave solutions E_x, $E_y \neq 0$ provided that $n^2 = \varepsilon$, whereas the last row leads to the *longitudinal* solution $E_z \neq 0$ provided that $\varepsilon(\omega) = 0$. In most metals this condition occurs at energies of the order of 15 - 20 eV, but their existence in the visible range gives rise to the yellow and red colors, respectively, of Au and Cu.

Another type of plasmon occurs at planar boundaries between media. These are excitations whose magnitudes decrease exponentially in both directions away from the interface, and in the literature are usually called either surface plasmons or surface electromagnetic waves. Their description can be obtained as follows. Let ε_a be the dielectric function of the ambient and ε_s that of the substrate. Then substitute the respective plane waves in the ambient and substrate dispersion equations, and apply the boundary conditions on tangential **H** and **E**. This leads to the reciprocal attenuation lengths k_{az} and k_{sz} into media a and s, respectively:

$$k_{az} = \frac{\omega}{c}\sqrt{\frac{-\varepsilon_a^2}{\varepsilon_s + \varepsilon_a}}; \tag{33}$$

$$k_{sz} = \frac{\omega}{c}\sqrt{\frac{-\varepsilon_s^2}{\varepsilon_s + \varepsilon_a}}.$$

The component k_x of the wave vector along the surface is given by

$$k_x = \frac{\omega}{c}\sqrt{\frac{\varepsilon_a \varepsilon_s}{\varepsilon_s + \varepsilon_a}}. \tag{34}$$

It is clear from the above expressions that interface plasmons can exist only if the real parts of ε_a and ε_s have opposite signs, and if $Re(\varepsilon_s) < -Re(\varepsilon_a)$. Hence the substrate must either be a metal or exhibit a resonance such that the above inequality is satisfied. The typical situation is that s is a metal, in which case the plasmon consists of electric field components parallel and perpendicular to the surface and a magnetic field parallel to the surface (transverse-magnetic wave). Associated with these fields is a surface charge density σ, but no surface current **K**.

It is also clear from these expressions that the magnitude of k_{az} is greater than k_a, where k_a is the wave vector of the plane wave in the ambient. Hence it is not possible to excite surface plasmons directly by plane waves incident from the ambient side of the interface. Methods must be used to increase k_a so that its projection along the surface can match k_z. This can be accomplished in two ways. In the Otto configuration the incoming plane wave is passed through a prism near the metal film, which is separated from the prism by a small air gap so that the dielectric directly adjacent to the film is 1 [18]. In the more widely used Kretschmann configuration the metal film is deposited directly on the prism [19]. In both cases coupling is done by evanescent waves, so it is necessary to keep either air gap or metal film thin enough so that the wave is not completely attenuated by the time it reaches the air-metal boundary. In the Kretschmann configuration an interface plasmon also exists at the prism-metal interface, but this cannot be excited by a plane wave propagating in the prism for the reasons mentioned above.

Interface plasmons have been used for many years as a diagnostic tool for electrolyte studies, generally involving Ag films [20]. As with the plasmons associated with nanoscopically inhomogeneous materials, they are built into the Fresnel equations. As a result, structures related to these resonances automatically appear when the Fresnel equations are evaluated to obtain reflectances of these configurations. Again, no additional equations are required.

Negative-index metamaterials

In the above we have focused our attention on ε rather than the magnetic permeability μ, because as noted above the dominant interaction of an electromagnetic wave with a material occurs via **E**. We typically approximate μ as 1 and write n as $\sqrt{\varepsilon}$. However, a rigorous derivation shows that $n = \sqrt{\varepsilon\mu}$. Hence if both ε and μ are negative, n itself will be negative, which in physical terms means that the group and phase velocities of a wave will move in opposite directions.

Materials with negative n exhibit characteristics that are quite different from those with which we are familiar with materials with positive n [14, 21]. Aside from details such as refraction that bends light rays in the direction opposite to what we expect from Snell's Law for positive n, aspects of particular interest are the focusing of light at sub-wavelength resolution with planar sheets of negative-n material, and the bending of light around an object so in effect it becomes invisible. The focusing properties depend on the reverse nature of refraction: rays emanating from a point source arriving at a planar surface of a negative-n material will be refracted back toward a focus internal to the material. If the material is sufficiently thick, these will come to a point focus within the material then emerge from the back side where they are again refracted toward what is now an external focus. If the thickness is properly chosen the size of the second (external) image can be much reduced relative to the original image, although the smaller it is the closer it must also be to the back side of the material. Despite this complication it can be appreciated that this phenomenon is potentially of great potential interest to lithography.

The difficulty in realizing negative-index behavior lies in realizing negative values of μ. While negative values of ε are easily obtained in the optical-frequency range through the use of metals, no homogeneous natural or synthetic material exists for which μ is negative in the same frequency range. However, Pendry noted that split-ring resonators shaped like the letter C with a very small gap behave as LC resonant circuits in the appropriate frequency range, and hence could simulate negative-μ behavior through resonances [21]. Although the frequency range of such resonances is highly limited, such resonances can be obtained with realizable structures in the microwave spectral region, solving the problem of obtaining negative values of μ.

The next challenge was that of finding structures that would give useful negative values of ε. Pendry solved this by noting that structures of isolated parallel short wires would behave as energy-storage devices with large effective values of mass, leading to plasma frequencies orders of magnitude less than those typically encountered with bulk metals and therefore also realizable in the microwave region. Calculations for 2 μm diameter Al wires on a 5 mm grid showed that such structures behaved as if they were free- electron-like with a mass sufficiently enhanced to produce a plasma frequency at 8 GHz. With an effective metal giving a reasonable negative value of ε and a resonant circuit a negative value of μ, a negative-index metamaterial in the microwave spectral region was realized. Since that time negative-index behavior has been demonstrated at a wavelength of 2.0 μm [22]. However, the question of broadband negative-index behavior, necessary for wide-ranging applications in the visible, remains open.

CONCLUSION

Although the basic physics of the optical properties of macroscopically homogeneous but nanoscopically inhomogeneous materials has been well understood for years, we are now in a position where such structures can be created. As a result the relevant equations are currently being studied much more thoroughly to gain a better understanding of the properties of materials that are inhomogeneous on these length scales. However, negative-index metamaterials represent a genuinely new class whose properties are only now being explored. As capabilities for fabricating designer materials with nanoscale inhomogeneities continues to improve and advances continue to be made, this understanding becomes increasingly important.

ACKNOWLEDGEMENTS

It is a pleasure to acknowledge the invitation by S. Tomas, O. Miranda, and co-organizers to present this series of lectures at the Advanced Summer School organized by the Department of Physics of Cinvestav. Financial support from the Academia Mexicana de Ciencias and the United States - México Foundation for Science is also gratefully acknowledged.

REFERENCES

1. R. Clausius, *Die mechanische Behandlung der Elektricität*, Vieweg, Branschweig (1879).
2. O. F. Mossotti, *Mem. Mat. Fis. della Soc. Ital. di Sci. in Modena* **24**, 49-74 (1850).
3. P. P. Ewald, *Dissertation*, Munich, 1912; *Ann. Phys. (Leipzig)* **49**, 1-38 (1916).
4. C. W. Oseen, *Ann. Phys. (Leipzig)* **48**, 1-56 (1915).
5. J. C. M. Garnett, *Philos. Trans. Roy. Soc. (London)*, Ser. A, **203**, 385-420 (1904).
6. J. C. M. Garnett, *Philos. Trans. Roy. Soc. (London)*, Ser. A, **205**, 237-288 (1906).
7. P. N. Sen, C. Scala, and M. H. Cohen, *Geophysics* **46**, 781-795 (1981).
8. D. E. Aspnes and H. G. Craighead, *Appl. Opt.* **25**, 1299-1309 (1986).
9. D. A. G. Bruggeman, *Ann. Phys.* **24**, 665-679 (1935).
10. O. Wiener, *Ber. Verhandlungen König.-Sächsischen Gesells. wisseschaften Leipzig*, 256-277 (1910).
11. Z. Hashin and S. Shtrikman, *J. Appl. Phys.* **33**, 3125-3131 (1962).
12. D. J. Bergman, *Phys. Rev. Lett.* **44**, 1285-1287 (1980).
13. G. W. Milton, *Appl. Phys. Lett.* **37**, 300-302 (1980).
14. J. B. Pendry, A. J. Holden, D. J. Robbins, and W. J. Stewart, *IEEE Trans. Microwave Theory Tech.* **47**, 2075-2084 (1999).
15. See, e.g., G. D. Powell, J.-F. Wang, and D. E. Aspnes, *Phys. Rev. B* **65**, 205320 (2002).
16. D. E. Aspnes, *Phys. Rev. Lett.* **48**, 1629-1632 (1982).
17. D. E. Aspnes, E. Kinsbron, and D. D. Bacon, *Phys. Rev. B* **21**, 3290-3299 (1980).
18. A. Otto, *Z. Phys.* **216**, 398-410 (1968).
19. E. Kretschmann, *Z. Phys.* **241**, 313-324 (1971).
20. See, for example, F. Abeles, *Surface Sci.* **56**, 237-251 (1976).
21. J. B. Pendry, D. Schurig, and D. R. Smith, *Science* **312**, 1780-1782 (2006).
22. S. Zhang, W. Fan, N. C. Panoiu, K. J. Malloy, R. M. Osgood, and S. R. J. Brueck, *Phys. Rev. Lett.* **95**, 137404 (2005).

Optical Properties of InAs Quantum Dots Grown on Variable Stoichiometry $In_xGa_{1-x}As$ and $In_{0.53}Al_yGa_{0.43-y}As$ Layers

Julio G. Mendoza-Alvarez[1], Mauricio P. Pires[2], Sandra M. Landi[2], Patricia L. Souza[2], Jose M. Villas-Boas[3], and Nelson Studart[4]

[1]*Departamento de Física, Cinvestav-IPN, Apdo. Postal 14-740, Mexico DF 07000. Mexico*
[2]*Labsem, CETUC, PUC-Rio, Marques de Sao Vicente 225, Rio de Janeiro. Brazil*
[3]*Department of Physics and Astronomy, Ohio University, Athens, OH 45701-2979. USA*
[4]*Departamento de Fisica. Universidade Federal de Sao Carlos, SP, 13565-905. Brazil*

Abstract. The use of InP substrates has made possible to obtain InAs QD layers with room temperature photoluminescence (PL) in the range 2.0-2.2 μm. This last result was possible because of the shift to lower energies of the InAs QD energy bandgap due to the reduction in the strain field between the InAs and InP, as compared to the case for InAs and GaAs. To study the importance in the control of the strain field between the InAs dots and the beneath layer onto which they are grown, we have used two approximations: 1) Growth of InAs QDs layers on top of $In_xGa_{1-x}As$ layers with variable stoichiometry; that is, with variable In concentration.; and 2) Growth of InAs QDs layers on top of $In_xAl_yGa_{1-x-y}As$ layers with variable Al concentration. In both cases we have used the metal organic vapor phase epitaxy growth technique to grow four-layer structures on top of InP substrates: the first layer being an InP buffer layer, followed by an $In_xGa_{1-x}As$ or $In_xAl_yGa_{1-x-y}As$ layer, then a layer of InAs dot material, and finally covered by a InP layer. The control in the parameters of the structure will enable us to control the separation between the first electronic levels in the conduction band of the QD; and so, to control the range of photodetection in the medium infrared between 6 and 8 microns. We show measurements of the low temperature PL spectra for both series of different InAs QDs samples, and through the analysis of these spectra using a 1D effective-mass model for the four-layer structure, we have been able to study the influence of the strain effects for the case of the $In_xGa_{1-x}As$ sublayers, and of the variable potential well for the case of the $In_xAl_yGa_{1-x-y}As$ layers.

INTRODUCTION

The appearance of new and enhanced optical, electrical, and structural properties due to quantum confinement (QC) effects was theoretically shown since the early paper of Tsu and Esaki in 1973 on the transport properties of a finite superlattice [1] and the demonstration of quantum-size effects in one-dimensional confinement by Dingle, Wiegman and Henry [2]. These QC effects arise when the motion of carriers is restricted in one, two or three dimensions. We know form the basic courses on quantum mechanics that if we have a three dimensional potential well of size L with infinite potential energy barriers, then the solution to Schrodinger's equation give us quantized eigenstates with eigenvalues given by the expression:

CP885, *Advanced Summer School in Physics 2006, Frontiers in Contemporary Physics—EAV06,* edited by O. Miranda, M. Carbajal, L. M. Montaño, O. Rosas-Ortiz, and S. A. Tomás Velázquez
 978-0-7354-0385-7/07/$23.00

$$E_n = E_g + \frac{9\,\hbar^2 \pi^2}{2\,m^* L^2}\left(n_1^2 + n_2^2 + n_3^2\right) \quad ,n=1,2,3,..... \tag{1}$$

so, the energy states are quantized and the separation between those energy levels depend on the carrier mass and on the size of the cube.

From the solid state point of view, the physical situation of the confinement takes place when the carriers, electrons and holes, inside the crystal experience a constraint in their movement along an specific direction due to some kind of artificial "barrier" in that direction. The physical parameter involved in the confinement is the so-called exciton Bohr radius that corresponds to the size of an exciton, that is, an electron-hole pair. This radius is expressed as:

$$a_B = \frac{4\pi\varepsilon\hbar^2}{m^* e^2} \tag{2}$$

where ε is the dielectric constant, and m* is the exciton reduced mass which depends on the electron and hole effective masses: $(m^*)^{-1} = m_e^{-1} + m_h^{-1}$. It can then be observed that the properties of the material will enter through its dielectric constant and their carrier effective masses. This exciton Bohr radius is the key parameter to compare with the crystal dimensions and determine if quantum confinement effects could be present; if a_B is of the order or smaller than the physical dimension, L, which we have reduced in the crystal, then the exciton wavefunction will be affected by the boundaries, and quantized energy levels will arise in the conduction and valence bands in solving the Schodinger's equation for the electron and hole motion inside the crystal. Usually, three regimes are distinguished for carrier confinement:

1. Strong quantum confinement if: $a_B \gg L$
2. Intermediate quantum confinement if: $a_B \sim L$
3. Weak quantum confinement if: $a_B \ll L$

For semiconductors there will be a large variation in the exciton Bohr radius owing to the variation in their electron and hole effective masses. In these cases the formula for a_B can be expressed as:

$$a_B = 0.053\left(\frac{\varepsilon}{m^*}\right) \tag{3}$$

where ε is the semiconductor dielectric constant and m* is the exciton reduce mass; a_B will be given in nanometers. In Table 1 we show values for m_e, m_h, ε and a_B for some of the most important III-V, II-VI and IV semiconductors.

TABLE 1. Values of the dielectric constant, **ε**; electron effective mass, $\mathbf{m_e}$; heavy hole effective mass, $\mathbf{m_h}$; and the exciton Bohr radius, $\mathbf{a_B}$, for important III-V, II-VI and IV semiconductors.

Semiconductor	ε	m_e	m_h	a_B (nm)
GaAs	13.2	0.067	0.5	11.9
AlAs	10.06	0.19	0.409	4.1
InAs	15.1	0.022	0.41	38.1
InP	12.6	0.077	0.6	9.8
InSb	17.7	0.014	0.45	69.0
GaSb	15.7	0.041	0.28	23.2
GaN	10.4	0.27	0.8	2.7
CdTe	10.2	0.09	0.72	6.8
ZnSe	7.6	0.13	0.57	3.8
CdS	9.1	0.20	0.7	3.1
CdSe	10.1	0.12	0.45	5.7
ZnO	8.8	0.24	0.59	2.7
Si	11.9	0.1905	0.537	4.5

From the values for a_B in this table we can see that in order to get reasonable quantum confinement effects in GaAs nanocrystals, their radius should be smaller or of the order of 100 Å; whereas for InAs nanocrystals the radius should be around 380 Å.

When the confinement is in one direction we have a quantum well where carriers are restricted to move in a 2D plane; for confinement in two directions we have a quantum wire in which carriers move along a 1D line; and for confinement in the three directions we obtain the so-called quantum dots in which carriers are confined in the three dimensions, a 0D case. The density of states function, $\rho(E)$, expresses the number of allowed energy states per volume unit and per energy unit and so, it depends strongly on the dimensionality of the semiconductor system. In Fig. 1 we show the behavior of $\rho(E)$ for the four cases: bulk material (3D), where $\rho(E) \sim E^{1/2}$; quantum wells (2D), where $\rho(E) \sim \sigma(E-E_n)$ (step function); quantum wires (1D), where $\rho(E) \sim (E-E_n)^{-1/2}\,\sigma(E-E_n)$; and quantum dots (0D), where $\rho(E) \sim \delta(E-E_n)$ (delta function). This behavior of $\rho(E)$ is the main responsible for the new optical and electrical properties which present low dimensional semiconductors.

In particular, for quantum dots (QDs) the position of the quantized energy levels will be determined by the reduced effective mass and by the height of the potential barrier at their boundaries.

Semiconductor quantum dots, also called nanocrystalline or nanoparticle semiconductors have been grown by several techniques such as CdSe nanoparticles grown by chemical bath deposition emitting at energies between 2.05 to 2.28 eV for nanocrystalline sizes in the range 50-39 Å respectively [3]; CdNiTe nanostructures thin films grown by r.f. magnetron sputtering with grain sizes around 30 nm [4]; and colloidal III-V semiconductors [5]. However, for application of III-V semiconductor QDs in optoelectronic devices, the molecular beam epitaxy (MBE) and metal organic chemical vapor deposition (MOCVD) layer deposition methods have been the more reliable growth techniques.

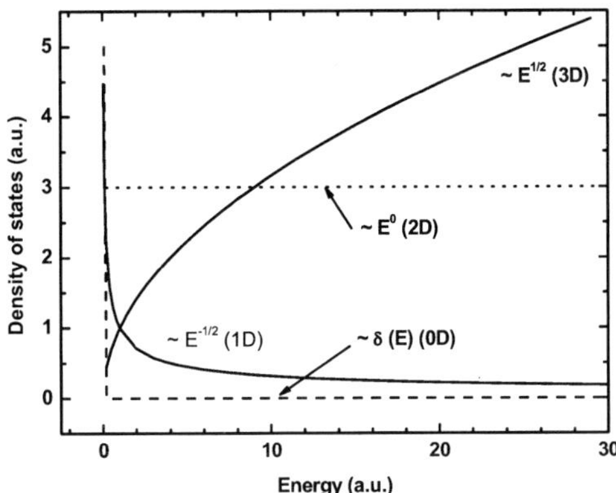

FIGURE 1. Behaviour of the density of states function for 3-dimensional (bulk) semiconductors; 2-dimensional (quantum wells) semiconductors; 1-dimensional (quantum wires) semiconductors; and 0-dimensional (quantum dots) semiconductors.

Since 1990, it has been known that three-dimensional islands of $In_xGa_{1-x}As$ could be spontaneously grown on top of a GaAs substrate in a highly strained semiconductor epitaxy [6]. This method to grow such 3D dimensional islands of nanometric size, or QDs, is actually known as the Stranski-Krastanow (S-K) growth mode and it is based on the relaxation of the build-up strain, that takes place when a semiconductor layer with a quite different lattice parameter is grown on a crystalline substrate, through the formation of defect-free 3D islands which grow on top a very thin wetting layer of the same material as the QDs. In the S-K mode, first takes place the growth of a thin 2D layer until a critical thickness is reached where the material on the surface reorganizes forming 3D islands relaxing the strain due to the mismatch of the crystalline lattices. The size and shape of the QDs will depend on the substrate temperature and the growth rate. The formation of the 3D islands can be seen as a phase transition, starting with a metastable, supercritical 2D wetting layer and ending with a more stable configuration of 3D islands surrounded by a thin wetting layer.

The key parameter which takes into account the difference between the lattice parameters of the layer and substrate is the so-called lattice mismatch defined as:

$$\frac{\Delta a}{a} = \frac{a_l - a_s}{a_s} \tag{4}$$

For small lattice mismatch values, let's say around 1-3 % the layers grow under strain with respect to the substrate until this strain is relieved through the formation of dislocations which are undesirable for application of these layers in optoelectronic devices. When the mismatch reaches larger values, the strain relief is not through the formation of dislocations but instead, it is relieved by forming the 3D islands known as QDs. This transition between a 2D to a 3D growth mode has been experimentally shown by S. Guha et al [6] showing that when a $In_{0.11}Ga_{0.89}As$ layer was grown on a GaAs substrate, dislocations were formed to relief the strain due to the difference in lattice parameters; whereas when a $In_{0.5}Ga_{0.5}As$ layer was grown on a GaAs substrate, then defect-free 3D islands were formed to release the strain without dislocations production. In Fig. 2 it is shown a very important and useful graphic where several II-VI and III-V binary and ternary semiconductors are plotted showing their bandgap energy in the y-axis and their corresponding lattice parameter in the x-axis. In this figure we see that there is a big difference in lattice parameters between InAs (6.0584Å) and GaAs (5.6232Å) or InP (5.8687Å); the lattice mismatch for the heterojunction InAs/GaAs is 8.2 %, whereas for the heterojunction InAs/InP the mismatch is 3.23%.

The first system developed was the growth of InAs QDs on a GaAs substrate with good control on fundamental properties such as the QD density and size, through studies on the dependence of these properties on the substrate temperature and on the deposition rate [7-9]. Semiconductor lasers based on InAs QDs layers grown on GaAs have been reported with emission wavelengths in the range: 0.9-1.3 μm [10], but emission wavelengths larger than 1.55 microns are required for applications in optical communications systems. This type of emission can be obtained by using the so-called strain relieved layers (SRL) to grow on top of them the QDs; the physics behind these SRL's is to use a layer with a larger lattice constant such that the strain between this layer and the InAs QD layer could be reduced, and then it would reduce the separation between the ground state for electrons in the conduction band and the ground state for heavy holes in the valence band; that is, it would reduce the effective bandgap energy for the InAs QDs and thus the emission and absorption edge would shift to lower energies. Following this idea, emission at 1.55 microns has been obtained growing a layer of InAs QDs on a SRL of InGaAs [11], or by the growth of InAs QDs on top of a SRL of GaNAs [12].

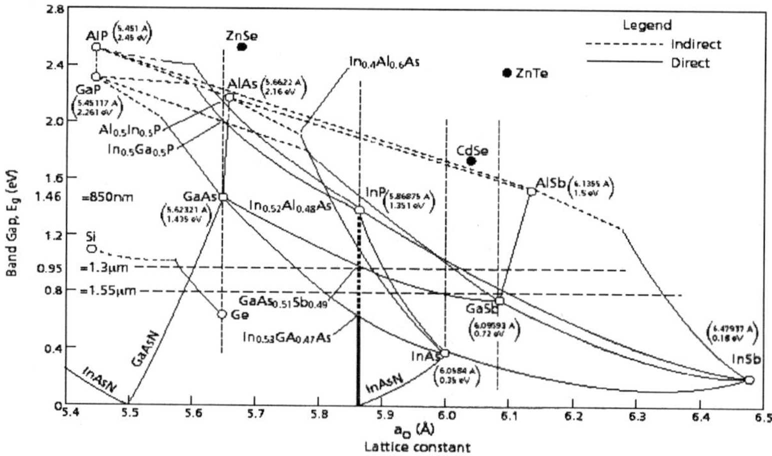

FIGURE 2. Graphic representation of the room temperature bandgap energy and the lattice parameters for some of the most relevant III-V and II-VI semiconductors.

In this paper we report on the growth of layers of InAs QDs grown on different SRL following two approximations: 1) Changing slightly the stoichiometry of an $In_xGa_{1-x}As$ layer around the value of x=0.53 that correspond to a lattice-matched situation ($In_{0.53}Ga_{0.47}As$); and 2) Using as SRL a layer of the quaternary alloy $In_{0.53}Al_yGa_{0.47-y}As$ with variable aluminum concentration. The InAs QDs layers have been characterized by low temperature photoluminescence spectroscopy to study the influence of the beneath layers of either InGaAs or InAlGaAs, on the InAs QDs formation.

EXPERIMENTAL METHODS

Samples were grown by MOVPE with phosphine (PH3), arsine (AsH3), trimethyl-indium (TMI), trimethyl-gallium (TMG), and trimethylaluminium (TMA) as precursors, and hydrogen as carrier gas. We used Fe-doped (SI) InP (001) ''epi-ready'' wafers with a 2^0 misorientation towards the nearest <011> direction for our experiments. A 150nm-thick InP buffer layer was grown, initially at a growth temperature of 630 ^{0}C, after which the reactor was cooled down to 600 ^{0}C. Then a 10 nm-thick $In_xGa_{1-x}As$ layer was grown with a ratio V/III of 30; or a 500 nm-thick $In_{0.53}Al_yGa_{0.47-y}As$ layer was grown with the same V/III ratio. The growth rates for InP, $In_xGa_{1-x}As$ and $In_{0.53}Al_yGa_{0.47-y}As$ were 1.4 and 2.8 ML/s, respectively. The temperature was ramped down to 500 ^{0}C during a 5 min period without growth, to keep the well-defined terrace structure before dot deposition. 2.5ML InAs dot material was deposited on the surface at a growth rate of 0.5ML/s and then the surface was annealed for 12 s under an arsine flow. At this point, the samples were either cooled down under arsine containing atmosphere, or capped with 50nm InP during which the temperature was ramped up to 600 ^{0}C. For capping the samples, the arsine was switched off and replaced by phosphine. Then TMI was reintroduced into the reactor, after a delay time of 1 s in order to reduce carry over effects.

The dot densities and their height were measured by using an atomic force microscope (AFM) in the tapping mode, and from the AFM images over an area of 2x2 μm^2, the data histograms were made. The error on the height measurement is given by the standard deviation.

For the study of InAs QDs grown on a slightly mismatched InGaAs layer, the degree of mismatch was determined from X-ray diffraction measurements for the (004) reflection.

Low temperature photoluminescence (PL) measurements were done placing the sample on a closed-cycle helium cryostat with a variable temperature controller in the range 10-300 K. The 488-nm line of an Ar ion laser was used as the exciting source for PL, and the laser power was changed in the range 40-200 mW. The radiative emission from the InAs QDs samples was detected either by a thermoelectrically-cooled InGaAs photodetector, or by a liquid nitrogen-cooled Ge detector, coupled to a lock-in amplifier.

RESULTS AND DISCUSSION

First, a layer of InAs QDs was grown on top of an InGaAs layer with a stoichiometry slightly departed from the composition corresponding to a lattice-matched situation: $In_{0.53}Ga_{0.47}As$, in order to study the influence of the strain at the InP/InGaAs interface on the InAs QDs properties. In Fig. 3 we show a schematics of the multilayer structure grown on an InP substrate. Sample labeled 840 was grown under lattice-matched conditions; samples 859 and 860 were grown with a negative lattice mismatch (layer lattice parameter < substrate lattice parameter, tensile strain) of 900 and 2200 ppm (In concentrations of 51.8% and 48.4%,) respectively; and samples 861 and 862 were grown with a positive lattice mismatch (layer lattice parameter > substrate lattice parameter, compressive strain) of 900 and 2000 ppm (In concentration of 55.5% and 57.8%), respectively.

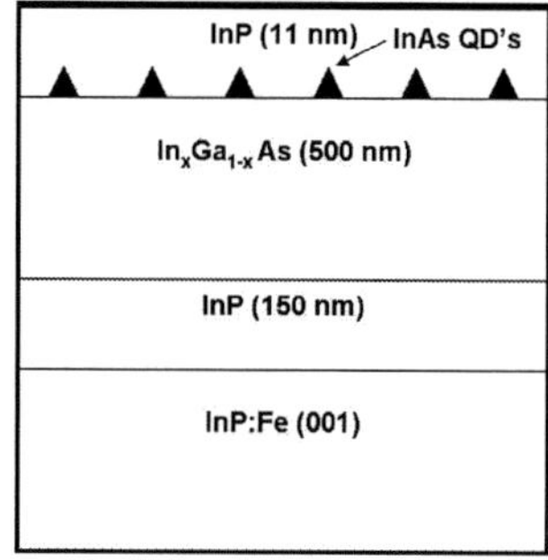

FIGURE 3. Multilayer structure for the InAs QDs layer grown on top of a InGaAs layer slightly mismatched to the InP substrate. The InP capping layer is used to cover the QD layer for photoluminescence measurements.

In Fig. 4 we show the AFM micrograph for sample 830, which was grown under the same conditions as sample 840 but without covering it with the InP layer, corresponding to the growth of InAs QDs grown on a $In_{0.53}Ga_{0.47}As$ layer lattice-matched to an InP substrate. From the histogram resulting of the analysis of the AFM micrograph, a dot density of 1.48×10^{10} cm^{-2} is obtained, with an average dot height of 8.2 ± 1.5 nm. As observed, the curve is slightly asymmetric showing the presence of more dots with larger heights. For the different samples measured, a variation between 7 and 10 nm were found for the QD heights, the larger value was measured for sample 859, and the smaller for sample 861.

The PL spectra for the set of five samples of InAs QDs grown on InGaAs layers of variable stoichiometry are shown in Fig. 5. These spectra were measured at a temperature of 14 K and for a laser power of 10 mW; the apparent minimum at 0.66 eV is due to a band of atmospheric water vapor absorption. All the spectra show an asymmetric PL band towards higher energies that is an evidence of the presence of two superimposed emission bands around 0.64 and 0.67 eV for these samples. It is observed that the peak in the PL emission has its higher value for sample 840 which correspond to the QDs grown on the InGaAs lattice-matched layer. When the InGaAs layer deviates from the matching conditions, the PL band shifts to lower energies, being this shift larger for those samples grown on the positive mismatched InGaAs layers.

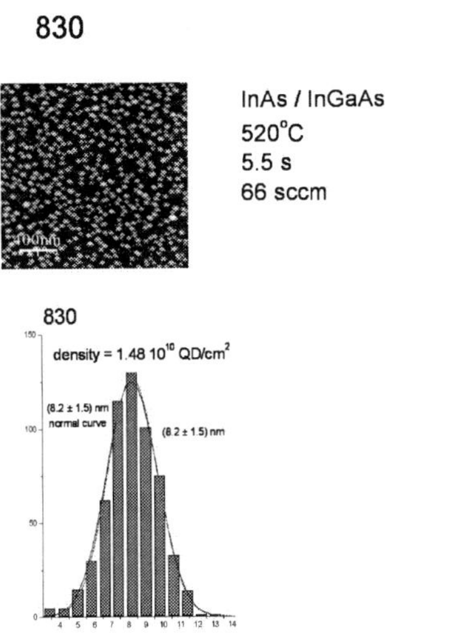

FIGURE 4. AFM micrograph for the InAs QDs layer grown on top of a lattice-matched layer of $In_{0.53}Ga_{0.47}As$. A dot density of 1.48 x 10^{10} cm^{-2} with average heights of 8.2 nm were obtained from the histogram shown.

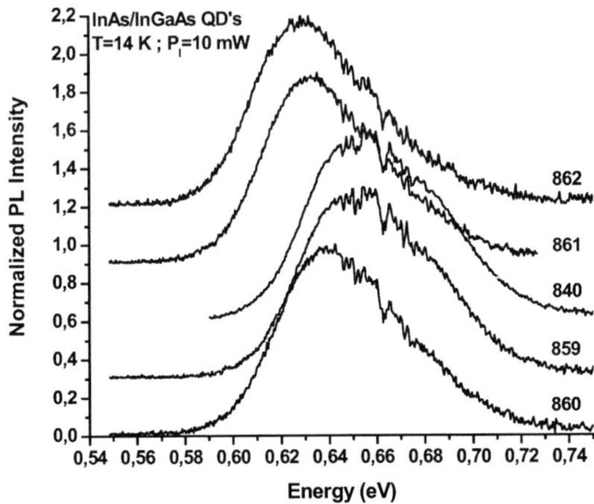

FIGURE 5. PL spectra for the set of InAs QDs samples grown on $In_xGa_{1-x}As$ layers of different stoichiometry, measured at a temperature of 14 K. Sample 840 correspond to a lattice-matched $In_{0.53}Ga_{0.47}As$ layer, samples 861 and 862 were grown on positive mismatched InGaAs layers, and samples 859 and 860 correspond to growth on negative mismatched InGaAs layers.

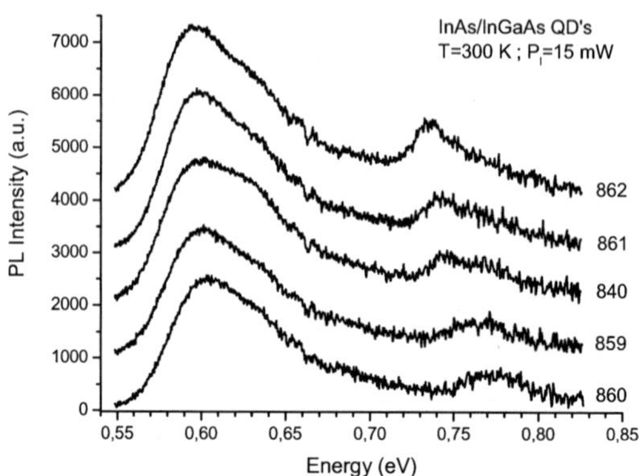

FIGURE 6. Room temperature PL spectra for the set of InAs QDs samples grown on $In_xGa_{1-x}As$ layers of different stoichiometry.

In Fig. 6, the PL spectra for the same set of samples, but measured at room temperature, are shown. The fact that we can measure PL at room temperature is an indication of deep confinement in the quantum dots that blocks out the possibility of electron escape from the dots due to a temperature-activated ionization process. In this figure we observe that the spectra are composed of the asymmetric band with peaks around 0.6 and 0.63 eV, and a weaker PL band at about 0.75-0.77 eV; this last band is due to radiative emission from the InGaAs layer.

Using a two Gaussian fit to the PL spectra measured at low temperature shown in Fig. 5, we were able to follow the behavior of the energy of both peaks as a function of the degree of lattice mismatch, and the results are shown in Fig. 7. It can be clearly observed in the figure that both peaks present the same behavior, showing a larger red shift when the InAs QDs are grown on positive mismatched InGaAs layers than for QDs grown on negative mismatched InGaAs layers.

The two peaks shown in Fig. 7 have been identified as resulting from radiative transitions between: 1) The ground electron state in the conduction band (CB), 1e, and the ground heavy hole state in the valence band (VB), 1hh; and 2) The ground electron state in the CB and the ground light hole state in the VB, 1lh. We have used a one-dimensional model which takes into account the layer structure used, the layer thicknesses, layer compositions and the strains between the layers, in order to have an estimation of the energy levels positions in the CB and VB, as shown in Fig. 8. In this figure we show the results for the theoretical model assuming an InGaAs layer thickness of 10 nm, and a quantum dot height of 6 nm. The predicted energy separation between the transitions 1e-1hh and 1e-1lh is around 50 meV, that compares favorably with the experimental separation measured of around 40-45 meV shown in Fig. 7.

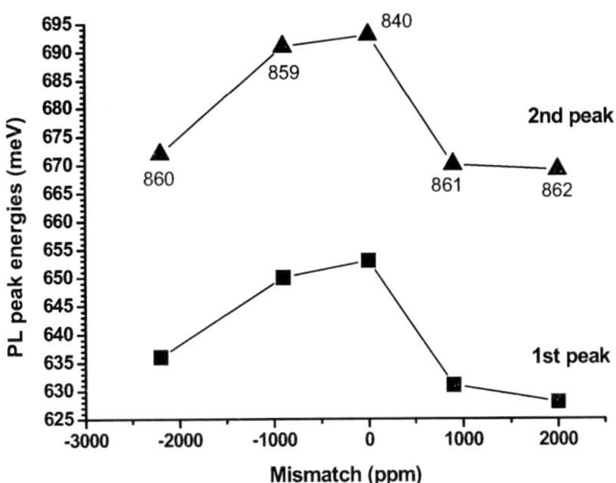

FIGURE 7. Behaviour of the energy of the two emission bands present in the PL spectra of the InAs QDs layers measured at a temperature of 14 K, as a function of the degree of lattice mismatch between the InGaAs layer and the InP substrate.

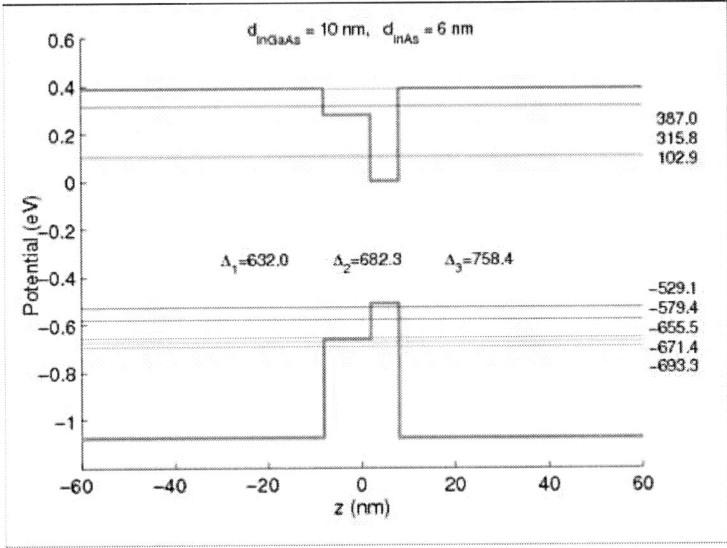

FIGURE 8. Theoretical calculation of the energy levels distribution in the InAs QDs conduction and valence bands, using a one-dimensional effective mass model, for the structure InP/InGaAs/InAs QDs/InP.

The behavior shown in Fig. 7 for the energy peak position as a function of the mismatch degree can be understood qualitatively as due to the following four effects [13]:

1. A variation in the InGaAs bandgap energy as the In concentration changes; increasing the In content will decrease the bandgap energy, as it is clear from Fig. 2.

2. A change in the InAs QD bandgap energy as a result of the change in the QD size.

3. Intermixing between the InGaAs layer and the InAs QDs, Ga diffusion into the dot will be larger as the Ga content in the InGaAs layer is higher; this effect will increase the bandgap energy of the dot as the In concentration in the InGaAs layer increases.

4. Variation in the strain fields around the InAs QDs when the InGaAs stoichiometry changes; this variation will induce a change in the InAs QD bandgap energy.

The second approximation we used to study the variation in the properties of the InAs QD layer, was to use a quaternary alloy of the type InAlGaAs to grow on top of it the layer of InAs QDs. The physical idea was to choose lattice-matching growth conditions to grow the quaternary alloy on an InP substrate, and to change the stoichiometry of the InAlGaAs layer in order to control the barrier height between the InAs QD and the InAlGaAs layer, and in this way to control the emission energy of the QDs. Also, a larger barrier height will result in better carrier confinement and then in a better performance at higher temperatures. Room temperature emission wavelengths around 2.1 microns have been reported for this system [14]

For this purpose, a set of InAs QDs/$In_{0.53}Al_yGa_{0.47-y}As$ heterostructures were grown by MOVPE varying the Al concentration with the values: y = 0, 0.058, 0.11, and 0.165; with sample labeling of 594, 596, 588 and 599, respectively. The grown heterostructure is shown in Fig. 9.

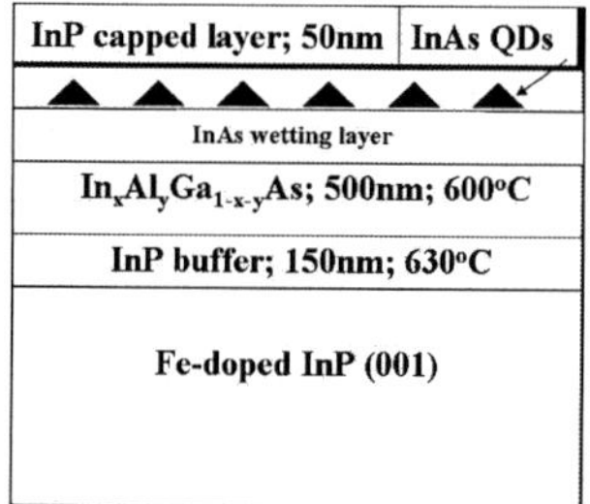

FIGURE 9. Multilayer structure grown by MOVPE for the growth of InAs QDs layers on top of lattice-matched $In_{0.53}Al_yGa_{0.43-y}As$ layers with variable Al concentration.

The AFM micrographs for the four samples are shown in Fig. 10 where a highly uniform quantum dot density can be observed; from the histograms resulting from the micrographs for these InAs QDs samples we obtained QD densities in the range 1.5-2.0 x 10^{10} cm^{-2}, and QD heights in the range 5.2-6.1 nm. There is a slight tendency to have larger QD densities and larger QD heights as the Al concentration increases in the quaternary InAlGaAs alloy.

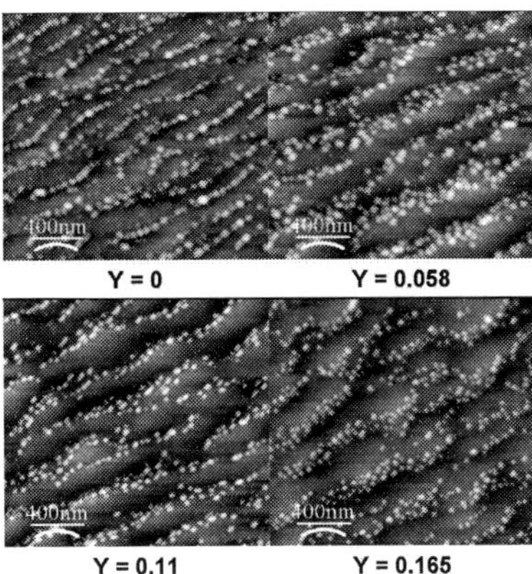

FIGURE 10. AFM micrographs showing the distribution of the InAs QDs for the set of four QD samples grown on top of $In_{0.53}Al_yGa_{0.47-y}As$ for Al concentrations of: y = 0, 0.058, 0.11, and 0.165.

The PL spectra measured at 14 K for the four InAs QDs samples are shown in Fig. 11. For the QDs grown on a lattice-matched $In_{0.53}Ga_{0.47}As$ layer, the PL spectra show a peak at 0.64 eV and a shoulder at 0.673 eV; again, features observed at 0.645 and 0.66 eV are due to water vapor absorption bands. It can be observed that when the QDs are grown on layers with Al, there is a shift to higher energies, to around 0.68 eV, as a result of an increase in the barrier height between the InAs QD and the InAlGaAs layer which shifts the levels in the CB and VB to higher energies. For those samples grown on the InAlGaAs layer, a broad and low intensity PL band is also observed at energies of about 0.81-0.82 eV which we associate to emission from the quaternary InAlGaAs alloy.

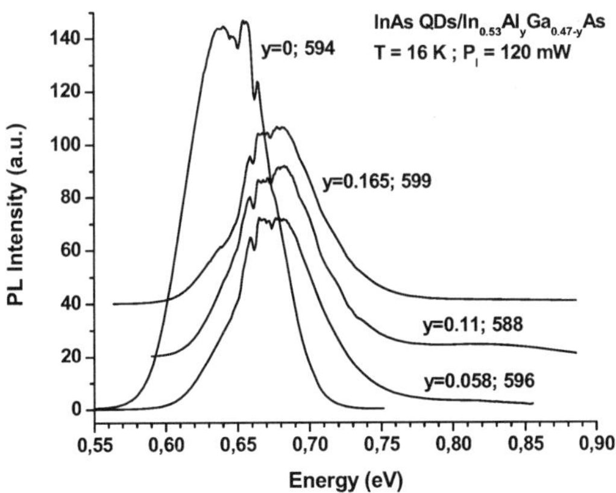

FIGURE 11. PL spectra for the set of InAs QDs samples grown on $In_{0.53}Al_yGa_{0.47-y}As$ layers with different Al concentration. The spectra were measured at a temperature of 14 K.

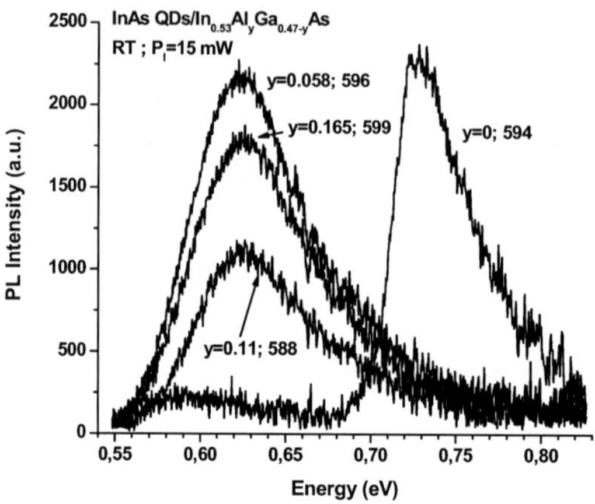

FIGURE 12. Room temperature PL spectra for the InAs QDs samples grown on $In_{0.53}Al_yGa_{0.47-y}As$ layers with different Al concentrations.

In Fig. 12 are shown the PL spectra for the set of four samples measured at room temperature. For the sample of InAs QDs grown on the lattice-matched InGaAs layer, a low intensity PL band is observed peaking at 0.69 eV, and a very intense band with its peak at about 0.72 eV; this last band comes from emission of the $In_{0.53}Ga_{0.47}As$ layer. The PL emission for the QDs samples, grown on InAlGaAs layers with variable Al concentration, shift to higher energies with an energy peak in the range 0.62-0.63 eV, and a shoulder around 0.67-0.68 eV. The PL intensity for those samples grown on InAlGaAs layers is larger than for the sample grown on the InGaAs lattice-matched layer, and this is due to a better carrier confinement between the InAs QD and the InAlGaAs layer.

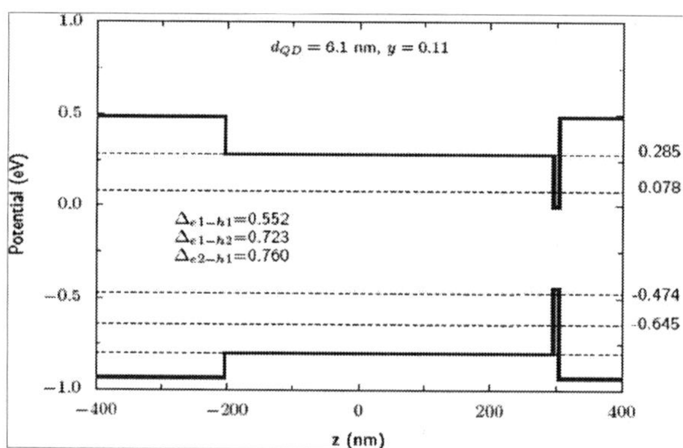

FIGURE 13. Theoretical calculation of the energy levels distribution in the InAs QDs conduction and valence bands, using a one-dimensional effective mass model, for the structure InP/InAlGaAs/InAs QDs/InP.

Using the same one-dimensional effective mass model for the InP/InAlGaAs/InAs/InP structure, we have calculated the energy levels distribution in the CB and the VB as shown in Fig. 13. These energy levels correspond to an Al concentration of 11% in the quaternary alloy, and to an InAs QD height of 6.1 nm. In this figure are shown only the energy levels corresponding to the heavy hole states in the VB.

CONCLUSIONS

In summary, in this paper we have shown results on the optical characterization, using the low temperature photoluminescence spectroscopy, of InAs QDs grown on two different types of layers: slightly strained InGaAs layers with In concentrations around 53%, and quaternary $In_{0.53}Al_yGa_{0.47-y}As$ lattice-matched alloys. We show that using the MOVPE growth technique we were able to grow InAs QDs with acceptable large densities in the range 1-2 x 10^{10} cm^{-2}, and dot heights around 6 to 8 nm. For samples of InAs QDs grown on InGaAs layers slightly mismatched to the InP substrate, we observed red shifts on the PL spectra which could be qualitatively understood by taking into account four different physical processes: the InAs QD bandgap variation due to a change in the InGaAs stoichiometry, the bandgap variation due to dot size variation, bandgap variation due to intermixing between the InGaAs layer and the InAs QDs, and bandgap variation due to changes in the strain fields when the InGaAs layer stoichiometry changes. For the InAs QDs grown on lattice-matched InAlGaAs with variable Al concentration we observed in the PL spectra a shift to higher energies as a result of the increase in the barrier height between the InAs QD and the InAlGaAs layer, and also faster decrease in the carrier confinement with the temperature when the QDs are grown on a lattice-matched InGaAs layer without Al. For both sets of InAs QDs samples, a one-dimensional effective mass model which took into account the multilayer structure, the layer thicknesses, the layer stoichiometries, and the strain fields around the InAs QDs due to the beneath layer and the cap layer, predicted the distribution of energy levels in the CB and VB of the dots, in good agreement with the results obtained from the fitting to the measured PL spectra.

REFERENCES

1. R. Tsu and L. Esaki, *Appl. Phys. Lett.* **22**, 562-564 (1973).
2. R. Dingle, W. Wiegman, and C.H. Henry, *Phys. Rev. Lett.* **33**, 827 (1974)
3. S.A. Empedocles, D.J. Norris, and M.G. Bawendi, *Phys. Rev. Lett.* **77**, 3873 (1996)
4. O. Alvarez-Fregoso, J.G. Mendoza-Alvarez, O. Zelaya-Angel, F. Morales, *J. Appl. Phys.* **80**, 2834 (1996)
5. U. Banin, C.J. Lee, A.A. Guzelian, A.V. Kadavanich, A.P. Alivisatos, W. Jaskolki, G.W. Bryant, Al. L. Efros, and M. Rosen, *J. Chem. Phys.* **109**, 2306 (1998)
6. S. Guha, A. Madhukar, and K.C. Rajkumar, *Appl. Phys. Lett.* **57**, 2110 (1990)
7. O. Suekane et al, "Growth Temperature Dependence of lnAs Islands Grown on GaAs (001) Substrates", 2001 International Conference on Indium Phosophide and Related Materials Conference Proceedings, Nara, Japan, IEEE, NY, 2001, pp.288-291.
8. A. Madhukar et al, *Appl. Surf. Sci.* **123/124**, 266 (1998)

9. S. Fafard, Z.R. Wasilewsky, C.Ni. Allen, D. Picard, M. Spanner, J.P. McCaffrey, P.G. Piva, *Phys. Rev. B* **59**, 15368 (1999)
10. S. Fafard, Z.R. Wasilewsky, C.Ni. Allen, K. Hinzer, J.P. McCaffrey, Y. Feng, *Appl. Phys. Lett.* **75**, 986 (1999)
11. Jun Tatebayashi, Masao Nishioka, and Yasuhiko Arakawa, *Appl. Phys. Lett.* **78**, 3469 (2001)
12. X.Q. Zhang, Sasikala Ganapathy, Hidekazu Kumano, Kasturi Uesugi, and Ikuo Suemune, *J. Appl. Phys.* **92**, 6813 (2002)
13. J.G. Mendoza-Alvarez, M.P. Pires, S.M. Landi, A.S. Lopes, P.L. Souza, J.M. Villas-Boas, N. Studart, *Physica E* **32**, 85 (2006)
14 M. Borgstrom, M.P. Pires, T. Bryllert, S. Landi, W. Seifert, P.L. Souza, *J. Cryst. Growth* **252**, 481 (2003)

Photoluminescence Emission from Heteroestructures $SiO_2/Si/SiO_2$ Growth by RF Reactive Sputtering

E. Mota-Pineda and M. Meléndez Lira

Cinvestav, Departamento de Física, Apdo. 17-740 México D.F. 07000, Mexico

Abstract. We prepared heterostructures SiO2/Si/SiO2 by RF sputtering on Si (100) and glass. We employed a polycrystalline Si target with Ar and O_2 as working gases. We investigate the effect of the partial pressure of oxygen and the thickness of the Si interlayer on the electronic properties of the heterostructures. A broad luminescent band (around 1.7 eV) appears by effect of the Si interlayer and its intensity increases with the Si layer thickness. The results were discussed in terms of a model of quantum confinement of Si embedded in a SiO_2 matrix.

Since the discovery of light emission from porous silicon [1], an intense investigation of a material compatible with silicon technology and with excellent detector and emitter optical properties has been developed. This investigation has led to a great variety of materials with these characteristics [2,3,4]. In recent years, a great deal of research on silicon nanocrystals embedded in a silicon oxide matrix has been conducted because of their potential for applications in silicon-based optoelectronic devices.[5,6]. As an indirect band gap material, bulk Si is known to be inefficient in light emission. Besides, the low band gap of bulk Si at about 1.1 eV allows only for infrared emission, instead of visible light. Therefore, optoelectronic devices are currently built on compound semiconductors due to their high efficiency in light emission. Compound semiconductors, however, are hard to integrate into cheap and versatile silicon circuits due to lattice mismatch problems. Once Si devices can be made exhibiting high electroluminescence (EL) efficiency, the low cost and powerful optoelectronic integrated circuit Si devices will become feasible.

We prepared heterostructures SiO2/Si/SiO2 by sputter reactive magnetron deposition on Si (100) and corning glass substrates at 400°C. We employed a polycrystalline Si target with Ar and O2 as working gases. The sputtering chamber was evacuated to 10^{-7} Torr before argon gas was introduced. The sputtering pressure was 15 mTorr, and the RF power was 200 W. The samples were prepared under continuous flow by pumping off the sputter gas with a turbomolecular pump. The target-substrate distance was of 5 cm. We investigate the effect of the partial pressure of oxygen (25-75%) and the thickness (5 and 10 minutes of grow) of the Si interlayer on the properties of the heterostructures. In order to observe the evolution of the PL bands, we deposited three types of samples varying the partial pressure of oxygen of 25 to 75 percent during the growth. We observed no luminescence in the SiO2 film.

CP885, *Advanced Summer School in Physics 2006, Frontiers in Contemporary Physics—EAV06*, edited by O. Miranda, M. Carbajal, L. M. Montaño, O. Rosas-Ortiz, and S. A. Tomás Velázquez

The thickness of each one of the samples is present in figure 1A. The curve a) show the thickness of SiO2 samples. Curves b) and c) show the thickness of the films of Si deposited during 5 and 10 minutes respectively between the SiO2 matrixes. The thickness of the deposited films of SiO_2 are reduced with the increase of the partial pressure of oxygen in the camera because the average free way of the silicon is reduced with the increment in the pressure in the chamber.

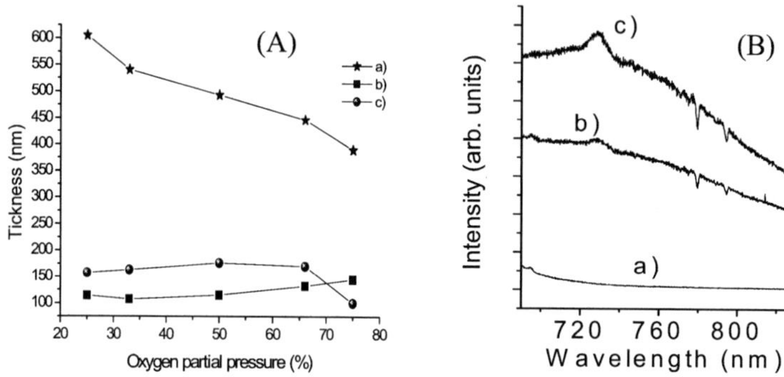

FIGURE 1. (A) Thickness of film in function of oxygen partial pressure (B) The PL spectra functions of time of grow of Si interlayer for 75% of oxygen partial pressure samples. a) Without Interlayer, b) 5 minutes interlayer and c) 10 minutes interlayer.

Figure 1B shows the PL response from the heterostructures corresponding to different time of grow of the Si interlayer and 75% of oxygen partial pressure. The oxygen partial pressure appear not has a direct influence in the luminescence of the samples. The weaker band are observed in the sample prepared without intermediate layer of Si, whereas the greater spectra appear for those samples with a growth of 10 minutes of Si interlayer. PL is obtained at room temperature and increases with the time of grow of Si interlayer. We found that a luminescence band (around 1.7 eV) appear by effect of the Si interlayer and the intensity of this band increase with the time of growing of the Si layer.

In the samples growth on glass we made the UV-Vis transmission analysis in order to studied the absorption edge of the material. Figure 2A show the curves for the samples grow with 75% of oxygen partial pressure. We can observe a landslide in the absorption edge of 4 to 2.35 eV. We also performed Raman experiments. Crystalline silicon is characterized by a strong and thin band at 520.5 cm^{-1} and the amorphous silicon is characterized by two bands at 150 and 480 cm^{-1}. Raman experiments were first performed for samples deposited on glass substrates. We found that this samples are polycrystalline. Raman experiments were also performed on silicon substrates. In the SiO2 samples we no found evidence Si cluster. These results are in agreement with the PL results. In the samples with 5 minutes Si interlayer we found a linear increment of the signal of Si cluster when increasing the pressure. Finally in the samples with 10 minutes of grow of Si interlayer we found an intense signal of Si amorphous independently of the partial pressure of O2.

The FTIR absorption spectra show the absorption peaks associated with stretching ($1084 cm^{-1}$), bending ($812 cm^{-1}$) and rocking ($458 cm^{-1}$) vibration modes of the Si-O-Si bonds in SiO2 [7,8,9]. In addition all the samples present a tip in 666 cm^{-1} corresponding to Si-Si modes [10]. In the figure 2B are present the intensity of the tip of infrared in 1080 cm-1 of all the samples with the purpose of obtaining an idea of the tendency in the oxidation of each film. We found that samples with 50% and 66% of oxygen partial pressure are those for which the deposited SiO2 are respectively more and less stequiometric. The physical structure of the films was also studied by cross sectional transmission electron microscopy using a JEOL-2010 TEM operated to 300 kV. The figure 3 show two micrograph that confirms the growth of the SiO2/Si/SiO2 heterostructures. The SiO2 and Si bands appear as white and dark bands respectively; this contrast is caused by the large Si atomic density difference between the SiO2 (2.2×10^{22} atoms cm^{-3}) and Si (5×10^{22} atoms cm^{-3}).

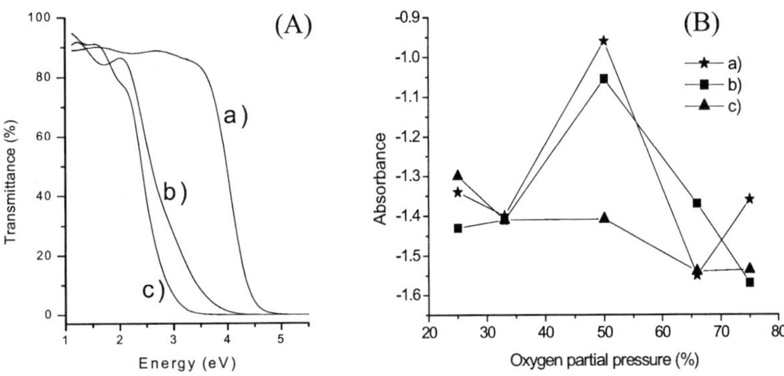

FIGURE 2. (A) The UV-Vis spectra functions of time of grow of Si interlayer. (B) Intensity of the tip of infrared in 1080 cm^{-1} of all the samples. a) Without Interlayer, b) 5 minutes interlayer and c) 10 minutes interlayer.

One of the important of quantum confinement is the modification of the energy band gap in these sandwiches Si films [11]. The change of band gap can occur at both the conduction band minimum (CBM) and the valence band maximum (VBM). According to effective mass theory [12] and assuming an infinite potential barrier at the interfaces, which should be a reasonable approximation because the band off-set at the SiO2/Si interfaces is several electrons volts, the energy gap variation, ΔE, should be determinate by $\Delta E= \Delta E_{CBM}+\Delta E_{VBM} = (\alpha+\beta)/d_{Si}^2$ where α and β are constants determined by the electron and hole effective masses. P. Nae Man, C. Chel-Jong, et al. [13] reported on the quantum confinement effect and light emission in the a-Si/SiN quantum dot samples prepared by plasma enhanced chemical vapor deposition. In their work, the photoluminescence peak energy E was expressed as E(eV)=1.56 + 2.40/d2 where d is the dot diameter in nm. With this model, we have Si quantum dot of 4.7 nm inside the matrix of SiO2.

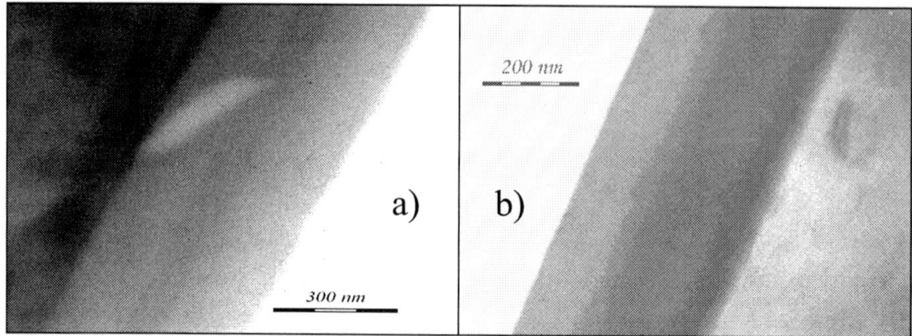

FIGURE 3. Cross-Sectional TEM micrograph taken of sample with a) 5 min. interlayer and 33% of Parcial pressure of O_2 and b) 10 min. interlayer and 75% of partial pressure of O_2.

In conclusion, SiO2/Si/SIO2 heterostructures were deposited on silicon (100) and glass substrates by RF reactive sputter magnetron. We investigate the effect of the partial pressure of oxygen and the thickness of the Si interlayer on the electronic properties of the heterostructures. The oxygen partial pressure not has a direct influence in the luminescence of the samples. A luminescence band around 1.7 eV appear by effect of the Si interlayer and the intensity of this band increase with the time of growing of the Si layer. The behavior of PL band 1.7 eV band presents a behavior that indicates the presence of nanocrystals inside the film. The Si quantum dot are of 4.7 nm inside the matrix of SiO2. In the UV-Vis transmission analysis, we observe a landslide in the absorption edge.from 4 eV to 2.35 eV. The change of band gap can occur by effect of quantum confinement.

REFERENCES

1. L. T.Canham, Appl. Phys. Lett. **57**, 1046 (1990).
2. Z. Song and X.M. Bao, Phys. Rev. B **55**, 6988 (1997).
3. T. Inokuma, Y. Wakayama, *et. al.*, Appl. Phys. **83**, 2228 (1998).
4. F. Wisken, D. Amans, *et. al.*, New J. Phys. **5**, 10 (2003).
5. N. Lalic and J. Linnros, J. Lumin. **80**, 263 (1999).
6. S.-H. Choi and R. G. Elliman, Appl. Phys. Lett. **75**, 968 (1999).
7. F. Ay, A. Aydinly, Optical Materials **26**, 33 (2004)
8. P. G. Pai, S. S. Chao, Y. Takagi, and G. Lucovsky, J. Vac. Sci. Technol. A **4**, 689 (1986).
9. A. Pereyra, Thin Solid Films **402**, 154 (2002).
10. D. B. Mawhinney, J.A. Glass and J.T. Yates, J. Phys. Chem. B. **101**, 1202 (1997).
11. Z. H. Lu and D. J. Lockwood, Nature **378**, 258 (1995).
12. Van Buuren, *et. al.* Appl. Phys. Lett. **63**, 2911 (1993).
13. P. Nae-Man, C. Chel-Jong Choi, *et. al.*, Phys. Rev. Lett. **86**, 1355 (2001).

Molecular Beam Epitaxial Growth of GaAs on (631) Oriented Substrates

Esteban Cruz Hernández[*], Alvaro Pulzara Mora[§], Juan-Salvador Rojas Ramírez[*], Rocío Contreras Hernández[*], Víctor H. Méndez García[¶], and Máximo López López[*]

[*]*Physics Department, Centro de Investigación y de Estudios Avanzados del IPN, Apartado Postal 14-740, México D.F., 07000, Mexico.*

[§]*Universidad Nacional de Colombia - Sede Manizales, A. A. 127, Colombia.*
[¶]*Instituto de Investigación en Comunicación Óptica, Universidad Autónoma de San Luis Potosí, Av. Karakorum 1470, Lomas 4a Sección, C.P. 78210, San Luis Potosí, Mexico.*

Abstract. In this work, we report the study of the homoepitaxial growth of GaAs on (631) oriented substrates by molecular beam epitaxy (MBE). We observed the spontaneous formation of a high density of large scale features on the surface. The hilly like features are elongated towards the [-5, 9, 3] direction. We show the dependence of these structures with the growth conditions and we present the possibility of to create quantum wires structures on this surface.

INTRODUCTION

In the last years one of the most imperative goals in the semiconductor research field has been the development and optimization of low-dimensional structures. MBE growth of III-V compound semiconductors on high index substrates is an interesting and promising way to induce self-organized phenomena to produce low dimensional structures, such as quantum dots and wires. Some of the high index surfaces where the MBE growth has been reported are: (11n), (122), (133), (012), and (2, 5, 11). The (631) plane is interesting because facets of this orientation spontaneously appear during the MBE growth of III-V semiconductor under diverse circumstances [1-3]. We describe here a review of some of the results we have found in our work group [4-6].

EXPERIMENTAL, THE (631) PLANE

We employed GaAs(631)A undoped wafers; some crystallographic directions relevant to the discussion are indicated in the schematic drawing of a wafer in the inset of Fig. 1(a). They were grown in a Riber C21 MBE system under usual conditions [3] and were in situ monitored by reflection high-energy electron diffraction (RHEED). We handled three principal parameters: growth temperature (T_g), As_4/Ga ratio and

CP885, *Advanced Summer School in Physics 2006, Frontiers in Contemporary Physics—EAV06*, edited by O. Miranda, M. Carbajal, L. M. Montaño, O. Rosas-Ortiz, and S. A. Tomás Velázquez

growth time (t_g). Then, we have three different groups of samples, each group with two fixed parameters and the other one variable. The samples surface was analyzed by atomic force microscopy (AFM) and high-resolution scanning electron microscopy (HRSEM). The optical properties of the samples were studied by photoreflectance spectroscopy (PR) at room temperature employing a standard experimental setup with a He-Ne laser (λ=632.8 nm).

In Fig. 1(a) is show a ball-and-stick model of the GaAs(631) unreconstructed surface. In this figure the topmost Ga and As atoms are illustrated by the larger balls with grey and black color, respectively. The bulk-terminated surface unit cell is defined by the vectors $\boldsymbol{a}$ and $\boldsymbol{b}$, with $\boldsymbol{a} = a_{GaAs}$ [-1,2,0], and $\boldsymbol{b}$=1/2a_{GaAs}[0,1,-3], where a_{GaAs} is the GaAs lattice constant. Along the [0,1,-3] direction we can observe an alignment of atoms (marked with dashed lines) formed by the topmost and second topmost Ga atoms.

In Fig. 1(b) we show a HRSEM image of the surface of a 500nm thick GaAs layer grown at a substrate temperature of 490 °C. We observe that the surface is covered by elongated hillock structures with a density of 28x10^5 /cm^2. The hillocks are oriented roughly along the [-5,9,3] direction, that is; perpendicular to the surface unit vector $\boldsymbol{b}$. The hillocks are limited by four sidewalls close to planes of the type: (211), (311), (6 3 -1), and (11 7 1).

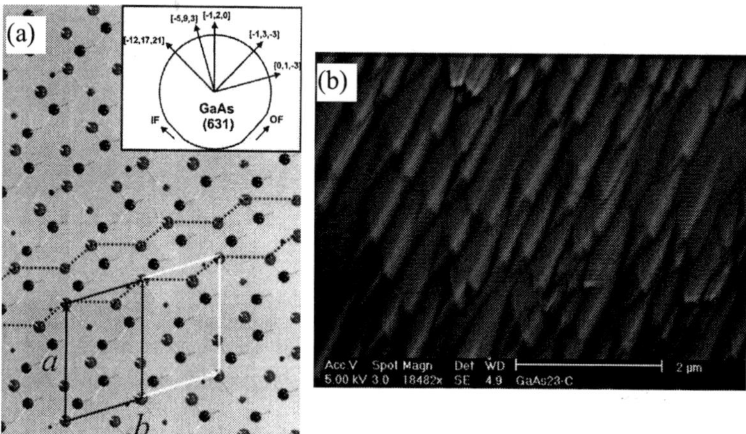

Figure 1. (a) ball-and-stick model of the unreconstructed GaAs(631)A surface. The inset shows a schematic figure of a (631) wafer with some crystallographic directions indicated. (b) A HRSEM image of a sample at T_g = 490 °C, the surface presents a high density of hillocks.

RESULTS AND DISCUSSION

First, we go to discuss the dependence of the hilly like features (HLF) with the growth temperature. In Fig. 2 shows two AFM images of samples grew to different temperatures (490 and 510 °C) in which we can appreciate the very small hillock dimensions. From these images we can to highlight several general behaviors of this series: a) We observed the formation of a high density of hillocks with a preferential elongation towards the [-5,9,3] direction; b) the length, width and height of the hillocks increase when T_g is increased and, c) the hillock density decrease when T_g increases. We have explained this behavior in terms of adatom diffusion anisotropy and diffusion barriers [6].

The variation of the hillock dimensions as a function of the As_4/Ga ratio at fixed T_g=540 °C shows a small variation of the hillock width and height. In contrast, the hillock length rapidly increases when the As_4/Ga ratio is increased due to the fact that, at high As_4/Ga ratios, longer surface structures are formed by the coalescence of individual hillocks along the [-5 9 3] direction.

In the third series, varying growth time from 3 to 30 minutes, we have understood about the way the hillocks develop from its first stages into a high density of mountainous like structures. Figure 3 shows sequential AFM images where we can appreciate the different stages of the hillocks formation. The process begins with a spontaneous formation of very tiny mound like features at ~ t_g= 3 min, then, from t_g= 6 to 12 min these mounds are jointed together in groups of 2 or 3. At 30 min of growth, the mounds have been jointed into individual hillocks limited by 4 well defined facets.

PR spectroscopy at room temperature was employed to study the optical properties of the samples. In order to obtain the energy values of the transitions observed in the PR spectra we used the third-derivative functional form developed by Aspnes [7],

$$\frac{\Delta R}{R} = \mathrm{Re}\left[Ce^{i\theta}\left(\hbar\omega - E_i + i\Gamma\right)^{-5/2}\right]$$

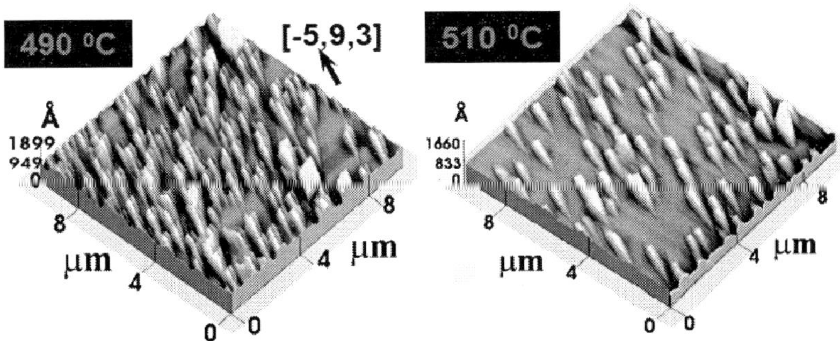

Figure 2. AFM images of two samples at T_g= 490 and 510 °C where we can see the high order hillocks orientation and the marked difference in hillock density.

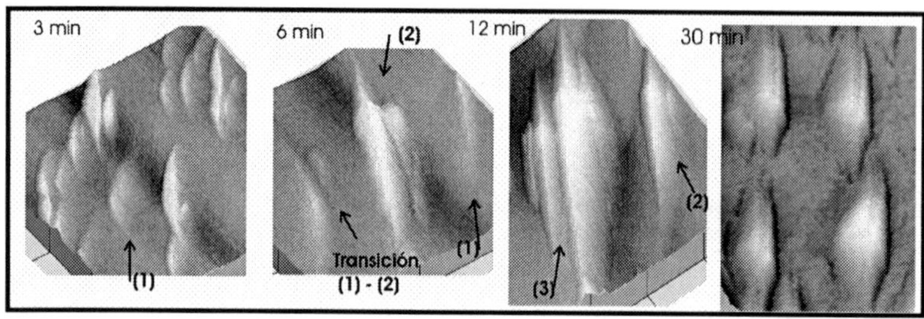

Figure 3. AFM images at different growth times showing the progressive formation of the well defined hillocks. The formation is a coalescence process between adjacent small mounds.

where C is the amplitude, θ the phase angle, ħω the incident photon energy, Γ the line width, and E_i the energy of the transition. Figure 4 shows PR spectra from the set of samples grown at different temperatures. The fittings to the Aspnes equation are shown by the solid lines and the energy transitions are marked by arrows.

The inset of Fig. 4 shows $\Delta E = E_2 - E_1$ in function of the hillock density. RS is a reference sample without overgrowth in which we have only an energy transition as it should be for a simple substrate. On the other hand, we found two different energy transitions for our grown samples. We have explained this second transition energy as due to a possible generation of a strain field in the GaAs epilayers induced by the hillock structures, which would split the band gap energy transition.

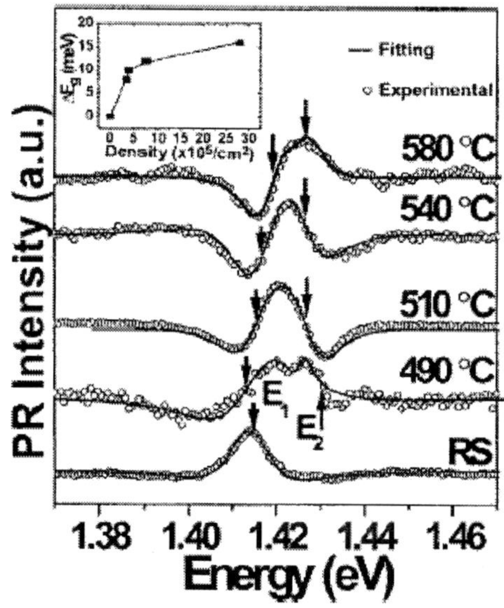

Figure 4. PR spectra from the set of samples grown at different temperatures. The spectrum RS is from a desorbed GaAs(631) substrate. The solid lines show the fitting to the Aspnes third-derivative functional form. The arrows indicate the energy transitions E_1 and E_2 found by the fits. The inset shows the variation of $\Delta E = E_2 - E_1$ as a function of the hillock density.

CONCLUSIONS AND PERSPECTIVES

In summary, we have studied the homoepitaxial growth of GaAs on (631) oriented substrates. We observed the formation of a high density of hillocks with a preferential elongation towards the [-5 9 3] direction. By PR spectroscopy an interesting splitting of the band gap transition was found. We think GaAs(631) substrates are good candidates to look for the growth conditions necessary to synthesize quantum wires. Studies are underway to synthesize self-assembled low dimensional structures on the (631) plane by varying other growth parameters to reduce the surface features size.

ACKNOWLEDGMENTS

This work was partially supported by CONACyT-Mexico, and UC-MEXUS. The authors would like to thank the technical assistance of R. Fragoso, M. Guerrero, E. Gomez, Z. Rivera, and A. Guillen. One of the authors (A.P.M.) thanks the support by the program CUAUTHEMOC III SRE-Mexico.

REFERENCES

1. M. Lopez, T. Ishikawa, and Y. Nomura, *Electron. Lett.* **29**, 2225 (1993).
2. H. Lee, R. Lowe-Webb, W. Yang, and P. C. Sercel, *Appl. Phys. Lett.* **72**, 812 (1998).
3. Y. Suzuki, T. Kaizu, and K. Yamaguchi, *Physica E* (Amsterdam) **21**, 555 (2004).
4. E. Cruz-Hernandez, A. Pulzara-Mora, F. Ramirez-Arenas, J. S. Rojas-Ramirez, V. H. Mendez Garcia and M. Lopez Lopez, *Jpn. J. Appl. Phys.*, **44**, L1556 (2005).
5. V.H. Mendez-García, F.J. Ramirez-Arenas, A. Lastras-Martínez, E. Cruz-Hernandez, A. Pulzara-Mora, J. S. Rojas-Ramírez, M. Lopez Lopez, *Appl. Surf. Sci*, **252** 5530 (2006).
6. E. Cruz-Hernández, J. S. Rojas-Ramírez, A. Pulzara-Mora, C. Vázquez-López, V. H. Méndez-García, and M. Lopez-Lopez, *J. Vac. Sci. Technol. B*, **24**(3), (2006).
7. F. H. Pollak, *Surf. Interface Anal.* **31**, 938 (2001).

PARTICIPANTS

Aguilar Ayala, Yareni (UAP, Puebla)
Alvarez, Abigail (UV , Veracruz)
Ávila Avendaño, Jesús Alberto (IPN)
Berumen Torres, Javier Alejandro (UAZ, Zacatecas)
Bolaños, Azucena (Cinvestav)
Bringas, Guadalupe (UAEM, Toluca)
Briones Jurado, Claudia (UNAM-FES, Cuautitlán)
Bustamante, Mauricio (PUC, Perú)
Cadena Arenas, Antonio (ESFM-IPN)
Camacho, Abraham (ESFM-IPN)
Carrión Ruiz, Francisco Javier (ESFM-IPN)
Castañeda Hernández, Alfredo (IF-UASLP, San Luis Potosí)
Castañeda Miranda, Elizabeth (Cinvestav)
Castellanos Diéguez, Pablo (USAC, Guatemala)
Castellanos Durán, Fidel Roberto (ESFM-IPN)
Cázares Ramírez, José Manuel (ESFM-IPN)
Cervantes, Aldrin (Cinvestav)
Contreras Astorga, Alonso (Cinvestav)
Corona Oran, Juan Carlos (Cinvestav)
Cuesta Sánchez, Vladimir (Cinvestav)
Esparza Moreno, Karina Patricia (UABC, Baja California)
Espinosa Olivares, María Guadalupe (Cinvestav)
Flores Cano, Leonardo (ESFM-IPN)
Flores Fuentes, Naria Adriana (ESFM-IPN)
Flores Monterrosas, Imelda Nalleli (ESIME-IPN)
Frías Hernández, Daniel (IT Orizaba, Veracruz)
Gaitán, Ricardo (UNAM-FES, Cuautitlán)
García, Edgar (USAC, Guatemala)
García Hernández, Luis Abraham (UPIITA-IPN)
García Molina, Nohemí (UAEM, Toluca)
Goiz, Klara (IFUNAM)
Gómez Izquierdo, Juan Carlos (Cinvestav)
González Alcudia, Michel (ESFM-IPN)
Guadarrama Lezama, Issai (UAEM, Toluca)
Heredia de la Cruz, Ivan (Cinvestav)
Hernández Becerril, Mario (UAEM, Toluca)
Hernández Chacón, Roberto (U del Valle, Guatemala)
Hernández Cruz, Iliana (ESFM-IPN)
Hernández Hernández, David (UAEM, Toluca)
Hernández Pinto, Roger José (Cinvestav)
Hernández Ruiz, Ligia (ESFM-IPN)
Herrera González, Baldomero (UAEM, Toluca)
Huerta Montiel, Grissel (UV, Veracruz)
Hurtado Torres, Ernesto Raul (ESIME-IPN)

Jardón Flores, Juan Luis (UAEM, Toluca)
Jiménez García, Mónica Noemí (UPIITA-IPN)
Jiménez Gutiérrez, Alberto (BUAP, Puebla)
Leyte González, Roy (UAEM, Toluca)
López Luis, Alberto (Cinvestav)
López Estrada, Omar (ESFM-IPN)
Luévano, José Rubén (UAM-A)
Lueza Martínez, Francisco Israel (UAM-A)
Luis Pineda, Loyda (ESIME-IPN)
Magaña, Ricardo (Cinvestav)
Martínez, Jorge (Cinvestav)
Martínez, Isaac (Cinvestav)
Martínez Dorantes, Miguel (ESFM-IPN)
Martínez Elena, Asunción Zeferino (ESFM-IPN)
Martínez Merino, Aldo (Cinvestav)
Mijangos Delfin, Zoé Emmanuel (ESFM-IPN)
de Moure Flores, Francisco Javier (UAM-A)
Muñoz García, Jorge Enrique (UMSNH, Michoacán)
Navarro López, Ninfa (UAZ, Zacatecas)
Nieto Rodriguez, Miriam Alicia (ESIME-IPN)
Nuñez Magaña, Tania Gudelia (UJAT, Tabasco)
Nuñez Ruedas, Nayeli (UAEM, Toluca)
Olivares Alonso, Oscar (ESFM-IPN)
Olivos Lagunes, Emilia (UV, Veracruz)
Ortega Sigala, José Juan (UAZ, Zacatecas)
Ortiz Madrigal, Nestor (UMSNH, Michoacán)
Pacheco Ortin, Sandy Maria (UNAM-FES, Cuautitlán)
Paniagua, Pablo (Cinvestav)
Pelayo Cárdenas, José de Jesús (U de G, Guadalajara)
Pérez Lara, Carlos Eugenio (PUC, Perú)
Pérez Martínez, Ricardo (Cinvestav)
Ramírez Reyes, Abdiel (ESFM-IPN)
Ramírez, Manolo (Cinvestav)
Rivera Martínez, Israel (ESFM-IPN)
Rodríguez, Patricia (Cinvestav)
Rojas Ramírez, Juan Salvador (ESFM-IPN)
Rueda Becerril, Jesús Misráyim (UAEM, Toluca)
Sacahui, Rodrigo (USAC, Guatemala)
Sánchez, Joel (UAEM, Toluca)
Sánchez Cruz, Patricia (ESFM-IPN)
Sandoval Fuentes, Eduardo (ESIME-IPN)
Sauceda, Huziel (UAS, Sinaloa)
Váldez Rojas, José Ernesto (UNAM-FES, Cuautitlán)
Vargas Calderón, Alejandra (UV, Veracruz)
Véliz Osorio, Alvaro Roberto (U del Valle, Guatemala)
Villada Balbuena, Mario (UAEM, Toluca)

Viveros Méndez, Perla Xochitl (Cinvestav)
Yañez Roldán, Etna (IFUG, Guanajuato)
Yee Rendón, Bruce (UAS, Sinaloa)
Zamora Cruz, Manuel (ESIME-IPN)

INVITED SPEAKERS

Arauz, José Luis (Cinvestav and UASLP)
Herrera, Gerardo (Cinvestav)
Marat, Luis (Cinvestav)
Matos, Tonatiuh (Cinvestav)
Méndez, José (Cinvestav)
Mendoza, Julio (Cinvestav)
Pérez, Miguel Angel (Cinvestav)
Rojas, Luis Fernando (Cinvestav)
Zelaya, Orlando (Cinvestav)

CONTRIBUTOR SPEAKERS

Alfaro Martínez, Adrián (Cinvestav)
Marín Satibañez, Benjamín Marcos (ESFM-IPN)
Barranco, Juan (Cinvestav)
Bernal Bautista, Argelia (Cinvestav)
Cruz Hernández, Esteban (Cinvestav)
Cruz y Cruz, Sara (Cinvestav and UPIITA-IPN)
Díaz, Celia (Cinvestav)
Espinoza, Catalina (Universitat de Valencia, España)
Fernández, Nicolás (Cinvestav)
Flores Amado, Abel (Cinvestav)
García Salcedo, Ricardo (CINVESTAV-Unidad Monterrey, Monterrey, México)
Hernández, Martín (Cinvestav)
López Carrera, Benjamín (ESFM-IPN)
Méndez Chávez, Francisco Javier (ESFM-IPN)
Mendoza Castrejón, Arturo (ESFM-IPN)
Miguel Pilar, Zelin (ESFM-IPN)
Montes de Oca Yemha, José Halim (ESFM-IPN)
Moreno González, Claudia (Universidad de Guadalajara, Jalisco, México)
Mota Pineda, Esteban (Cinvestav)
Torres Vega Gabino (Cinvestav)

SUPPORT TEAM

Juan Barranco (Cinvestav)
Luz del Carmen Cortés (ESIME-IPN)
Sara Cruz (Cinvestav and UPIITA-IPN)
Nicolás Fernández (Cinvestav)
Elvia Lomelí (Cinvestav)
Israel Lomelí (ESMC-IPN)
Miriam Lomelí (Secretary)

Author Index